energy science

Constants, symbols, and conversion factors

kilo (k) = 10^3; mega (M) = 10^6; giga (G) = 10^9;

tera (T) = 10^{12}; peta (P) = 10^{15}; exa (E) = 10^{18};

femto (f) = 10^{-15}; pico (p) = 10^{-12}; nano (n) = 10^{-9};

micro (μ) = 10^{-6}; milli (m) = 10^{-3}; centi (c) = 10^{-2};

deci (d) = 10^{-1}; 1 billion = 1000 million

1 kilowatt-hour (kWh) = 3.6 MJ

1 British thermal unit (Btu) $\approx$ 1.055 kJ

1 therm = 10^5 Btu $\approx$ 105.5 MJ

1 quad = 10^{15} Btu $\approx$ 1.055 EJ

1 barrel of oil equivalent (boe) $\approx$ 6 GJ

1 tonne of oil equivalent (toe) = 41.868 GJ

1 calorie (cal) $\approx$ 4.18 J

1 Calorie (food) $\approx$ 4.18 kJ

1 megaelectron volt (MeV) = 10^6 eV

$\approx 1.602 \times 10^{-13}$ J

1 horsepower (hp) $\approx$ 0.746 kilowatts (kW)

1 TWh per year $\approx$ 114 MW

1 metric tonne (t) = 1000 kg $\approx$ 2205 lb

1 US ton = 2000 lb

1 imperial ton = 2240 lb

1 mile $\approx$ 1.609 km

1 mph $\approx$ 0.447 m s^{-1}

0 °C = 32 °F $\approx$ 273 K

100 °C = 212 °F $\approx$ 373 K

1 barn (b) = 10^{-28} m^2

1 hectare (ha) = 10^4 m^2 $\approx$ 2.47 acres

1 acre $\approx$ 0.405 ha

1 square kilometre = 10^6 m^2 = 100 ha

$\approx$ 0.386 square miles

1 US ton per acre $\approx$ 2.24 t ha^{-1}

1 MW per square kilometer = 10 kW ha^{-1}

1 kW m^{-2} = 100 mW cm^{-2}

1 m^3 = 1000 litres = 1000 dm^3 $\approx$ 35.3 ft^3

1 US gallon $\approx$ 3.79 litres

1 imperial gallon $\approx$ 1.20 US gallons $\approx$ 4.55 litres

1 barrel of oil = 42 US gallons of oil

1 mile per gallon (US) $\approx$ 0.425 km litre^{-1}

1 pascal (Pa) = 1 N m^{-2}

1 bar = 10^5 Pa

1 atmosphere (atm) $\approx$ 1.013 bar

1 US dollar ($) $\approx$ 0.63 UK pound (£)

$\approx$ 0.78 euro (€) (May 2012)

Densities:

air $\approx$ 1.2 kg m^{-3} (20 °C); water $\approx$ 1000 kg m^{-3};

petrol (gasoline) $\approx$ 730 kg m^{-3};

ethanol $\approx$ 790 kg m^{-3}; diesel $\approx$ 840 kg m^{-3};

biodiesel $\approx$ 880 kg m^{-3}

Energy densities (LHV):

carbohydrates $\approx$ 15 MJ kg^{-1}; ethanol $\approx$ 27 MJ kg^{-1};

biodiesel $\approx$ 38 MJ kg^{-1}; diesel $\approx$ 43 MJ kg^{-1};

petrol (gasoline) $\approx$ 43.5 MJ kg^{-1};

coal (bituminous) $\approx$ 27 MJ kg^{-1};

natural gas $\approx$ 34.6 MJ m^{-3}

Avogadro number $N_A \approx 6.022 \times 10^{23}$ mol^{-1}

Gas constant $R \approx 8.314$ J mol^{-1} K^{-1}

Boltzmann constant $k \approx 1.381 \times 10^{-23}$ J K^{-1}

Stefan–Boltzmann constant σ

$\approx 5.67 \times 10^{-8}$ J K^{-4} m^{-2} s^{-1}

speed of light $c \approx 2.998 \times 10^8$ m s^{-1}

Planck's constant $h \approx 6.626 \times 10^{-34}$ J s

$hc \approx 1240$ eV nm

magnitude of electron charge $e \approx 1.602 \times 10^{-19}$ C

$e^2/4\pi\varepsilon_0 \approx 1.44$ MeV fm

Faraday constant $\approx$ 96 485 C mol^{-1}

gravitational constant $G \approx 6.674 \times 10^{-11}$ m^3 kg^{-1}s^{-2}

acceleration due to gravity $g \approx 9.81$ m s^{-2}

radius of Earth $\approx$ 6378 km

Solar constant $\approx$ 1.37 kW m^{-2}

mass of electron $m_e \approx 9.109 \times 10^{-31}$ kg

mass of proton $m_p \approx 1.673 \times 10^{-27}$ kg

mass of neutron $m_n \approx 1.675 \times 10^{-27}$ kg

permittivity of vacuum $\varepsilon_0 \approx 8.854 \times 10^{-12}$ C V^{-1} m^{-1}

permeability of vacuum $\mu_0 = 4\pi \times 10^{-7}$ V s^2 C^{-1} m^{-1}

energy science

principles, technologies, and impacts

second edition

John Andrews and Nick Jelley

OXFORD
UNIVERSITY PRESS

OXFORD

UNIVERSITY PRESS

Great Clarendon Street, Oxford, OX2 6DP,
United Kingdom

Oxford University Press is a department of the University of Oxford.
It furthers the University's objective of excellence in research, scholarship,
and education by publishing worldwide. Oxford is a registered trade mark of
Oxford University Press in the UK and in certain other countries

British Library Cataloguing in Publication Data

Data available

ISBN 978-0-19-959237-1

Printed in Great Britain by
Ashford Colour Press Ltd

Preface

Harnessing the Earth's energy resources has been a source of inspiration since ancient times. Energy devices have transformed civilization beyond the wildest imagination of our predecessors. These days, energy is always in the news. Why is this so? How is the global demand for energy going to be satisfied in the future? Can we avoid global warming becoming an insurmountable crisis? What are the options?

These questions have to be answered. What the present civilization does with the remaining energy resources will have a profound effect on the lives of future generations and the state of the planet. One approach is to do nothing and to assume that market forces and governments will sort it all out. The energy field is not short of uninformed or politically motivated opinions and commercial interests! This book is for those who prefer to make up their own minds, through an understanding of the science involved and the impact of the various technologies on society.

The idea for writing this book originated from undergraduate lecture courses given by the authors in Bristol and Oxford. The main focus of the book is to explain the physical principles underlying each technology and to discuss each of the technologies and their environmental, economic, and social impacts. It describes all the key areas of energy science, covering fossil fuels, nuclear energy, alternative energy, and the emerging energy technologies. Energy is a broad subject that crosses the boundaries between the traditional scientific and engineering disciplines. It is not essential to have a background in any particular discipline in order to use this book, apart from a general knowledge of science and mathematics to about high school standard. The aim is to enable students, professionals, and lay-readers to make quantitative estimates and form sound judgements. The material is presented in such a way that it can be understood on different levels. All the important results are described qualitatively using straightforward mathematical methods, with numerical **Examples**. More difficult mathematical **Derivations** and items of supplementary information are contained in **Boxes**. These boxes can be by-passed by those who do not wish to consider such detail. Readers are encouraged to work through the **Exercises** at the end of each chapter. The exercises are designed to be informative and thought-provoking, but not intimidating! For those who want to stretch their cerebral muscles and gain a deeper understanding of some of the more difficult points, some starred exercises have been added in each chapter.

Chapter 1 presents a brief history of energy technology, together with a review of the long-term energy trends and the evidence for global warming. Chapters 2 and 3 describe the essentials of thermal physics and the exploitation of fossil fuels in thermal power plants; many readers will already be aware of the basic concepts but may not be familiar with the details of the thermodynamic cycles, the greenhouse effect, the nature of fossil fuels, and geothermal power. Chapter 4 gives a brief account of the fluid mechanics needed to understand the exploitation of fluid-based devices, including hydropower, tidal power and wave power (Chapter 5), and wind power (Chapter 6). Chapters 7–10 are self-contained and can be read in any order, covering solar power, biomass and nuclear power (both fission and fusion). Chapter 11 gives a brief account of the principles of electricity generation and transmission, and electrical energy storage. Finally, Chapter 12 highlights the central issues concerning the impact of energy on society, with particular attention to the dangers of global warming and how it may be combated, emphasizes the importance of low carbon technologies, and gives a current appraisal of the likely energy scene of the future.

New to this edition

The layout of the second edition of the book is similar to that in the first edition, but a significant amount of new material has been added on energy conservation, the nature of fossil fuels and carbon capture and sequestration (in a separate chapter), advances in solar cell technology, a reassessment of the potential of biomass, the impact of the Fukushima disaster on the outlook for nuclear power, the development of smart grids, energy storage devices and electric cars, life-cycle analysis, and a reappraisal of the likelihood that the world will be successful in combating the challenge of global warming in the coming decades. In addition, the second edition contains numerous Case studies of energy projects taken from around the world of general interest, and includes many more Exercises for students to test their understanding of the subject.

Online Resource Centre

The book is accompanied by an Online Resource Centre at www.oxfordtextbooks.co.uk/orc/andrews_jelley2e/, which features additional materials for both students and lecturers.

For students:

• Multiple-choice questions to check your understanding as you progress through the text.

For lecturers:

• Figures from the book in electronic format.
• Full solutions to end-of-chapter exercises.

Acknowledgements

The authors would like to express their appreciation to their families, to their editors, Jonathan Crowe, Dewi Jackson, and production editor Philippa Hendry, and to their friends and colleagues for numerous ideas, criticisms, and helpful suggestions, in particular to David Andrews, Katherine Blundell, Niel Bowerman, David Cherns, Peter Cook, Steve Cowley, George Doucas, Nigel Dowrick, Mahieddine Emziane, Nick Eyre, Kieran Finan, Godfrey Gardner, Wina Graus, Nick Green, John Hannay, David Howey, Oliver Inderwildi, William Ingram, Tegid Jones, Chris Llewellyn-Smith, James Loach, Helen O'Keeffe, Robert Orzanna, Robert Paynter, Bruce Pilsworth, John Pye, Rachel Quarrel, Alex Schekochihin, Margaret Stevens, Robert Taylor, Andrew Tindal, Peter Wakefield, and Justin Wark.

Contents

6 Wind power

1 An introduction to *Energy Science*

→ Introduction

For millions of years the impact of life on Earth was energy-neutral: animals and plants coexisted in a continuous cycle. When primitive human beings discovered how to make fire, the balance of nature began to shift irreversibly. With fire, it was possible to cook meat, deter predators, and fashion metals into tools and deadly weapons. In the last two centuries, humankind has discovered how to convert heat into electricity, the most versatile and convenient form of energy. Electricity has enabled astonishing advances to be made in science and engineering, transforming civilization and making life far more comfortable than that of our predecessors. However, in the process it has created a consumer society that treats electricity and other forms of energy as commodities that should be available on demand.

In the relatively short period since the start of the Industrial Revolution (in the second half of the eighteenth century), a significant fraction of the fossil reserves of the planet that took hundreds of millions of years to evolve have been significantly depleted—in one year we consume what took about a million years to lay down. The emission of carbon dioxide and other products of combustion is now having a noticeable impact on the global climate. The threat to life in the relatively near future could be dire unless humankind can rise to the greatest challenge it has faced since its emergence as the dominant species on Earth.

Energy conversion is a disparate subject, but it is possible to obtain a good understanding of the essentials by applying basic physical principles. Energy issues tend to be open-ended and controversial. Addressing them with an independent mind is a rewarding and intellectually stimulating exercise. It is important to be objective, to pay attention to fact rather than opinion, to challenge assumptions, and always to look for constructive solutions.

1.1 A brief history of energy technology

Throughout history the harnessing of energy in its various forms has presented great intellectual challenges and stimulated scientific discovery. The energy technologies of today are

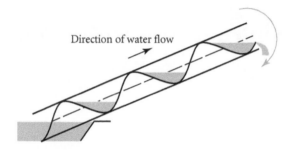

Direction of water flow

Fig. 1.1 Archimedes' screw.

the result of advances in scientific understanding, inspiration, and gradual improvements in engineering design over many centuries. By the time of the Roman Empire, water engineering was already a well-established technology. Thousands of years earlier, irrigation systems had greatly enlarged the area of farming land around the River Nile and increased the prosperity of ancient Egypt. An illustration of the ingenuity of the early engineers is demonstrated by Archimedes' screw (Fig. 1.1), a device used to extract water from rivers, empty grain from the holds of ships, and clear water from flooded mines (e.g. Rio Tinto in Spain). There are records showing that the device was used to irrigate the Hanging Gardens of Babylon (before the birth of Archimedes!), and it is still widely used today by the water and chemical industries.

In order to raise water, Archimedes' screw is encased in a hollow cylinder. One end is then immersed in water, and the water trapped in the hollows of the screw is transported upwards by rotating the device about its axis (see Exercises 1.6 and 1.18).

Waterwheels (see Chapter 5) existed in the ancient world and by AD 1000 were common throughout western Europe. The early waterwheels were very inefficient but designs gradually improved over the centuries (see Fig. 5.1). A technological breakthrough was made in 1832 with the invention of the Fourneyron turbine (see Section 5.1), which used fixed guide vanes to direct water between the blades of a rotating runner. The design of the vanes and the blades enabled most of the kinetic energy of the incident flow to be captured. The Fourneyron turbine pioneered the development of modern turbines, leading to the emergence of hydropower as one of the major providers of electricity today.

The earliest recorded steam engine was a toy device, invented by Hero of Alexandria, in the first century AD (Fig. 1.2). It essentially comprised a hollow metal sphere filled with steam, supported by two pivots. The steam is allowed to escape through two bent spouts and the momentum of the steam jets produces a reaction in the opposite direction on the spouts, which causes the sphere to rotate. The Industrial Revolution was made possible by the emergence of steam power. However, before the first commercial steam engines appeared on the scene, there were serious misunderstandings about the nature of vacuum and air pressure that needed to be resolved. In 1644 Torricelli (a follower of Galileo) invented the mercury barometer. He proved that the rise of the column of mercury was due to the difference in pressure between the atmosphere and the vacuum above the mercury inside the column. The next major breakthrough came about through the invention of the piston air pump, by von Guericke in 1650. In one spectacular demonstration, von Guericke took two identical metal hemispheres (known as the 'Magdeburg hemispheres', after the town where von Guericke was

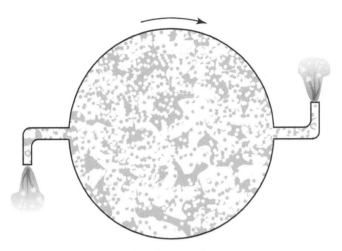

Fig. 1.2 Hero's steam engine.

burgomaster and a military engineer during the Thirty Years' War) and placed them together to form a complete sphere. The hemispheres were in touching contact along a flange, but there was no method of fixing. Von Guericke pumped out the air inside the sphere and invited two teams of eight horses to pull the hemispheres apart; they were unsuccessful, because the force exerted by the air was too great (see Exercise 1.4).

In 1666, the Académie des Sciences was established in Paris and the Dutch scientist, Christiaan Huygens, was made its first president. Huygens appointed two young assistants, Gottfried Leibniz and Denis Papin, who modified a von Guericke air pump to conduct experiments with gunpowder. A small charge of gunpowder was exploded inside a cylindrical chamber containing a tightly fitting piston. The air was expelled through two leather valves, thereby creating a partial vacuum. The piston then collapsed because of the imbalance in air pressure across it, and useful work was done in the process. Huygens realized the enormous potential of this discovery, from new forms of transport to powerful engines that could revolutionize industry.

Not surprisingly, gunpowder explosions proved to be too difficult to control! Papin, with help from Leibniz, tried using water instead of gunpowder, exploiting the fact that water expands to 160 times its original volume when converted to steam. Papin placed a quantity of water inside a piston chamber and heated it from the outside by a flame. The pressure of the steam raised the piston against the pressure of the air (Fig. 1.3(a)). In order to return the piston to its original position, he poured cold water over the outside of the cylinder so that the steam condensed back to water and the space inside the cylinder dropped to below atmospheric pressure.

Papin was forced to flee France in 1685 as a result of religious persecution. He settled in London and worked at the Royal Society, where he continued to develop his steam engine. In 1690 he applied for a patent to build commercial steam engines, but the patent was awarded instead to Thomas Savery, a military engineer, who had been strongly influenced by Papin's ideas. Papin's friendship with Leibniz meant that he did not get the support of Isaac Newton, who was locked in a bitter dispute with Leibniz as to which of them was the inventor of calculus. Papin died a pauper in 1712.

Savery's steam engines consumed huge amounts of coal and proved to be uneconomic. The first successful steam engine was built in 1712 by Thomas Newcomen, a blacksmith. Newcomen had discovered by accident that the steam inside a Savery steam engine condensed suddenly after some cold water had leaked into the steam chamber. Newcomen exploited this effect by installing a pipe to squirt a jet of cold water directly into the steam chamber (Fig. 1.3(b)). The Newcomen steam engine was a large structure with a long horizontal beam that rocked to and fro, with the rise and fall of the piston. Though it could only perform about five or six strokes a minute, it was capable of lifting large volumes of water from flooded mines. In the early Newcomen engines, a boy attendant was employed to open and shut two taps, one that allowed steam into the cylinder and the other that turned on a jet of cold water to condense the steam. This was a very monotonous job but, one day, a young lad called Humphrey Potter had a bright idea. He wanted to play with his friends and decided to connect a cord from each of the taps to the beam, so that the taps opened and closed at just the right moments in the cycle. It worked, and his invention was incorporated into all Newcomen steam engines.

A much more efficient steam engine was patented by James Watt in 1769. Watt was an instrument-maker and was working as an assistant to Professor Black, who had discovered the latent heat of steam (see Section 2.1). Watt took a keen interest in steam engines and the properties of steam. When asked one day to repair a malfunctioning model of a Newcomen steam engine, he calculated that about 80% of the heat was lost in heating the walls of the steam cylinder. Watt deduced that it would be much better to condense the steam in a separate chamber (known as the condenser), so that the temperature of the walls in the piston chamber could be maintained and thereby conserve heat (Fig. 1.3(c)).

Watt needed money to exploit his idea and formed a partnership with a wealthy iron foundry owner, Matthew Boulton. This gave Watt the finances to develop a commercial steam engine, but he soon ran into a major technical hitch: the cylinder castings were distorted and allowed too much steam to escape. Fortunately for Watt, a breakthrough in the manufacture of cannons provided him with the solution he needed. Cannons were constructed as thick-walled cylindrical tubes, but irregularities in the casting process meant that cannonballs often missed their target. In 1775 John 'Iron Mad' Wilkinson produced a solid cast iron block, from which he bored a smooth cylindrical hole of exactly the right shape and size. This improved the ballistics of cannonballs and enabled Watt to build leak-tight steam engines. The first Boulton and Watt steam engines were sold in 1776 and by 1824 they had produced 1164 machines.

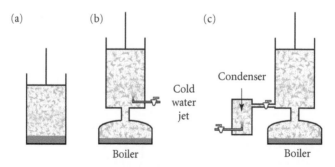

Fig. 1.3 Evolution of design of early steam engines: (a) Papin; (b) Newcomen; (c) Watt.

Surprisingly, perhaps, it was not until the middle of the nineteenth century that the concept that heat is a form of energy was finally accepted. Heat was originally thought to be a fluid, known as caloric, that was deemed to flow from hot bodies to cold bodies and could neither be created nor destroyed. The caloric theory was a remnant of the science of the ancient Greeks, who believed that all matter consisted of four basic substances: air, fire, earth, and water. The caloric theory was shown to be erroneous by Benjamin Rumford, an American scientist who had worked as a spy for the British in the American War of Independence. Rumford fled to Europe, where he married the widow of Antoine Lavoisier, one of the joint discovers of oxygen (and guillotined during the French Revolution in 1794). Rumford found occupation in Bavaria, where he improved the manufacture of cannons and was made a count of the Holy Roman Empire. Cannons were bored under water and the boring process made the water boil. Rumford observed that the water boiled for only as long as the boring process was continued. He deduced that caloric was apparently being produced by friction, in contradiction to the belief that it was uncreatable. Later, in the 1840s, an amateur scientist, James Joule showed that heat and mechanical energy are equivalent, and that energy is conserved.

In 1824, a young French scientist, Nicholas Carnot, proved that the maximum possible efficiency of an ideal heat engine depends only on the values of the hot and cold temperatures between which it operates. Carnot proved that the result was independent of the nature of the working fluid, but his explanation assumed the validity of the caloric theory: 'The motive power of heat is independent of the agents employed to realize it; its quantity is fixed solely by the temperatures of the bodies between which is effected, finally, the transfer of caloric' (see Exercise 1.7).

Steam power continued to advance through the nineteenth century with the development of steam trains and ships, powered by reciprocating steam engines. In 1884, Charles Parsons invented the rotary steam turbine. His great innovation was to realize that the power in a high pressure jet of steam could be exploited more efficiently if the pressure was dropped in stages across sets of turbine blades, rather than in a single step, resulting in a very compact and powerful engine. In order to demonstrate the superiority of his invention to a sceptical British Admiralty, he fitted his vessel *Turbinia* with three rotary steam turbines. He had the audacity to appear with *Turbinia*, uninvited, at the display of Her Majesty's fleet for Queen Victoria's Diamond Jubilee at Spithead. He was chased by the Navy's fastest patrol boats but they were no match for *Turbinia*. Parsons had made his point and the rotary steam turbine was accepted.

Earlier in the nineteenth century Michael Faraday, a self-taught scientist, was making exciting discoveries in electromagnetism. Faraday had begun his career at the age of 12 as a bookbinder. This gave him the opportunity to read learned books on science and he was eventually appointed by Sir Humphry Davy as an Assistant at the Royal Institution, in London. During the 1820s, Faraday was intrigued by two recent discoveries: one by Oersted, that the needle of a compass is deflected at right angles to the direction of flow of an electric current in a wire, and the other by Ampère, that two current-carrying wires exert a force on each other. In 1831, Faraday published his laws of electromagnetic induction, based on his own discoveries that a current is set up in a closed circuit by a changing magnetic field and also in a loop of wire when moved through a stationary magnetic field.

Faraday also showed that a steady current is induced across a rotating copper disc between the poles of a strong magnet (Fig. 1.4). This result led to the invention of the dynamo, which

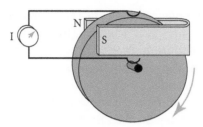

Fig. 1.4 Faraday's rotating disc.

helped the introduction of electric lighting. The early dynamos produced very spiky outputs; the first device to produce a smooth current was the Gramme dynamo, using a continuous loop of wire wrapped around a rotating iron ring. Electric telegraphy and electroplating were two of the first useful applications of electricity, followed by arc-lighting for public service. The first patent for an incandescent lamp was awarded to the American inventor and entrepreneur, Thomas Edison, in 1879. He used a loop of carbonized cotton thread that glowed in a vacuum for over 40 hours. Earlier, in 1860, Joseph Swan in England had patented the world's first light bulb, but it had a short lifetime and was inefficient. During the 1870s, Swan improved the vacuum inside his bulbs and formed the Swan Electric Light Company in 1881. The Edison and Swan companies were rivals in the development of the incandescent lamp, and eventually merged in England as the Edi-Swan Company.

In order to capitalize on his invention of the incandescent lamp, Edison patented his electric distribution system in 1880. In 1881, he built the world's first power station at Holborn Viaduct in London, which produced 160 kW of power for lighting and electric motors. In the following year he built a similar power station in New York City, to provide electric lighting for Wall Street and the banking community. There was intense rivalry between Thomas Edison's direct current system and a system using alternating current, promoted by George Westinghouse. Edison staged public events to highlight the dangers of alternating current, with live electrocutions of dogs, cats, and even an elephant! However, the alternating current system became the one generally adopted worldwide.

The first large-scale hydroelectric power station was built in 1895 on the US side of Niagara Falls, using Fourneyron turbines. The first half of the twentieth century witnessed a massive construction programme of coal-fired power stations and hydroelectric plants. The most significant new development in the second half of the twentieth century was nuclear power. The idea of producing electricity from nuclear fission reactors was an afterthought of the Manhattan Project, a secret military enterprise by the Allied powers in the Second World War to build an atomic bomb. The early fission reactors were used for producing materials for nuclear weapons. Reactors solely for electricity generation did not appear until the latter half of the 1950s.

The early years of nuclear power were heralded as a new era of cheap, inexhaustible, and safe electricity. In the 1970s, the developed world became more dependent on nuclear power after a series of large jumps in oil prices following the Arab–Israeli War in 1973. The most striking example was in France, where the government decided in the interests of national energy security to commit the country to nuclear power as the principal means of generating electricity; over 75% of French electricity is currently generated from nuclear power.

A number of high-profile accidents at various nuclear plants around the world, and growing public concern about the disposal of nuclear waste, eroded public confidence in nuclear power during the late 1970s and 1980s. The worst incidents were a partial meltdown of a pressurized water reactor at Three Mile Island (USA) in 1979, and a complete meltdown of an RBMK reactor at Chernobyl (Ukraine) in 1986. Human error and design faults were found to be significant factors in both cases. As a result there was a general improvement in nuclear safety standards worldwide and greater international support for the effective regulation of civil nuclear installations. However, the impact of these accidents considerably slowed the building of new reactors. The recent accident in 2011 in Fukushima, Japan, which was the result of a series of tsunamis following a massive undersea earthquake that measured 9.0 on the Moment scale, has further set back the deployment of nuclear power, though it still has the potential to play a significant role in combating global warming.

Alternative energy was a neglected area until the oil price shocks of the 1970s. Western governments then began to sponsor research programmes into various alternative energy technologies with the aim of reducing their dependence on oil. Funding was, however, on a much smaller scale than that for nuclear power. Wind power was the first alternative energy technology to become commercially viable, benefiting from low capital costs and tax breaks. Some other alternative technologies are now becoming competitive in niche areas. However, large-scale power generation by alternative energy sources can have a significant environmental impact. In particular, wind power has aroused public opposition in areas of outstanding natural beauty. Nonetheless, alternative energy is establishing a foothold and is likely to gain more support as the effects of global warming are realized and become more pronounced, and the remaining fossil fuel reserves of the planet become uneconomic to extract.

To complete this brief historical overview, Table 1.1 compares the power scales involved in a small selection of energy-related devices, from antiquity to the present day. It is a tribute to the achievement of humankind in applying scientific knowledge for the benefit of society, but it demonstrates that modern civilization has become dependent on consuming vast amounts of energy in order to maintain a comfortable lifestyle.

Table 1.1 Power scales

Device	Power (kW)
Treadwheel (AD 0)	0.2
Tour de France cyclist (uphill)	0.5
Strong horse	0.7
Newcomen steam engine (1712)	4
Fourneyron water turbine (1832)	30
Parsons' steam turbine (1900)	10^3
Smith–Putnam wind turbine (1942)	1.3×10^3
Boeing 747 gas turbine (1969)	6×10^4
Sizewell B nuclear power station (1992)	1.2×10^6
Drax coal power station (1986)	3.9×10^6

1.2 **Global energy trends**

A measure of the standard of living in a country is the Human Development Index (HDI), which combines indicators of educational attainment, life expectancy, and income. There is a strong correlation for low values of the HDI of a country with the energy use per capita, as shown in Fig. 1.5. However, there is a large spread in energy consumption per capita between different highly developed countries, indicating some scope for reducing consumption by improvements in efficiency and changes in lifestyle.

It is natural that less-developed countries will seek to increase their GDP (gross domestic product) and thereby increase their energy consumption per capita. The global population is predicted to be 9.2 billion by 2050, compared with 7.0 billion in 2011 and 2.5 billion in 1950, and the demand for energy is projected to increase by 40% (BP2030) by 2030, based on present trends.

Figure 1.6 shows the global energy consumption in 2010. The primary energy demand (excluding biomass) is predicted to remain approximately constant in OECD countries over the period up to 2030 at ~5500 megatonnes of oil equivalent (Mtoe) while growing significantly in non-OECD countries from ~6500 Mtoe to ~11 000 Mtoe. During these next 20 years the GDP of OECD and non-OECD countries is predicted to rise by ~1.5 and ~2.5 times, respectively, while the population is expected to increase by 1.4 billion compared with 1.6 billion in the previous 20 years, with nearly all the increase occurring in non-OECD countries. The breakdown by source is shown in Fig. 1.7 (note that hydropower is plotted separately from renewables).

The energy intensity (the amount of energy per unit of GDP) is expected to decrease globally over the next 20 years and energy efficiency is expected to improve. As an agricultural society becomes industrialized and more developed its energy intensity increases, but as more service sector jobs grow and technology develops the energy intensity falls as more high value commodities that require less energy are manufactured. However, the total energy demand is still predicted to increase by ~40% from ~12 000 to ~16 500 Mtoe, with only a slow decrease in

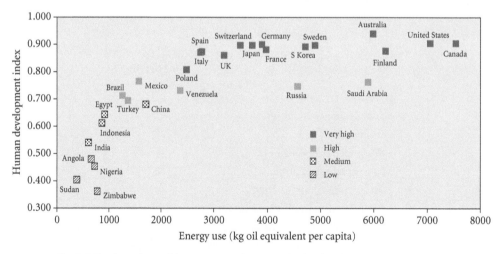

Fig. 1.5 The dependence of the Human Development Index (HDI) on the energy use per capita. *Source:* UNDP2009, WB2009.

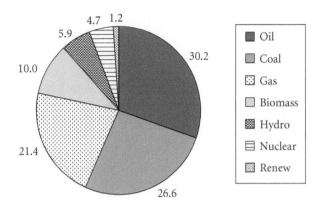

Fig. 1.6 Primary energy consumption in 2010 by fuel type (%), with WEC estimate for biomass. Total = 558 EJ = 17.7 TWy. *Source:* BP2011, WEC2010.

the growth rate by 2030: over this period fossil fuels up by ~30% to ~13 500 Mtoe, with a shift away from coal to gas, and renewables up by ~500% from ~150 to ~800 Mtoe.

Electricity generation is expected to increase by 65% from ~21 000 TWh in 2010 to ~35 000 TWh in 2030, with renewables generating ~10% globally compared with ~3% in 2008 (see Fig. 1.8).

The present (2010) distribution of fuel demand by region is shown in Fig. 1.9. China's economy is expected to continue to grow rapidly and to be the largest in the world by 2030.

In China, and globally, fuel switching to sources with lower carbon dioxide emissions per unit of energy (gas is about half that of coal, see Table 1.2) is expected to result in a percentage increase in Gt of CO_2 per annum of 27% in 2030 compared with that in 2010, when ~31 Gt of CO_2 was emitted, lower than the estimated 40% increase in energy demand. However, continuing such CO_2 emissions would put the world at risk of significant climate change owing to the associated global warming.

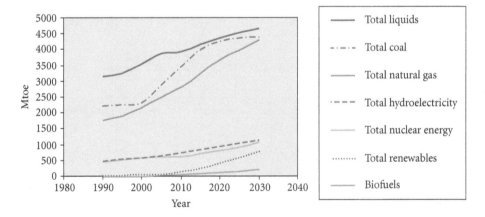

Fig. 1.7 Predicted consumption of energy from fossil and low-carbon sources out to 2030. *Source:* BP2030.

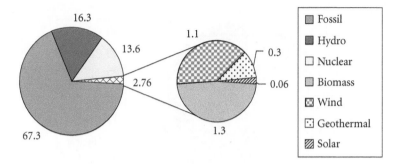

Fig. 1.8 Electricity generation (%) by fuel in 2008 (total 19 100 TWh = 2.18 TWy).
Source: EIA2008.

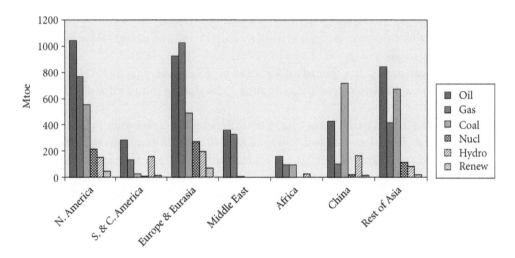

Fig. 1.9 The present distribution of energy fuel by region.
Source: BP2011.

Table 1.2 Average values for emissions (gCO_2eq/kWh) from life-cycle analyses of different sources of electricity (CCS = carbon capture and storage)

Source	gCO_2eq/kWh	Source	gCO_2eq/kWh
Hydro	8	CCS	133
Nuclear	10	Gas	550
Wind	14	Oil	800
Solar PV	57	Coal	1000
Biomass	70	Lignite	1100

Source: Weisser2007.

1.3 Global warming and the greenhouse effect

Since the Industrial Revolution there has been a sharp increase in the burning of fossil fuels. This is important for the global climate because CO_2 is what is called a greenhouse gas. The main greenhouse gases in the atmosphere are water vapour (and water droplets) and CO_2. These gases absorb infrared radiation and this affects the temperature of the Earth. We can see why this happens by first considering the temperature of the Earth with an atmosphere completely transparent to all radiation. The Sun's radiation would then impinge directly on the Earth's surface. This would heat the surface until it was at a temperature of around $-19\,°C$, at which temperature it would radiate energy back out into space at the same rate at which it was received from the Sun.

Now consider the effect of adding an atmosphere that is completely transparent to the Sun's radiation, which is mainly in the visible, but absorbs infrared radiation. Radiation emitted by a surface that is near room temperature is mainly in the infrared. Thus the atmosphere absorbs the Earth's radiation and heats up. As a result, the atmosphere radiates infrared radiation both out into space and back to the Earth's surface, which now receives more radiant energy than with a completely transparent atmosphere and gets hotter. The temperature rises until the Earth's surface emits energy at the same rate as it receives energy. This is an oversimplification of what happens in our atmosphere but illustrates the key process. The trapping of infrared radiation and the consequent temperature rise is called the greenhouse effect. It is actually only a small part of the temperature rise in greenhouses, which is mainly from reduced convection, but the name has stuck. We discuss a simple model of the greenhouse effect in Chapter 2 (Box 2.1).

In 1827 the great French mathematician Jean Fourier was the first scientist to realize that the atmosphere acts like a greenhouse. However, it was not until 1860 that the Irish scientist John Tyndall identified water vapour, carbon dioxide, and methane as the main gases in the atmosphere that absorb infrared radiation. In particular, he realized that without the greenhouse effect 'the Sun would rise upon an island held fast in the grip of frost'.

The greenhouse effect causes a temperature rise of about $35\,°C$, so it is very important in determining and maintaining our global temperature. However, substantially increasing the concentration of greenhouse gases by burning fossil fuels has already had a noticeable impact on the global climate. Over the twentieth century, the average global temperature rose by $0.6\pm0.2\,°C$ and the International Panel on Climate Change (IPCC) predicts that global temperatures will rise 1.1–$6.4\,°C$ between 1990 and 2100, primarily due to the release of greenhouse gases from fossil fuel emissions. The range in temperature reflects different scenarios and their uncertainties in the resulting temperature rise. These scenarios model the effect of differing reliance on fossil fuels, the global economic growth rate, and the speed of uptake of new low-carbon technologies.

The average global temperature since the mid nineteenth century (when accurate temperature measurements started to be recorded) is shown in Fig. 1.10. There has been a steep rise since the 1970s and this rise is called global warming. It has been explained by physical models of the Earth's temperature that take into account the emission of greenhouse gases, in particular CO_2, due to human activity. Including these anthropogenic emissions gives good

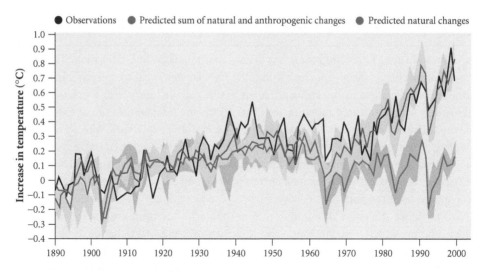

● Observations   ● Predicted sum of natural and anthropogenic changes   ● Predicted natural changes

Fig. 1.10 Curves showing global temperatures 1890–2000. Measured; predicted with only natural changes; and predicted with natural plus anthropogenic changes.
Source: Meehl2004.

agreement with observations; the temperature would have been expected to have remained about the same as in the 1970s if only natural causes were included.

The main greenhouse gas is water vapour (and water droplets). The effect of carbon dioxide, methane, CFCs, and other greenhouse gases is to enhance its effect by increasing the amount of water vapour in the atmosphere. Carbon dioxide concentrations have risen from about 280 parts per million in 1750, which is just before the Industrial Revolution, to about 390 parts per million in 2012. The characteristic timescale for an excess of water vapour in the atmosphere to disappear is a few days but, for other greenhouse gases and for the response of the interactions between the oceans and atmosphere, the characteristic timescales are typically 10–1000 years. Aerosols, such as those arising from the burning of fossil fuels (e.g. SO_2 aerosols which produce acid rain), tend to have a cooling effect on the climate and, at current levels, cancel out closely the warming effect of the non-CO_2 greenhouse gases such as methane. The net result is that the level of CO_2 equivalent gases (CO_2eq) is close to that of CO_2 alone.

1.3.1 Life-cycle analysis

The analysis of the amount of greenhouse gases emitted in a process is called a life-cycle analysis (LCA), which should include all stages of the process; e.g. for biomass, the emissions associated with any land change, the use of fertilizers, tractors, and processing. Table 1.2 shows the average values of emissions associated with various fossil and renewable sources of electricity.

It is important to note that it is not only energy-related emissions that matter: CO_2 given off as a result of land-use change (e.g. deforestation) is also a major concern. The percentage contributions from different sources of CO_2 and from other gases to global anthropogenic greenhouse gas emissions in CO_2eq in 2004 are shown in Fig. 1.11.

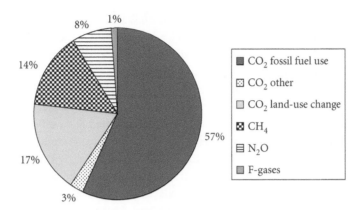

Fig. 1.11 The percentage contributions from different sources of CO_2 and from other gases to global anthropogenic greenhouse gas emissions in CO_2eq in 2004.
Source: IPCC2007.

1.4 Implications of global warming for energy supply

The different scenarios considered by the IPCC result in different levels of CO_2, with the more fossil fuel based or business-as-usual (BUA) scenario giving an approximate level of 600–800 ppmv (parts per million by volume) of CO_2 and a rise in temperature of ~3–4 °C by the end of the twenty-first century. As we discuss in Chapter 12, such a temperature rise would put the world at considerable risk of significant climate change. Although this would be only a small change in the global mean temperature, it would increase the chance of extreme weather conditions considerably (see Exercise 1.10). Limiting the temperature rise to ~2 °C, corresponding to a level of ~450 ppmv CO_2eq, which is equivalent roughly to holding CO_2 levels to less than 450 ppmv, would reduce the risk considerably.

To achieve 450ppmv we would need to reduce CO_2 emissions to effectively zero by 2100. Noting that the temperature rise to date since pre-industrial times is ~1 °C, during which time ~0.5 trillion tonnes of carbon have been emitted, reducing to zero carbon emission by 2100 would emit about the same mass of carbon, thereby leading to another ~1 °C rise. Since ~80% of our energy presently comes from fossil fuels, this means decarbonizing our energy supply quickly (since the resulting temperature rise only depends on the amount of carbon emitted and not when, the rate of reduction is lessened the sooner it is begun). This process would initially mean fuel switching to sources with lower carbon emissions per unit of energy, e.g. from coal to gas or renewables, and eventually satisfying our energy requirements from just renewable, nuclear, and carbon capture power plants.

To achieve this goal would also require reducing the global demand for energy but without denying the benefits to a society that are associated with energy (see Fig. 1.5). A significant reduction that meets such a condition can be made by improving end-use efficiency; e.g. by changes in building technology and in transport. Buildings account for about 30% of energy-related global CO_2 emissions and these could be reduced significantly through improvements in insulation and by the use of heat pumps and LEDs. Transport (i.e. mainly light vehicles) accounts for about 20% of energy-related emissions and significant savings would be made through the introduction of electric or initially plug-in hybrid cars. Another important measure is energy conservation, whereby energy savings are made through changes in behaviour; e.g. lowering thermostats, turning lights off, and using public transport rather than cars. These and other aspects of energy savings are discussed in more detail in Chapter 12.

Global warming is expected to become more pronounced over the coming decades. As we discuss in Chapter 12, the impact on weather systems, ocean currents, and sea levels, and on biodiversity, food production, and water availability, could be devastating in various parts of the world unless effective countermeasures are implemented and sustained over the next century. The scale of the problem requires a broad international commitment to stand any chance of success. The Kyoto Protocol was the first international agreement to commit countries to reduce or trade in emissions of carbon dioxide and other greenhouse gases, with the aim of a 5.2% reduction compared with 1990 levels for the period 2008–2012. However, there were some notable abstainers, including the USA and Australia, and there were concerns about the effectiveness of the Protocol and the likely degree of compliance of individual countries. Though the Kyoto Protocol was a step in the right direction, the world as a whole has yet to demonstrate the commitment required to combat the threat due to global warming.

1.5 **Energy units and capacity factors**

In order to appreciate the magnitudes of physical quantities and to be able to make comparisons, it is essential to have a sound grasp of physical units. To conclude this chapter, we define the most important units relating to energy. We also introduce the technique of dimensional analysis, which is employed throughout the book to derive basic physical relationships without the need for a full analysis of the system.

The SI system of units is used throughout this book, but a comprehensive list of units can be found at the front of the book, with conversions to the imperial system of units. Table 1.3 defines the key physical quantities related to energy.

Table 1.3 Energy-related units and conversion rates

Quantity	Unit	Definition
Force	newton (N)	Force required to accelerate 1 kg by $1\,\text{ms}^{-2}$
Energy	joule (J)	Work done by force of 1N in moving 1 kg by 1 m
Power	watt ($W = Js^{-1}$)	1 joule per second
Energy	kilowatt-hour (kWh)	$10^3 \times 60 \times 60 = 3.6 \times 10^6$ joules ≈ 3411 Btu ≈ 859.6 kilocalories
Energy	calorie	Energy required to heat 1 g of water by $1\,°C \approx 4.2\,$J
Energy	Btu	Energy required to heat 1lb of water by $1°F \approx 1.055\,\text{kJ} \approx 0.293\,\text{kWh}$
Volume	barrel	42 US gallons $\approx$ 35 imperial gallons $\approx$ 159 litres of oil $\approx 6\,$GJ
Fuel equivalence	1 tonne oil	≈ 1.5 tonnes hard coal ≈ 3.0 tonnes lignite $\approx 12\,000\,$kWh
Power	1 horsepower	550 ftlb per second or 0.746 kW

An important point to realize is that amounts of energy can refer to either thermal or electrical energy. In Fig. 1.6, which gives the contributions to the primary energy demand from different sources, the amounts refer to thermal energy. Conversion of oil, gas, or coal to electricity is indirect, since the first product is heat which is then converted to electricity, with a typical efficiency of 33%. Hydropower, nuclear, solar, wind, tidal, and wave all produce electricity. To make a fair comparison with the fossil fuel contribution, the electrical energy from these low-carbon sources has been multiplied by a factor of ~3 to give the equivalent amount of fossil fuel energy displaced.

When comparing the outputs of power plants using different sources of energy, it is also important to note that the maximum continuous or rated power output is usually quoted for a plant. For typical fossil fuel plants the average annual power output as a fraction of the rated power, called the capacity factor of the plant, is ~75% and for a nuclear power plant it is ~90%. For wind and solar farms, the capacity factor is much less: ~20–40% for wind and ~20% for solar, the lower values reflecting the variable nature of the source. The contribution that a technology can make by a certain time is termed the accessible potential and can be given in terms of the equivalent continuous power, e.g. an accessible potential of 600 GWe (where e refers to electrical output) would generate ~5000 TWh per year. How much of the accessible potential is actually deployed depends in particular on economic considerations.

As we will see in Chapter 2, when we consider heat pumps, an amount of electrical energy E can be used to move several times E as heat either from the environment to inside a building when heating, or vice versa when cooling. Electricity can also be used to power electric cars, and if the electricity supply were decarbonized then the emissions from transport as well as the energy used, could be significantly decreased.

1.5.1 Dimensional analysis

A useful development of the idea of units is dimensional analysis. This is a method for deriving an algebraic relationship between different physical quantities which relies on good physical intuition in choosing the appropriate physical variables.

Each variable is expressed in terms of its fundamental units of mass M, length L, time T, . . . , raised to some arbitrary index $a, b, c, \ldots$. The unknown indices $a, b, c, \ldots$ are then determined by equating the indices of like units using simple algebra. Care is needed in choosing the relevant physical variables and it is sometimes possible to obtain more than one solution to a given problem. An independent means of checking the result obtained is therefore desirable.

EXAMPLE 1.1

Check whether the physical dimensions of the expression for the hydrostatic pressure in a fluid, $\rho g h$, are consistent with the physical dimensions of pressure, p.

Replacing the individual symbols in terms of their fundamental physical units, we have

$$\rho g h \equiv (\text{kg m}^{-3})\,(\text{m s}^{-2})\,(\text{m}) = (\text{kg m s}^{-2})\,(\text{m}^{-2}) = \text{N m}^{-2}$$

Hence $\rho g h$ is consistent with the physical dimensions of pressure (force per unit area).

EXAMPLE 1.2

Use dimensional analysis to find an expression for the hydrostatic pressure in a fluid.

Suppose the hydrostatic pressure depends on the density ρ, the acceleration due to gravity g, and the depth h. We assume a general algebraic formula of the form

$$p = k \, \rho^a \, g^b \, h^c$$

where k is a dimensionless coefficient and the indices a, b, and c are numbers to be determined. Replacing each symbol by its fundamental physical units, we have

$$M^1 L^{-1} T^{-2} = (ML^{-3})^a (LT^{-2})^b (L)^c$$

or

$$M^1 L^{-1} T^{-2} = M^a L^{-3a+b+c} T^{-2b}$$

Since M, L, and T are independent physical quantities, we can equate the indices on both sides, yielding the equations

$$1 = a, \; -1 = -3a + b + c, \; -2 = -2b$$

so that $a = b = c = 1$. Hence, the expression for hydrostatic pressure is of the form $p = k\rho g h$. (The coefficient k cannot be determined from dimensional analysis since it is dimensionless.)

 SUMMARY

- Global energy production is expected to increase by around 40% between 2010 and 2030, with the increase virtually all in the developing countries.

- The greenhouse effect is a natural phenomenon due to absorption of solar radiation by the Earth's atmosphere, raising the temperature on the surface of Earth by about 35 °C.

- The main greenhouse gas is water vapour. Carbon dioxide, methane, CFCs, and other greenhouse gases enhance the effect of water vapour.

- The characteristic timescale for an excess of water vapour in the atmosphere to disappear is a few days but, for other greenhouse gases and for the response of the interactions between the oceans and atmosphere, the characteristic timescales are typically 10–1000 years.

- Carbon dioxide concentrations have risen from about 280 parts per million by volume in 1750 to about 390 parts per million by volume today.

- Carbon dioxide emissions need to fall to zero by ~2100 in order to limit the atmospheric concentration to ~450 ppmv and to restrict the temperature rise (compared to pre-industrial times) to 2 °C.

- Continuing to rely predominantly on fossil fuels for our energy—the business-as-usual scenario—could cause a temperature rise of ~4 °C and put the world at risk of significant climate change.

- Decarbonizing our electricity and energy supply, coupled with energy savings, is essential in order to combat climate change.

- Physical expressions should be checked to make sure that the physical units are consistent.

- Dimensional analysis is a useful technique for deriving algebraic relationships between physical quantities.

FURTHER READING

Borowitz, S. (1999). *Farewell fossil fuels*. Plenum, New York. Interesting book about the need for alternatives to fossil fuels.

Botkin, D.B. and Keller, E.A. (2003). *Environmental science*. Wiley, New York. Good discussion of issues about energy and the environment.

Boyle, G., Everett, R., and Ramage, J. (eds). (2003). *Energy systems and sustainability*. Oxford University Press, Oxford. Good qualitative discussion of issues.

Coopersmith, J. (2010). *Energy: the subtle concept*. Oxford University Press, Oxford. Excellent history of how the concept of energy evolved.

MacKay, D.J.C. (2009) *Sustainable energy: without the hot air*. UIT, Cambridge. Very readable introduction to alternative energy and energy policy.

Meehl, G.A. et al. (2004). Combinations of natural and anthropogenic forcings and 20th century climate. *J. Climate* **17**, 3721.

Richter, B. (2010). *Beyond smoke and mirrors*. Cambridge University Press, Cambridge. Good non-technical account of climate change science and on how we might reduce our reliance on fossil fuels.

Stein, R.S. and Powers, J. (2011). *The energy problem*. World Scientific Publishing, Singapore. Introduction to alternative energy sources for non-specialists.

Uglow, J. (2002). *The lunar men*. Faber, London. Interesting account of the friends who launched the Industrial Revolution.

Weisser, D. (2007). A guide to life-cycle greenhouse gas (GHG) emissions from electric supply technologies. *Energy* **32**, 1543–1559. (Weisser 2007).

WEB LINKS

www.energy.gov US Department of Energy website, appraisals of energy trends and energy technologies.

www.iea.org International Energy Agency, advisory body on energy policy.

www.metoffice.com/research/hadleycentre Research on climate change.

www.wikipedia.org Online encyclopaedia, articles on history of energy and energy-related topics.

www.worldenergy.org Neutral overview of global energy scene.

www.wri.org Environmental think-tank.

bp.com/statisticalreview (BP2011).

bp.com/energyoutlook (BP2030).

www.ipcc.ch/pdf/assessment-report/ar4/syr/ar4_syr.pdf Intergovernmental Panel on Climate Change (IPCC), AR-4, 2007 (IPCC2007).

www.cgd.ucar.edu/ccr/publications/meehl_additivity.pdf (Meehl2004).

hdr.undp.org/en/media/HDR_2011_EN_Table2.pdf (UNDP2009).

data.worldbank.org/indicator/EG.USE.PCAP.KG.OE (WB2009).

www.worldenergy.org/documents/ser2010exsumsept8.pdf (WEC2010).

www.eia.gov/cfapps/ipdbproject/IEDIndex3.cfm?tid=2&pid=2&aid=12 US Energy Information Administration (EIA2008).

EXERCISES

1.1 Can global warming be combated by energy conservation alone?

1.2 Assuming the volumes of the Greenland and Antarctic ice caps are 2.85×10^6 km^3 and 25.7×10^6 km^3, respectively, the radius of the Earth is 6378 km, and 70% of the Earth's surface is covered by sea, estimate the rise in sea level if both ice caps melted completely.

1.3 Describe the development of the steam engine.

1.4 Estimate the atmospheric force exerted on von Guericke's hemispheres. Diameter 51 cm; assume air pressure $= 10^5$ Nm^{-2}.

1.5 Is the quality of life of a society better or worse without electricity?

1.6 Construct a model of an Archimedes' screw using scrap materials.

1.7 Rewrite in modern terminology Carnot's statement, 'The motive power of heat is independent of the agents employed to realize it; its quantity is fixed solely by the temperatures of the bodies between which is effected, finally, the transfer of caloric'.

1.8 During the second phase of the Papin steam engine cycle the outer surface of the cylinder is cooled to condense the steam so that the piston returns to its original position. Using dimensional analysis, derive a relationship for the timescale for the inner surface of the cylinder to be significantly cooled, in terms of the thermal conductivity k, density ρ, specific heat capacity c, and wall thickness d. Why is an engine of this design very inefficient?

1.9 The number of children under 15 is about 2 billion (2012) and is expected to be the same in 2072. The global population is 7 billion in 2012; estimate the global population in 2072.

1.10 A region has a temperature variation that can be described by a Gaussian distribution with mean 25 °C and standard deviation 5 °C. The chance of the temperature being above 40.5 °C is 1 in 1000. Prolonged exposure to temperatures above 40 °C can be dangerous to health. If the mean temperature in this region rises by (a) 2 °C and (b) 4 °C and the standard deviation remains the same, calculate the chance that temperatures are above 40.5 °C. Discuss the relevance of your result to global warming.

1.11 Show that the physical dimensions of $\frac{1}{2}mv^2$ are consistent with the physical dimensions of work.

1.12 Use dimensional analysis to find an expression for the power in the wind incident on a wind turbine, in terms of the air density ρ, the wind speed v, and the cross-sectional area of the wind turbine A.

1.13 Convert (a) 1 MJ into Btu, (b) 800 kg of oil equivalent per year into kW.

1.14 Using a dimensional argument, show that the aerodynamic drag force F_D on an aeroplane travelling at a speed u is proportional to $\rho A u^2$, where ρ is the density of air and A is the frontal area of the aeroplane. How does the drag force depend on the altitude of the aeroplane?

1.15 Derive an expression for the characteristic time taken for viscous effects in a fluid to dissipate in terms of the distance x (L), density ρ (ML^{-3}), and the coefficient of viscosity μ (ML^{-1}T^{-1}).

1.16 Derive a dimensionless parameter based on specific heat c (L^{2}T$^{-2}\theta^{-1}$), coefficient of viscosity μ (ML^{-1}T^{-1}), and thermal conductivity k (MLT$^{-3}\theta^{-1}$), where θ represents the dimensions of absolute temperature.

1.17* Table 1.4 shows the rise in the average global near-surface temperature and the atmospheric concentration of carbon dioxide for each year over the period 1979–2005. Analyse whether a statistical correlation exists between the two sets of data.

Table 1.4 Carbon dioxide[1] concentrations and global temperature differences[2] ΔT (°C) for the period 1979–2005

	1979	1980	1981	1982	1983	1984	1985	1986	1987
CO_2	336.53	338.34	339.96	341.09	342.07	344.04	345.10	346.85	347.75
ΔT	0.06	0.10	0.13	0.12	0.19	−0.01	−0.02	0.02	0.17
	1988	1989	1990	1991	1992	1993	1994	1995	1996
CO_2	350.68	352.84	354.22	355.51	356.39	356.98	358.19	359.82	361.82
ΔT	0.16	0.10	0.25	0.20	0.06	0.11	0.17	0.27	0.13
	1997	1998	1999	2000	2001	2002	2003	2004	2005
CO_2	362.98	364.90	367.87	369.22	370.44	372.31	374.75	376.95	378.55
ΔT	0.36	0.52	0.27	0.24	0.40	0.45	0.45	0.44	0.47

[1] *Source:* Hadley Centre for Climate Prediction and Research (Crown copyright).
[2] *Source:* Mauna Loa Observatory, NOAA (temperature difference with respect to the average temperature over the period 1961–90).

1.18* Consider an Archimedes' screw consisting of a helical tube of wavelength λ wound around a cylinder of radius a. Prove that the angle of elevation of the axis of the cylinder, θ, for the device to be able to raise water is such that $\tan\theta \leq \dfrac{2\pi a}{\lambda}$.

2 Thermal energy

➔ **Introduction**

Thermal energy plays a vital role in modern civilization, both at a domestic level and in industrial processes, notably in the conversion of fossil fuel and other sources of heat, such as nuclear reactors, biomass convertors, solar collectors, and geothermal energy, into electrical energy. In this chapter we explain the basics of thermal physics, starting from the concepts of thermal energy and heat transfer, and show how they apply to domestic heating systems and the greenhouse effect: vital for maintaining the diversity of life on the planet. Next, we explain how the laws of thermodynamics relate to thermal power cycles and impose a fundamental limit on their thermodynamic efficiency. We then describe various thermodynamic quantities which are useful in explaining how particular energy devices function and in describing chemical reactions. Finally, we give a brief account of geothermal energy and mechanisms for its extraction and utilization.

2.1 Heat and temperature

Temperature is a characteristic of the thermal energy of a body due to the internal motion of molecules. Two bodies in mutual thermal contact are said to be in thermal equilibrium if they are both at the same temperature. Temperature was originally defined in terms of the freezing point and boiling point of water, but the modern definition of temperature is based on the efficiency of an ideal fluid working in a Carnot cycle (Section 2.7), independent of the properties of any particular material.

In general, apart from when a material changes phase (e.g. from solid to liquid), the temperature of any material increases as it absorbs heat. The heat ΔQ required to raise the temperature of unit mass of a material by an amount ΔT is given by

$$\Delta Q = c\Delta T \tag{2.1}$$

The coefficient c is roughly independent of temperature and is called the specific heat of the material. The original unit of thermal energy was the calorie, defined as the energy needed to change the temperature of one gram of liquid water by one degree Celsius, but thermal energy is now usually measured in joules. The energy equivalence of the two units is approximately given by

$$1 \text{ calorie} \approx 4.2 \text{ joules} \tag{2.2}$$

During a change of phase of a material, heat is absorbed and the temperature remains constant. The heat ΔQ required to change the phase of unit mass of material is called the latent heat L. Thus

$$\Delta Q = L \tag{2.3}$$

The latent heat of evaporation (from liquid to gas) is typically up to one order of magnitude larger than the latent heat of fusion (from solid to liquid).

2.2 Heat transfer

There are three basic forms of heat transfer: conduction, convection, and radiation.

Conduction is the transfer of thermal energy within a body due to the random motion of molecules. The average energy of the molecules is proportional to the temperature. Consider a bar of length d and cross-sectional area A, with one end at a fixed temperature T_1 and the other at a fixed temperature T_2, where $T_1 > T_2$. The more energetic molecules at the hot end transfer kinetic energy to the less energetic molecules at the cold end. In the steady state, the rate of flow of heat Q is constant along the length of the bar and is given by Fourier's law of heat conduction:

$$Q = kA\frac{(T_1 - T_2)}{d} \tag{2.4}$$

where k is called the thermal conductivity.

EXAMPLE 2.1

A steel bar of length $1\,\text{m}$ and cross-sectional area $1\,\text{cm}^2$ has one end at $1000\,°\text{C}$ and the other end at $0\,°\text{C}$. If the thermal conductivity of the bar is $50\,\text{W}\,\text{m}^{-1}°\text{C}^{-1}$, calculate the heat flow along the bar in the steady state, ignoring heat losses from the surface.

From eqn (2.4) the heat flow along the bar is given by

$$Q = kA\frac{(T_1 - T_2)}{d} \approx 50 \times 10^{-4} \times \frac{10^3}{1} = 5\,\text{W}$$

It should be noted that eqn (2.4) applies only in the steady state. In practice, it takes time for a solid body to establish a steady-state temperature distribution. For unsteady heat conduction, the characteristic time to establish a steady state is determined by the time t for an isotherm to diffuse a distance x, and is given by

$$t \approx \frac{x^2}{\kappa} \tag{2.5}$$

where $\kappa = \frac{k}{\rho c}$ (m^2 s^{-1}) is called the thermal diffusivity of the material. The validation of eqn (2.5) is outside the scope of this book but can be found in books on the mathematics of heat conduction (e.g. Carslaw and Jaeger, 1959), but its form can be derived by dimensional analysis (see Exercise 2.2).

EXAMPLE 2.2

Suppose the bar in Example 2.1 is initially at $0\,°C$ and that one end is raised to $1000\,°C$ at $t=0$. Estimate the characteristic time for the temperature in the bar to achieve the steady-state temperature at (a) $1\,cm$, (b) $10\,cm$ from the hot end. ($\rho \approx 8 \times 10^3$ kg m^{-3}, $k \approx 3.5 \times 10^2$ Wm^{-1}°C^{-1}, $c \approx 4 \times 10^2$ J kg^{-1}°C^{-1})

The thermal diffusivity of the material is given by $\kappa = \frac{k}{\rho c} \approx 1.1 \times 10^{-4}$ m^2 s^{-1}. Substituting in eqn (2.5) yields (a) $t \approx \frac{x^2}{\kappa} = \frac{(10^{-2})^2}{1.1 \times 10^{-4}} \approx 0.9$ s, (b) $t \approx \frac{(10^{-1})^2}{1.1 \times 10^{-4}} \approx 90$ s

---

Convection is the transport of heat due to the bulk motion of a fluid. Consider a fluid of density ρ and temperature T moving with velocity u. The mass flow per unit area per second is ρu and the thermal energy per unit mass is cT. The rate of flow of heat per unit area by convection is the product of ρu and cT, i.e.

$$\frac{Q}{A} = (\rho u)(cT) = \rho u cT \tag{2.6}$$

When a cold fluid flows over a hot surface (forced convection), the rate of heat transfer from the surface to the fluid is greater than for a stationary fluid. Initially a layer of fluid adjacent to the wall is heated by thermal conduction and this hot fluid is then transported into the body of the fluid away from the surface. As a result, the net heat transfer into the body of the fluid is much larger than that by heat conduction alone.

In forced convection, the rate of heat transfer per unit area is often expressed in the form

$$\frac{Q}{A} = Nu \frac{k(T_s - T_\infty)}{L} = h\Delta T \tag{2.7}$$

where T_s is temperature of the surface, T_∞ is the temperature in the body of the fluid ($\Delta T = T_s - T_\infty$), L is a characteristic length and Nu is a dimensionless parameter known as the Nusselt number. The heat transfer coefficient $h = Nu\,k/L$. The choice of L depends on the geometrical set-up; e.g. for heat transfer from a bar in a cross-flow, it is appropriate to take L to be the radius of the bar. The Nusselt number is the ratio of the convective to the conductive heat flux and is a function of two other non-dimensional parameters: the Prandtl number, $Pr=c\mu/k$, and the Reynolds number, $Re=\rho uL/\mu$, where μ is the coefficient of dynamic (or absolute) viscosity (see Section 4.5). The Prandtl number can be expressed as $Pr=v/\kappa$, where v is the kinematic viscosity ($v=\mu/\rho$, see Section 4.5) or momentum diffusivity and κ is the thermal diffusivity. Pr depends only on the properties of the material (i.e. specific heat c, thermal conductivity k, and coefficient of viscosity μ), whereas the Reynolds number Re also depends on the velocity of the fluid. Re is the ratio of the rate of transfer of momentum flux in the same direction and normal to the direction of flow and its value determines whether the flow is turbulent or laminar. For fluid flow in pipes, $Re=uD/v$, where D is the diameter of the pipe, and the flow is laminar or turbulent when $Re<2300$ or $Re>4000$, respectively, and is transitional for intermediary values. The Reynolds number can also be expressed as the ratio of inertial to viscous forces (see Section 4.5).

The numerical value of the Nusselt number for a given Pr and Re is usually obtained from empirical correlations (see Example 2.3). As an example, for fully developed turbulent flow in a smooth pipe, the Nusselt number Nu is related to the Reynolds number Re and the Prandtl number Pr by the empirical correlation

$$Nu=\frac{1/2\,f\,Re\,Pr}{1+2Re^{-1/8}(Pr-1)}$$

where $f\approx0.08Re^{-1/4}$ is called the 'friction factor'. For example, putting $Re=10^4$ and $Pr=1$ yields $Nu\approx0.04(10^4)^{3/4}=40$; ~40 times the heat transfer due to heat conduction alone.

EXAMPLE 2.3

Water is heated in the economizer (see Section 3.10) of the boiler in a power station from 35 °C to 165 °C. The tube wall is at 360 °C and its diameter D is 50 mm. Calculate the length L of tube required when the flow speed is 1 m s^{-1}.

Mean temperature is 100 °C, where $\rho=961\,\mathrm{kg\,m^{-3}}$, $v=2.93\times10^{-7}\,\mathrm{m^2\,s^{-1}}$, $Pr=1.74$, $k=0.68\,\mathrm{W\,m^{-1}\,°C^{-1}}$, $c=4216\,\mathrm{J\,kg^{-1}\,°C^{-1}}$.

$Re=1\times0.05/(2.93\times10^{-7})=1.71\times10^5$, so the flow is turbulent

$$Nu=\frac{0.04Re^{3/4}Pr}{1+2Re^{-1/8}(Pr-1)}=441,\ \text{so } h=Nu\,k/D=5998$$

The rate of heat input, $h\times$ surface area $\times\Delta T$, must equal the rate of increase of sensible heat in the water, $u\times\rho\times$ cross-sectional area $\times c\times$ increase in temperature, i.e.

$h\times\pi\,D\times L\times(360-100)=u\times\rho\times(\pi\,D^2/4)\times c\times(165-35)$, so

$L=1\times961\times0.05\times4216\times130/(4\times5998\times260)=4.22\,\mathrm{m}$

Radiative heat transfer is the transport of energy by electromagnetic waves. Unlike conduction and convection, heat can be transferred by radiation in a vacuum. The energy radiated per unit area per second (i.e. the power per unit area) from a surface at temperature T is given by the Stefan–Boltzmann law

$$P_e = \varepsilon \sigma T^4 \qquad (2.8)$$

where ε is the emissivity of the surface and $\sigma \approx 5.67 \times 10^{-8}\,\mathrm{W\,m^{-2}\,K^{-4}}$ is the Stefan–Boltzmann constant. ε is a dimensionless number and varies from 0 to 1, depending on the nature of the surface.

Opaque surfaces absorb radiation from the environment. The absorptivity of a surface is the same as its emissivity. The rate of absorption per unit area is $P_a = \varepsilon \sigma T_0^4$, where T_0 is the temperature of the environment. Hence the *net* rate of emission per unit area per second is given by

$$P = P_e - P_a = \varepsilon \sigma (T^4 - T_0^4) \qquad (2.9)$$

A surface that absorbs *all* incident radiation is known as a black body. A good practical approximation to a black body is a cavity with a small pinhole that connects it with the outside environment. Since nearly all of the radiation entering the pinhole is absorbed by the surface inside the cavity before it can escape out of the pinhole, the absorptivity is very close to unity and the energy distribution of the radiation emitted by the pinhole is effectively determined by the temperature of the surface inside the cavity.

The temperature of the outer surface of the Sun determines the flux of radiation incident on the upper atmosphere of the Earth (see Example 2.4 and Chapter 7). Also, radiation is the dominant mode of heat transfer in the furnace of a fossil-fuel power plant (see Section 2.11).

EXAMPLE 2.4

Estimate the solar power per square metre incident at the equator. Assume that the surface temperature of the outer surface of the Sun is 5800 K, the emissivity is unity, the radius of the Sun is 7×10^8 m and the distance from the Sun to the Earth is 1.5×10^8 km.

The total power emitted by the Sun is the power per unit area per second, eqn (2.8), multiplied by the surface area of the Sun, i.e.

$$P_S = \sigma T^4 \times 4\pi r_S^2 \approx (5.67 \times 10^{-8}) \times (5.8 \times 10^3)^4 \times 4 \times 3.14 \times (7 \times 10^8)^2 \approx 3.9 \times 10^{26}\,\mathrm{W}$$

The fraction of solar power incident on $1\,\mathrm{m}^2$ at the equator is the solid angle subtended from the Sun, $\Omega = \dfrac{1}{4\pi d^2} \approx \dfrac{1}{4 \times 3.14 \times (1.5 \times 10^{11})^2} \approx 3.5 \times 10^{-24}$, where d is the distance from the Earth to the Sun. Hence the incident solar power per unit area at the equator is $P_S \Omega \approx (3.9 \times 10^{26}) \times (3.5 \times 10^{-24}) \approx 1.37\,\mathrm{kW\,m^{-2}}$

2.3 **Thermal insulation**

Maintaining buildings at a comfortable temperature is one of the most significant uses of thermal energy, for heating in winter and cooling in summer. In industrialized countries, 35 to 40% of energy consumption is in buildings, and in Europe, 30% is for space and water heating alone. Insulation can save a lot of this energy. The typical heat losses from an old domestic house without good insulation, i.e. with solid walls two bricks in thickness, are shown in Fig. 2.1.

We express the rate of heat transfer through a material in the form

$$Q = UA\,\Delta T = \frac{1}{R}A\,\Delta T \qquad (2.10)$$

where A is the area, U is the thermal conductance or U-value ($\mathrm{Wm^{-2}K^{-1}}$), R is the thermal resistivity or R-value ($\mathrm{m^2\,K\,W^{-1}}$), and ΔT (K) is the temperature difference across the material. In the case of heat transfer by conduction (e.g. through glass or stationary air), it follows from eqn (2.4) that

$$U = \frac{1}{R} = \frac{k}{d} \qquad (2.11)$$

In the UK, the installation of central heating has raised the average temperature in older houses from $12\,^\circ\mathrm{C}$ to $18\,^\circ\mathrm{C}$, with about 60% of total domestic energy being used for space heating.

Domestic heating bills can be significantly reduced by

- installing a more efficient heating system (condensing boilers are over 90% efficient, compared with 70–80% for a conventional boilers, but ground- and air-source heat pumps (see Section 2.10) are generally even better);

- reducing the temperature of the internal environment (and wearing warmer clothing!);

- decreasing the U-value of the building elements, by installing thicker roof insulation, cavity wall insulation, double-glazing, etc.

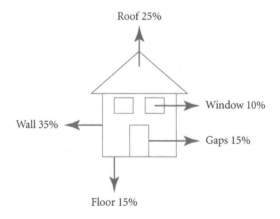

Fig. 2.1 Typical heat losses from an old domestic house.

Table 2.1 Maximum allowed U-values (Wm^{-2}K^{-1}) for building elements

Building element	New buildings	Existing buildings
Cavity wall	0.3	0.7
Pitched roof (insulation at ceiling level)	0.16	0.35
Floor	0.22	0.7
Window	1.8	3.3
Door (unglazed)	3.0	3.3

Table 2.1 gives maximum allowed U-values for building elements for existing and new buildings in the UK, which illustrates how governmental policy can lead to significant reductions in energy usage. Also see Example 2.5 and Exercise 2.4.

EXAMPLE 2.5

The U-value of a window is given by $1/U=1/h_1+1/h_o+d/k$, where $(1/h_1+1/h_o)$ is the thermal resistance of the inner and outer surface air layers ($\sim$0.16 W^{-1} m^2 K) and d/k is the thermal resistance of the window. For a double-glazed unit, with two panes of glass of thickness d_g separated by an air gap d_a, the thermal resistance of the window is given by $2d_g/k_g+d_a/k_a$, where k_g and k_a are the respective thermal conductivities (see Exercise 2.3). Calculate the U-value for the glass component (i.e. neglecting the frame) of (a) a single-glazed and (b) a double-glazed window with dimensions $d_g=d_a=5$ mm, $k_g=1.5$ W m^{-1} K^{-1}, and $k_a=2.1\times10^{-2}$ W m^{-1} K^{-1}.

(a) $1/U=0.16+0.005/1.5=0.163$ W^{-1} m^2 K so $U=6.1$ W m^{-2} K^{-1}

(b) $1/U=0.16+0.01/1.5+0.005/0.021=0.405$ W^{-1} m^2 K so $U=2.5$ W m^{-2} K^{-1}

2.4 Thermal mass

Another way of reducing the energy bill for heating or cooling a building is to utilize the effect of thermal mass. The essential idea is to use the thermal capacity of the building to keep the temperature inside the building between comfortable limits for the occupants.

Fig. 2.2 compares the internal temperatures of a building with low thermal mass (e.g. timber cladding and stud walls) and a building with large thermal mass (e.g. brick or concrete blocks), over a 24h period. The building with low thermal mass closely follows the external temperature, whereas the building with large thermal mass is less responsive and exhibits a significant time lag in peak temperature.

Maintaining a large office or school classroom with large windows at a reasonable temperature on hot summer days can be helped by means of a thick concrete floor, which acts as a large heat sink (and cools down overnight). Thermal mass is also important in winter, when the interior walls, floors, and furnishings absorb solar radiation through windows, and later radiate the heat into the room when the temperature falls.

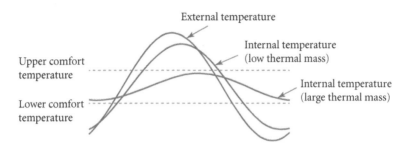

Fig. 2.2 Effect of thermal mass over a 24 h period.

In general, thermal mass is most significant in those parts of the world where there is a large difference between the daytime and night-time temperatures. Traditional houses in hot, dry climates tend to be built with very thick walls. During the day, solar heat is absorbed by the outer layer of the walls without allowing much heat transfer to the interior of the building. At night, when the outside temperature is much lower, the heat is radiated back to the environment and transferred by conduction to the interior.

2.5 Solar thermal energy

Solar thermal energy is the direct conversion of electromagnetic radiation from the Sun into heat, via a solar thermal collector. It is used especially in colder climates to provide supplementary space and water heating for domestic and commercial buildings, by mounting solar panels on Sun-facing surfaces.

The two most common designs of solar panels are the flat plate collector and the evacuated tube collector. In the flat plate collector (Fig. 2.3(a)), the incident solar radiation is absorbed by a flat plate with a black coating. Heat is conducted along the plate to pipes attached to the plate, and is transferred to a fluid (e.g. water or antifreeze) inside the pipes. Flat plate collectors lose a considerable amount of heat to the environment in cold and windy conditions and the efficiency is further reduced by condensation and pollution.

The evacuated tube collector (Fig. 2.3(b)) consists of an array of parallel tubes, with a reflecting surface on the rear of the panel. The incident solar radiation penetrates an outer glass tube and is absorbed on the blackened outer surface of a metal heat pipe, which is surrounded by a vacuum barrier to reduce heat loss to the environment. The heat pipe contains a fluid which evaporates in contact with the hot inner surface of the tube; the vapour rises because of buoyancy forces and the heat is transferred to a heat exchanger. The cooled vapour/liquid from the heat exchanger is then returned to the heat pipe, to complete the cycle.

The schematic design of a typical domestic solar thermal energy installation is shown in Fig. 2.4. The working fluid is contained in a closed circuit, with a controller to regulate the flow. Heat from the solar panel is stored in a thermally insulated tank, which provides hot water for radiators and hot water taps. It should be noted that a back-up heating system is usually required for cloudy days and periods of higher demand. However, the capital cost of

(a)

(b)

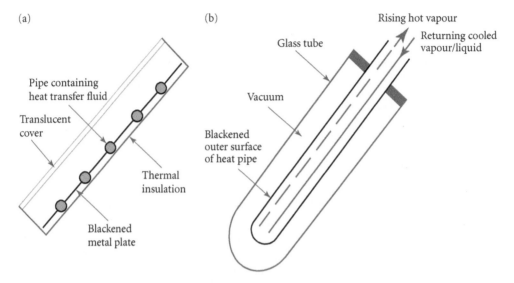

Pipe containing
heat transfer fluid

Translucent
cover

Thermal
insulation

Blackened
metal plate

Glass tube

Vacuum

Blackened
outer surface
of heat pipe

Rising hot vapour

Returning cooled
vapour/liquid

Fig. 2.3 (a) Flat plate collector. (b) Evacuated tube collector.

a domestic solar thermal installation is offset by a significant reduction in water heating bills, typically around 50–80%.

The temperature of the fluid leaving the solar panel is determined by energy conservation. Equating the power absorbed by the collector to the rate at which heat is absorbed by the fluid, we have

$$P_{\text{solar}} = \eta A G_n = \rho Q c T_{\text{out}} - \rho Q c T_{\text{in}} \tag{2.12}$$

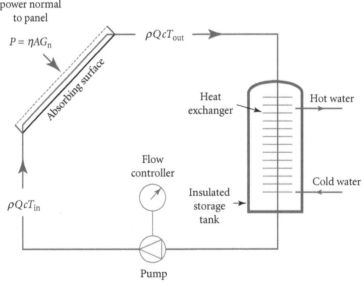

Incident solar
power normal
to panel

$P = \eta A G_n$

$\rho Q c T_{\text{out}}$

Absorbing surface

Heat
exchanger

Hot water

Flow
controller

Insulated
storage
tank

Cold water

$\rho Q c T_{\text{in}}$

Pump

Fig. 2.4 Solar thermal energy system for domestic hot water.

where η is the efficiency of the collector, $A(\text{m}^2)$ is the area of the collector, $G_n(\text{Wm}^{-2})$ is the solar energy per unit area per second normal to the surface of the collector, ρ (kg m^{-3}) and c ($\text{J kg}^{-1}\text{K}^{-1}$) are the density and specific heat of the fluid, and Q (m^3s^{-1}) is the volume flow rate.

Rearranging eqn (2.12), we obtain the output temperature as

$$T_{out} = T_{in} + \frac{\eta A G_n}{\rho Q c} \qquad (2.13)$$

EXAMPLE 2.6

Calculate the time taken to heat 1000 litres of water to $T_{out} = 333\,\text{K}$, given that $T_{in} = 293\,\text{K}$, $\eta = 0.8$, $A = 20\,\text{m}^2$, $G_n = 500\,\text{Wm}^{-2}$, $\rho = 10^3\,\text{kg m}^{-3}$, $c = 4 \times 10^3\,\text{J kg}^{-1}\text{K}^{-1}$.

From eqn (2.12), we have

$$Q = \frac{\eta A G_n}{\rho c (T_{out} - T_{in})} = \frac{(0.8)(20\,\text{m}^2)(5 \times 10^2\ \text{Wm}^{-2})}{(10^3\,\text{kg m}^{-3})(4 \times 10^3\,\text{J kg}^{-1}\text{K}^{-1})(40\,\text{K})} \approx 5 \times 10^{-5}\,\text{m}^3\,\text{s}^{-1} \qquad (2.14)$$

Hence, the time taken to produce 1000 litres of hot water is

$$t \approx \frac{1000 \times 10^{-3}\ \text{m}^3}{(5 \times 10^{-5}\,\text{m}^3\text{s}^{-1})} = 20\,000\ \text{s} = 5\ \text{hours}\ 33\ \text{minutes}$$

There are numerous other applications of solar thermal energy, notably

- solar air thermal collectors, popular in the USA and Canada in buildings where there is an existing ducting system for heating and cooling;
- evaporation ponds, for concentrating dissolved solids (e.g. salt);
- transpired collectors, for heating air to dry food crops, wood, and wood fuels;
- solar cookers, for cooking, drying, and the pasteurization of milk.

2.6 The greenhouse effect

The effect of the atmosphere on the transmission of radiation is very important in determining the temperature at the surface of the Earth. As discussed qualitatively in Chapter 1, the absorption of infrared radiation by the atmosphere gives rise to the greenhouse effect, which raises the Earth's temperature by about 35 °C. In Box 2.1 we describe a simple model of the greenhouse effect.

Box 2.1 The greenhouse effect

As shown in Example 2.4, the solar radiation incident on the Earth's atmosphere has an intensity of $1.37\,\mathrm{kW\,m^{-2}}$; this number is known as the solar constant. A fraction of this radiation, called the albedo, is reflected by clouds in the atmosphere and by the surface of the Earth back into outer space. The albedo is close to 30%. The radiation is absorbed by an area πR^2 of the Earth's surface (i.e. the cross section facing the Sun), where R is the radius of the Earth. As a result, the Earth heats up until it emits as much radiation as it receives. The radiation emitted from a surface at room temperature is in the infrared. We ignore the geothermal heat flux at the surface of the Earth, which is much smaller than the incident solar flux.

We first consider what would happen if the Earth's atmosphere did not absorb any of the incident solar radiation (which is mainly in the visible part of the spectrum), nor any infrared radiation from the Earth's surface. Let A be the value of the albedo and S be the incident solar intensity, and assume the Earth's surface has an emissivity of 1 (i.e. it acts like a black body). In equilibrium, when the Earth's surface is at a temperature T, we have

$$(1-A)S\pi R^2 = 4\pi R^2 \sigma T^4$$

noting that radiation is emitted by the whole of the Earth's surface (area$=4\pi R^2$). Putting $A=0.3$ and $S=1.37\,\mathrm{kW\,m^{-2}}$ gives $T=255\,\mathrm{K}=-18\,°\mathrm{C}$.

We now consider what would happen if the atmosphere absorbed all the infrared radiation emitted by the Earth, but still transmitted all the incident solar radiation.

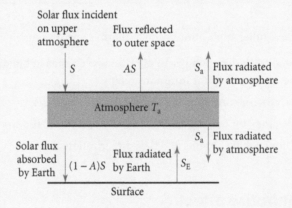

Fig. 2.5 Greenhouse effect.

The atmosphere would then heat up to some temperature T_a such that the energy radiated into space was equal to what it received. Assuming the same albedo as before, and assuming that the emissivity of the atmosphere is $\varepsilon=1$, then T_a is given by

$$(1-A)S\pi R^2 = 4\pi R^2 \sigma T_a^4$$

which yields $T_a = T = 255$ K.

The atmosphere radiates into outer space and down towards the Earth's surface in equal amounts, as shown in Fig. 2.5.

The Earth's surface therefore receives more radiation and heats up to a temperature T_E. For steady-state energy equilibrium, we require

$$(1-A)S\pi R^2 + 4\pi R^2 \sigma T_a^4 = 4\pi R^2 \sigma T_E^4$$

Substituting for $4\pi R^2 \sigma T_a^4$ from above, we obtain $T_E^4 = 2T^4$, so $T_E = 303$ K $= 30\,^\circ$C.

The rise in surface temperature due to the absorption of infrared radiation is called the greenhouse effect. While this model is an oversimplification, it illustrates the importance of the absorption of infrared radiation by the atmosphere. We consider the effect of partial absorption of the visible radiation by the Earth's atmosphere in Exercise 2.12. A better model of the atmosphere is obtained from considering the effects of both convection and the absorption of radiation.

Radiative–convective equilibrium

At a height of about 10 km the density of air is sufficiently low that radiation from that region can escape to space. This is the level, the tropopause, where we therefore expect the temperature to be ~255 K. Below a height of ~10 km (the troposphere), where the pressure of the atmosphere is above ~0.2 bar, convection dominates over radiation and conduction as the method of heat transfer within the atmosphere. A small 'parcel' of air in contact with the ground that is heated by conduction and radiation will rise, because its density is lower than that of the surrounding air. As the parcel rises it expands, since the pressure falls, and it does work. There is no significant heat loss by conduction or radiation so the motion is adiabatic, and the internal energy and hence the temperature of the air falls. For a dry atmosphere the rate of change of temperature with height, the lapse rate, is given by $dT/dz = -g/c_p$, where c_p is the specific heat at constant pressure and g is the Earth's gravitational acceleration. Substituting the value for air at STP (standard temperature and pressure, which is 0°C and 1 bar) gives a lapse rate of ~10 °C km^{-1}. The atmosphere is generally moist and a typical lapse rate is ~5 °C km^{-1}; the smaller rate is caused by the latent heat given out as the moisture condenses when the parcel of air cools on rising.

The effect of increasing the level of greenhouse gases (GHGs) in the atmosphere is to raise the tropopause to offset the increase in absorption of radiation in the atmosphere. The consequence of this is to raise the surface temperature, as illustrated in Fig. 2.6, as the lapse rate stays the same.

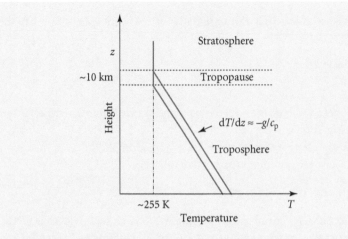

Fig. 2.6 A schematic plot of the change in temperature of the atmosphere with height, illustrating the effect of a change in the height of the tropopause caused by a change in the level of greenhouse gases.

2.7 Laws of thermodynamics and the efficiency of a Carnot cycle

The maximum possible efficiency of a thermal power plant can be obtained from the laws of thermodynamics without needing to consider the details of the fluid flow and heat transfer processes involved in the various stages of the plant.

The first law of thermodynamics is a statement of energy conservation taking thermal energy into account. Consider the system enclosed by the control volume V shown in Fig. 2.7. By energy conservation, the difference between the heat input to the system Q and the work done by the system W is equal to the change in the internal energy ΔU of the system, i.e.

$$Q - W = \Delta U \tag{2.15}$$

(Note the convention that Q and W are positive if heat flows into the system and work is done by the system.)

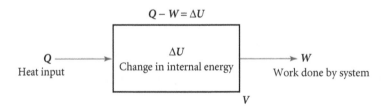

Fig. 2.7 First law of thermodynamics.

In order to convert heat into useful work in a steam power plant, the working fluid undergoes a change of phase at different stages in a closed cycle (i.e. the working fluid is reused), from liquid water, to a two-phase mixture of water and steam, to dry steam, and back to liquid water (Fig. 2.8). The key stages in the cycle are:

1. compressor (also known as the boiler feed pump) work W_{com} done on the system to compress cold water from sub-atmospheric pressure to high pressure;
2. boiler   heat Q_1 added to the system to convert cold water into steam;
3. turbine   work W_t done by the system (i.e. by steam) on the turbine blades;
4. condenser   heat Q_2 lost from the system to the environment in converting steam back to cold water.

After each complete cycle, the working fluid has the same internal energy U, so the net change in internal energy is zero, or $\Delta U = 0$. By the first law of thermodynamics (eqn (2.15)) we have

$$(Q_1 - Q_2) - (W_t - W_{com}) = 0$$

Hence the efficiency of the process is given by

$$\eta = \frac{[\text{net work output}]}{[\text{heat input}]} = \frac{W_t - W_{com}}{Q_1} = \frac{Q_1 - Q_2}{Q_1} = 1 - \frac{Q_2}{Q_1} \qquad (2.16)$$

Thus the efficiency is unity minus the ratio of the heat output in the condenser and the heat input in the boiler. For a perfect system there would be no heat loss in the condenser and all the heat supplied in the boiler would be used to do useful work. However, this heat increases the disorder (i.e. entropy, see Section 2.8) of the steam, which is unchanged when the steam does work passing through the turbine. As a result, some heat must be lost by the working

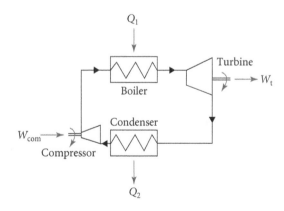

Fig. 2.8 Layout of thermal power plant.

fluid to the environment, to reduce the disorder of the fluid back to its original value, and the amount depends on the temperature of the condenser. Hence $Q_2 > 0$ and $\eta < 1$. It follows that there is an upper limit to the efficiency of a thermal power plant, and the thermal energy wasted heats the external environment.

From a thermodynamic point of view, the fact that the system is less than 100% efficient is a consequence of the second law of thermodynamics: no system operating in a closed cycle can convert all the heat absorbed from a heat reservoir into the same amount of work. Carnot proved that the maximum possible efficiency of a heat engine operating in a closed cycle between two heat reservoirs depends only on the ratio of the absolute temperatures of the reservoirs, i.e.

$$\eta_C = 1 - \frac{T_2}{T_1} \qquad (2.17)$$

where T_1 and T_2 are the absolute temperatures of the upper and lower reservoirs, respectively, measured in kelvins. (See Derivation 2.1.)

Derivation 2.1 Efficiency of a Carnot cycle

Consider an ideal gas with an equation of state $pV = nRT$ operating in the closed cycle $abcda$ shown in Fig. 2.9. Sections ab and cd are isotherms at T_1 and T_2, respectively; bc and da are reversible adiabatics (i.e. no heat transfer with the surroundings).
By the first law of thermodynamics (2.15) we have

$$đQ = dU + pdV = C_v\, dT + pdV \qquad (2.18)$$

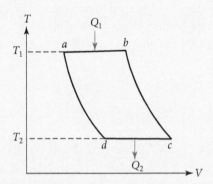

Fig. 2.9 Carnot cycle for perfect gas.

where C_v is the heat capacity at constant volume and $đ$ means an inexact differential as its value depends on the path of integration. Along isotherm ab we put $dT = 0$ in eqn (2.18), so that $đQ = pdV$. The total heat input is

$$Q_1 = \int_{V_a}^{V_b} p\,dV = nRT_1 \int_{V_a}^{V_b} \frac{dV}{V} = nRT_1 \ln(V_b/V_a)$$

Likewise, along isotherm cd the total heat output is $Q_2 = nRT_2 \ln(V_c/V_d)$. Hence

$$\frac{Q_2}{Q_1} = \frac{T_2 \ln(V_c/V_d)}{T_1 \ln(V_b/V_a)} \tag{2.19}$$

Along adiabatic da we put $đQ=0$ in eqn (2.18). Hence $C_v\,dT = -p\,dV$, or

$$C_v \frac{dT}{T} = -nR\frac{dV}{V}, \text{ since } pV = RT$$

Integrating along adiabatic da we obtain $C_v \ln(T_1/T_2) = -nR \ln(V_a/V_d)$.
Likewise, integrating along adiabatic bc gives $C_v \ln(T_1/T_2) = -nR \ln(V_b/V_c)$.
By inspection, $V_a/V_d = V_b/V_c$, or $V_c/V_d = V_b/V_a$, so that eqn (2.19) reduces to $Q_2/Q_1 = T_2/T_1$.
Replacing Q_2/Q_1 by T_2/T_1 in eqn (2.16) yields the efficiency of the Carnot cycle as $\eta_C = 1 - T_2/T_1$.

2.8 Useful thermodynamic quantities

There are six key quantities that are useful in describing the thermodynamics of a thermal power plant: temperature T, pressure p, specific volume v (volume per unit mass, i.e. the reciprocal of density), specific internal energy u, specific enthalpy h, and specific entropy s. In general, only two thermodynamic quantities are needed to completely specify the thermal state of a system.

Note: Unless otherwise stated, we assume that u, h, and s are all per unit mass and use lower case symbols, and drop the prefix *specific* in specific internal energy, etc.

In addition, the Gibbs free energy is very important for determining the conditions under which a chemical reaction takes place spontaneously and the maximum amount of work available, as from a battery. For chemical changes, the Gibbs free energy per mole is normally used.

The concept of internal energy, u, has already been introduced from the first law of thermodynamics (eqn (2.15)). Enthalpy is defined as

$$h = u + pv \tag{2.20}$$

Enthalpy is useful for describing

1. heat transfer at constant pressure (e.g. in boilers and condensers), where the change in enthalpy $h_2 - h_1$ is equal to the heat input Q.

$$h_2 - h_1 = Q \tag{2.21}$$

2. adiabatic ($Q=0$) compression or expansion (e.g. in compressors and turbines), where the net work done by the system is equal to minus the change in enthalpy.

$$W=h_1-h_2 \tag{2.22}$$

(See Derivation 2.2.)

Enthalpy is also useful in chemical reactions where the change in enthalpy is equal to the heat absorbed by the reacting system from its surroundings. The enthalpy change includes the internal energy change and the work associated with changes in volume. For example, the combustion of carbon is an exothermic reaction, i.e. gives out heat, so the change in enthalpy is negative:

$$C(graphite)+O_2(g) \rightarrow CO_2(g) \qquad \Delta H=-393\,kJ$$

The concept of entropy arises from the second law of thermodynamics and is a measure of the degree of disorder of a system. Essentially, there are two types of process whereby a system can change from one state to another: reversible processes and irreversible processes. In a reversible process, both the system and the surroundings can recover their original states. This can be achieved by changing the system so slowly that it remains in quasi-static thermal equilibrium throughout the process. In an irreversible process, however, the system and the surroundings are changed in such a way that they are unable to return to their original states (e.g. a scrambled egg). The change in entropy is defined as

$$\Delta s=\frac{\Delta Q_{rev}}{T} \tag{2.23}$$

where ΔQ_{rev} is the heat supplied reversibly to a system at an absolute temperature T. There is therefore no change in entropy in a reversible adiabatic process.

In a reversible process the total change in entropy of a system and its surroundings is zero, whereas in an irreversible process there is a net increase in entropy. It should be realized that the concept of reversibility is an idealization that is unachievable in practice.

Derivation 2.2

(a) Expansion of a gas at constant pressure

Consider unit mass of gas contained in a thermally conducting piston tube at constant pressure p (Fig. 2.10). Heat is added slowly to the gas (by heating the outside of the tube), so that the gas remains in approximate thermal equilibrium with the surroundings.

Suppose that an elemental amount of heat $đQ$ is required to expand the volume from v to $v+dv$. The work done by the gas on the surroundings is $đW=pdv$. By the first law of thermodynamics (eqn (2.15)), the change in internal energy is given by $du=đQ-pdv$. Since $p=$const., we can put $pdv=d(pv)$ and rewrite du in the form

$$du=đQ-d(pv)$$

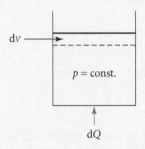

Fig. 2.10 Expansion of gas at constant pressure.

Integrating both sides we obtain the total change in internal energy as

$$u_2 - u_1 = Q - (p_2 v_2 - p_1 v_1)$$

Thus $(u_2 + p_2 v_2) - (u_1 + p_1 v_1) = Q$, or

$$h_2 - h_1 = Q$$

(b) Adiabatic compression or expansion

Figure 2.11 shows a process in which fluid moves from the inlet A to the outlet B and does work by rotating a shaft immersed in the fluid. Suppose the pressure is p_1 at A and p_2 at B. The work done in moving unit mass of fluid through a volume v_1 at A is $p_1 v_1$ and through a volume v_2 at B is $-p_2 v_2$. Thus the net work done by the fluid is $(p_2 v_2 - p_1 v_1)$. If the amount of work done by the shaft is W_s then the total work done by the system is

$$W = W_s + (p_2 v_2 - p_1 v_1)$$

From the first law of thermodynamics (eqn (2.15)), $Q = W + \Delta u$, where Q is the heat flow into the system. Hence, we have $Q = W_s + (p_2 v_2 - p_1 v_1) + u_2 - u_1 = W_s + \Delta h$.

For an infinitesimal process $đQ = đW_s + dh$. The change in enthalpy $dh = du + p\,dv + v\,dp = đQ + v\,dp$, so the work done by the shaft is given by $đW_s = -v\,dp$, or, when the process is adiabatic ($đQ = 0$),

$$W_s = h_1 - h_2$$

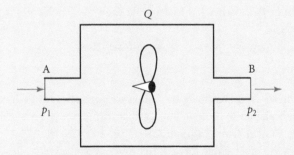

Fig. 2.11 Compression or expansion of a fluid.

2.9 **Chemical reactions**

The statement of the second law of thermodynamics that all practical processes proceed in a way that increases the total entropy means that in chemical processes a reaction will take place if the change in total entropy, ΔS_t, which equals the change in entropy of the reacting system, ΔS, plus that of the surroundings, ΔS_s, increases; i.e.

$$\Delta S_t = \Delta S + \Delta S_s > 0 \tag{2.24}$$

The heat absorbed by the reacting system from its surroundings at constant pressure is the change in enthalpy ΔH in the reaction. This flow of heat causes the entropy of the surroundings to change by $\Delta S_s = -\Delta H / T$, where T is the temperature of the system and surroundings. The condition expressed by eqn (2.24) is therefore equivalent to

$$T\Delta S - \Delta H > 0 \tag{2.25}$$

The Gibbs free energy, G, of a system is defined as

$$G = H - TS \tag{2.26}$$

so, under conditions of constant pressure and temperature, the condition for a reaction to occur spontaneously is given by (since T the absolute temperature is positive):

$$\Delta G = \Delta H - T\Delta S < 0 \tag{2.27}$$

As a first example, consider the change in phase of water from a liquid to a gas at a pressure of 1 bar and a temperature of 298 K. It takes energy to break apart the bonds that hold water molecules together in the liquid phase, and this latent heat of vaporization is given by the change in enthalpy: at 1 bar and 298 K, $\Delta H° = 44.1\,\text{kJ mol}^{-1}$. The entropy change is positive, as water is more disordered in the gas than in the liquid phase, and $\Delta S° = 118.7\,\text{J K}^{-1}\text{mol}^{-1}$. The change in the Gibbs free energy $\Delta G° = 8.58\,\text{kJ mol}^{-1}$, so at 1 bar and 298 K water does not spontaneously change from its liquid to its gas phase. (The standard state symbol ° refers to standard conditions, i.e. 1 bar pressure, but does not imply anything about temperature or phase, which have to be specified.)

We can, however, estimate water's boiling point by assuming that the values for $\Delta S°$ and $\Delta H°$ do not change significantly with temperature, in which case the boiling point is given by the value of T that makes $\Delta H° - T\Delta S° = 0$; i.e. $T = \Delta H° / \Delta S° = 44\,100 / 118.7 = 371.5\,\text{K}$, which is a good estimate. Above the boiling point, water molecules in the liquid absorb more energy from the surroundings than is needed to separate the water molecules, so the vaporization proceeds spontaneously.

As another example, consider the water–gas shift reaction:

$$CO + H_2O \leftrightarrow CO_2 + H_2 \tag{2.28}$$

The value of $\Delta G°$ can be estimated from the values for formation of each component at 1 bar and 298 K from their elements in their most stable forms; these standard values are called $\Delta G_f°$. For example, $CO_2(g)$ from $C(\text{graphite}) + O_2(g)$ has $\Delta G_f° = -394.4\,\text{kJ mol}^{-1}$, while H_2 has $\Delta G_f° = 0$, as H_2 is the most stable form of hydrogen. For the water–gas shift reaction, eqn (2.28), $\Delta G° = -28.6\,\text{kJ mol}^{-1}$ at 1 bar and 298 K, with the enthalpy change $\Delta H° = -41.1\,\text{kJ mol}^{-1}$

and the entropy change $\Delta S° = -42.1\,\text{J}\,\text{K}^{-1}\,\text{mol}^{-1}$. Therefore the forward reaction to produce H_2 does occur spontaneously at 1 bar and 298 K.

We can estimate the temperature at 1 bar when the forward and background reaction rates in eqn (2.28) are equal, i.e. the reaction is in equilibrium, for equal quantities of reactants and products by neglecting the change with temperature of $\Delta H°$ and $\Delta S°$ and finding T such that $\Delta G = \Delta H° - T\Delta S° = 0$. This gives $T = 950\,\text{K}$, close to the observed temperature of ~1080 K. Above this temperature, the reaction will produce an increasing amount of CO and H_2O with a corresponding decrease in the amount of H_2 and CO_2. This causes ΔG to decrease until the change in concentration of the gases corresponds to $\Delta G = 0$ and equilibrium is reached.

The decrease in the Gibbs free energy $-\Delta G$ in a chemical reaction also gives the maximum amount of work that can be extracted from the reaction. In a hydrogen fuel cell (see Section 11.16) the overall reaction is

$$H_2 + \tfrac{1}{2}O_2 \rightarrow H_2O, \quad \Delta H° = -285.8\,\text{kJ}\,\text{mol}^{-1} \tag{2.29}$$

The reaction occurs at constant pressure, so $-\Delta H$ gives the energy released when hydrogen and oxygen combine plus the work done when the volume of the gases decreases. The entropy of the gases decreases in this process, since the number of moles is reduced and the product is a liquid. In a reversible process the entropy of the entire system, reactants and surroundings, remains constant. As a result, an amount of heat equal to $-T\Delta S$, where ΔS is the change in the specific entropy of the gases, is transferred to the surroundings. The amount of energy available as electrical energy is minus the change in the Gibbs free energy $-\Delta G°$, where

$$\Delta G° = \Delta H° - T\Delta S° = (-285.8 + 48.7) = -237.1\,\text{kJ}\,\text{mol}^{-1} \tag{2.30}$$

The Gibbs free energy is therefore a very important quantity as it determines both the maximum amount of work available from a chemical reaction and the conditions under which a reaction occurs spontaneously.

2.9.1 Rate of chemical reactions

A chemical reaction will occur spontaneously, under conditions of constant pressure and temperature, if the change in Gibbs free energy is negative. But its rate will depend on the temperature and pressure of the reactants and on the reaction mechanism. An example of a reaction with a simple mechanism is

$$NO_2 + CO \leftrightarrow NO + CO_2$$

For the forward reaction, the probability that NO_2 and CO collide is proportional to the product of their concentrations $[NO_2][CO]$. As the NO_2 and CO molecules come closer, the interaction between them is at first repulsive and then attractive, i.e. there is a potential barrier that has to be overcome for the reaction to occur. The height of this barrier is called the activation energy E_{fA}. An estimate for the probability that the molecules interact is the probability that they have sufficient kinetic energy to overcome the barrier, which depends on the temperature of the gas as $\exp(-E_{fA}/k_B T)$, where k_B is the Boltzmann constant. The overall forward reaction rate dR_f/dt is then given by

$$dR_f/dT = k[NO_2][CO]$$

where k is the rate constant and is given in this model by $k = A \exp(-E_{fA}/k_B T)$, with A determined by the diameter of the molecules and their relative speed, proportional to $T^{1/2}$. In the transition-state model, the reaction proceeds by the two molecules forming a transient state at the activation energy E_{fA} (i.e. at the top of the potential barrier). The activation energy for the reverse reaction E_{rA} of $NO + CO_2$ forming $NO_2 + CO$ is higher and the difference equals the change in enthalpy in the reaction, i.e. $E_{fA} - E_{rA} = \Delta H$.

Reactions are often more complex, with the reaction proceeding through intermediary stages. An example is the reaction

$$2NO + O_2 \rightarrow 2NO_2$$

Two-molecule collisions are much more frequent than simultaneous three-molecule ones and a possible reaction mechanism is that two NO molecules collide to form the unstable N_2O_2 molecule and this dissociates very quickly back to two NO molecules. At equilibrium, the rate of dissociation of N_2O_2 equals the rate of formation of N_2O_2:

$$2NO \leftrightarrow N_2O_2$$

$[N_2O_2]_e$, the equilibrium value of the concentration $[N_2O_2]$, depends on the concentration of the reacting molecules, i.e. $[NO]^2$. The rate of the slower reaction of N_2O_2 and O_2 will then depend on $[N_2O_2]_e$ and the concentration $[O_2]$, i.e.

$$dR_f/dT = k[NO]^2[O_2]$$

A reaction can be speeded up by using a catalyst to bring together the reactants and to provide a route with a lower activation energy. An example is the use of a metal, such as copper, to facilitate the formation of methanol via the reaction

$$3H_2 + CO_2 \rightarrow CH_3OH + H_2O$$

The interaction of hydrogen molecules with the metal atoms on the surface of the catalyst gives rise to hydrogen atoms bonded to metal atoms, i.e. $H_2 + 2Cu \rightarrow 2Cu\text{-}H$. These reactive hydrogen atoms then combine with carbon dioxide molecules that bind to adjacent metal atoms on the surface to form (via a series of intermediary reactions) methanol.

The activation energy in the reaction of hydrogen and carbon dioxide when both are bonded to the catalyst is lower than when isolated hydrogen and carbon dioxide molecules interact. The catalyst provides an alternative route for the reaction to occur with a lower activation energy. The catalyst is not consumed nor chemically changed in the process.

2.10 Ground- and air-source heat pumps

An important application of a heat engine is to supply work to pump heat either into or out of a building; such an engine is called a heat pump. When used to pump heat to or from the ground or air, the process takes advantage of the fact that the ground, and to a lesser extent the air, is maintained at a relatively constant temperature by heat from the Sun (there is a very small contribution from geothermal energy when using the ground as a source). The temperature of the ground is reasonably constant below a depth of ~10 m (~10 °C in the UK). The amount of work required to pump a quantity of heat across temperature differences of 20–40 °C between the inside of the house and the ground or air is typically a factor of 3–5 times less than the heat transferred. Ground-source pumps draw heat from the ground during the winter, so it can be important to pump heat into the ground during the summer, which occurs when a heat pump is used to both air-condition and heat a building, in order to avoid the ground freezing. The operation of a heat pump is based on the same principle as a refrigerator.

In a refrigerator a fluid at high pressure and at near room temperature expands and cools. The cold fluid is piped through the refrigerator where it cools the contents. The fluid is then compressed to high pressure during which process it heats up. The fluid is cooled back to near room temperature in a heat exchanger, an air-cooled coil on the back of the refrigerator, and the cycle is repeated.

We can approximate the process by a Carnot cycle where heat Q_2 is extracted at a temperature T_2, heat Q_1 is expelled at a higher temperature T_1 (near room temperature), and work W is done by the compressor. $W = Q_1 - Q_2$, and for a Carnot cycle $Q_1/T_1 = Q_2/T_2$, so

$$Q_2 = T_2 W/(T_1 - T_2) \tag{2.31}$$

Thus a factor $T_2/(T_1 - T_2)$ more heat is extracted than the work done by the compressor.

A heat pump works on the same principle. Heat Q is either extracted from or transferred to the building. The ratio Q/W for a heat pump is called the coefficient of performance (COP), and, for an ideal heat pump heating a building, the $COP = T_1/(T_1 - T_2)$; e.g. for $\Delta T \equiv (T_1 - T_2) = 31°C$ and a ground temperature of $T_2 = 6°C = 279 K$, we have $COP = 10$. The actual COPs for heating or cooling units for buildings typically lie between 3 and 4.5. Even allowing for the inefficiency of the heat pump, much less energy is required to heat or cool buildings this way, compared with using direct heating units or cooling units that exhaust heat to the hot air rather than to the colder ground.

Ground-source heat pumps tend to be more efficient than air-source, because the ground is at a more stable and suitable temperature, being generally warmer in the winter and colder in the summer than the surface air temperature, but their capital cost is generally higher. For such pumps, water or a water–antifreeze mixture is circulated through pipes that are buried in the ground at a depth of typically 100–400 ft. There they extract or transfer heat to the ground depending on whether the pipe temperature is lower or higher than that of the ground. Inside the buildings ducts are used transport the hot or cool air throughout the building. The heat pump can also be used to provide domestic hot water.

In the United States nearly 500 000 geothermal units are used for heating and cooling. Over 40% of the electrical energy consumed and nearly 40% of CO_2 emissions in the USA are from space heating and cooling and for water heating in commercial and residential buildings. The US Environmental Protection Agency estimates that 100 000 domestic geothermal heat pump (GHP) units would save about 1.1 million tonnes of carbon over a 20-year period. In Europe there is a mature market for shallow geothermal technology in Sweden and Germany and a developing market in other European countries. The capacity of shallow geothermal installations in Europe grew from 5700 MW_{th} in 2007 to 11 500 MW_{th} in 2010, and is expected to rise to 30 000 MW_{th} by 2020—the capacity factor is ~0.2 (W_{th} = watts thermal). The principal countries using GHP are Sweden, USA, China, Denmark, Norway, Switzerland, Germany, and Canada, with China showing a significant expansion. On a global scale the rate of growth of the annual output of GHP has been very significant—~14.5 PJ y^{-1} in 1995; ~23.25 PJ y^{-1} in 2000; ~87.5 PJ y^{-1} in 2005—with an estimated ~4000 PJ y^{-1} by 2050.

2.11 Geothermal energy

The temperature in the interior of the Earth is around 4000 °C. It is maintained by the generation of heat produced by the radioactive decay of isotopes of heavy nuclei and the heat of crystallization due to the solidification of molten rock. Heat is conducted through the mantle and the average temperature gradient on the Earth's surface is typically around 30 °C km^{-1}. The average continental heat flux is about 65 kW km^{-2}, so an estimate of the sustainable geothermal resource is ~1–2 TW$_{th}$, taking 10–20% of the flux through the world's land area of ~150 million km^2.

Geologically active parts of the world such as Iceland, California, Italy, and New Zealand are close to the interfaces between tectonic plates. Naturally occurring steam jets (geysers) and hot springs up to 350 °C provide a ready source of thermal energy. Even in geologically stable regions of the world, geothermal energy can still be extracted by drilling boreholes to depths of a few kilometres and flushing water through hot rock formations. Up to about 180 °C, geothermal energy is primarily used for district heating, industry, and agricultural purposes, but above ~180 °C it can be used as feed-water heating for electric power plants. In the last two decades, plants with binary cycles, in which a low boiling point fluid (e.g. ammonia) is heated via a heat exchanger and the vapour produced is used to drive a turbine to generate electricity, have been developed that can use sources in the range 100–180 °C.

There are two basic types of rock formation that are suitable for 'mining' geothermal heat: aquifers and hot dry rocks. An aquifer is a layer of porous rock trapped between layers of impermeable rock, e.g. a layer of sandstone tens of metres in thickness. (Aquifers close to the surface provide vast reservoirs of rainwater, which are extracted by water authorities.) Aquifers at depths of 2–3 km are typically at 60–90 °C. Cold water is injected at some point in the aquifer through a borehole. The water flows through the aquifer and absorbs heat from the porous rock (see Section 2.11.1). The hot water is removed through a second borehole. In hot dry rock heat extraction, water is pumped at high pressure through narrow gaps in hot rock formations (see Section 2.11.2). A particularly suitable type of rock is granite, which is found in large blocks, typically 10–100 m in dimension. Granite contains isotopes of heavy nuclei, which release radioactive decay heat and raise the temperature of the rock. As a result, the temperature gradient in granite is higher than that in non-radioactive rock; it is possible to reach high temperatures at lower depths and thereby reduce drilling costs.

Geothermal power is a relatively harmless technology with little environmental impact. There is no carbon dioxide emission (except during the drilling process), although there is sometimes a release of H$_2$S gas and the water already trapped in aquifers can contain toxic heavy metals. As a consequence, it is usual to employ heat exchangers to keep the extracted water separate from that used for district heating or electricity generation. However, there is the possibility of induced seismic activity, which led to the cancellation of a geothermal project in Basel in 2009.

2.11.1 Heat extraction from an aquifer

In aquifer extraction, heat is removed from porous rock situated between layers of impermeable rock. Heat conduction from the impermeable rock above and below the aquifer is usually negligible over the timescale for heat extraction from an aquifer. For simplicity we consider a simple one-dimensional fluid flow model for heat removal from an aquifer (see Fig. 2.12). The actual flow field in the plane of the aquifer is two-dimensional (see Ex. 2.21) but a one-dimensional model gives a useful first approximation and illustrates the main physical features.

Cold water is injected at the inlet borehole $x=0$ and hot water is extracted at the outlet borehole $x=L$. Suppose that the aquifer is initially at temperature T_1 and the cold water at the inlet is at temperature T_0. Hence the heat available per unit volume from the rock is $\rho_r c_r(T_1-T_0)$. The power output of the system is the product of the heat per unit volume gained by the water, $\rho_w c_w(T_1-T_0)$, and the volume flow rate Q, i.e.

$$P=\rho_w c_w(T_1-T_0)Q \tag{2.32}$$

As cold water flows through the aquifer it absorbs heat from the hot porous rock. A 'cold front' moves from the inlet borehole to the outlet borehole over the lifetime of the system. The speed of the cold front v_f is given by

$$v_f=\lambda v_w \tag{2.33}$$

where $\lambda=\rho_w c_w/(1-\varphi)\rho_r c_r$ is a non-dimensional parameter (see Derivation 2.3) and $v_w=Q/A$ is the bulk velocity of the water in the aquifer, where A is its cross-sectional area. φ is the fraction by volume occupied by the pores, known as the porosity. The lifetime of the system is the time taken for the cold front to reach the outlet borehole and is given by

$$t_{life}=\frac{L}{v_f}=\frac{(1-\varphi)\rho_r c_r AL}{\rho_w c_w Q} \tag{2.34}$$

Hence, for a long lifetime it is desirable to have low porosity φ, large heat capacity per unit volume $\rho_r c_r$, large cross-sectional area of aquifer A, low volume flow rate Q, and large spacing

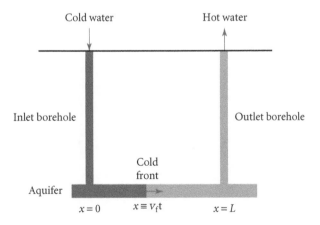

Fig. 2.12 Twin borehole system for heat extraction from aquifer.

L between the two boreholes. Since the total amount of thermal energy available from the system, $E=\rho_r c_r(T_1-T_0)AL$, is fixed, the choice of Q and the lifetime of the system are determined by the economics of the system.

Finally, in order to obtain a given volume flow rate Q it is necessary to apply a pressure drop Δp between the boreholes. By Darcy's law, the volume flow rate Q through a slab of porous rock of cross-sectional area A and thickness L is given by

$$Q=kA\frac{\Delta p}{L} \tag{2.35}$$

where k is a constant known as the permeability. Rearranging eqn (2.35), the pressure drop required for a given volume flow rate Q is given by

$$\Delta p=\frac{QL}{kA} \tag{2.36}$$

Derivation 2.3 Velocity of the cold front

Suppose the cold front moves a distance δx in a time interval δt (see Fig. 2.13). The volume of rock exposed is $\delta V_r=(1-\varphi)A\delta x$, where A is the cross-sectional area of the cold front. The amount of heat removed from the element is given by $\delta h_r=\rho_r c_r(T_1-T_0)(1-\varphi)A\delta t$. The volume of water passing through the element in a small time interval δt is $\delta V_w=Q\delta t$ and the heat gained per unit volume by the water is $\delta h_w=\rho_w c_w(T_1-T_0)Q\delta t$. By energy conservation, the heat lost by the rock is equal to the heat gained by the water, so that

$$\rho_r c_r(T_1-T_0)(1-\varphi)A\delta x=\rho_w c_w(T_1-T_0)Q\delta t$$

Rearranging, we obtain the velocity of the cold front as

$$v_f=\frac{dx}{dt}=\frac{\rho_w c_w}{(1-\varphi)\rho_r c_r}\frac{Q}{A}=\lambda v_w$$

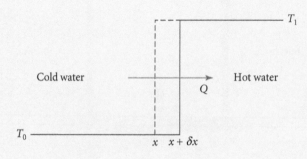

Fig. 2.13 Movement of cold front.

EXAMPLE 2.7 Heat extraction from an aquifer

A sandstone aquifer at 70 °C is 20 m thick and 100 m wide. The density, specific heat, porosity, and permeability are 2.3×10^3 kg m^{-3}, 1000 J kg^{-1} °C, 0.02, and 2×10^{-9} m^3 kg^{-1}, respectively. Estimate the volume flow rate needed to generate a power output of 1 MW, the lifetime of the system, and the pressure drop required for a borehole separation of 1 km. (Assume the water at inlet is at 10 °C, $\rho_w = 10^3$ kg m^{-3}, $c_w = 4000$ J kg^{-1} °C^{-1}.)

From eqns (2.32), (2.34), and (2.36) we have

$$Q = \frac{P}{\rho_w c_w (T_r - T_w)} = \frac{10^6}{(10^3)(4 \times 10^3)(70-10)} \approx 4 \times 10^{-3} \text{ m}^3 \text{s}^{-1}$$

$$t_{\text{life}} = \frac{(1-\varphi)\rho_r c_r A}{\rho_w c_w Q} L \approx \frac{(1-0.02)(2.3 \times 10^3)(10^3)(20 \times 100)}{(10^3)(4 \times 10^3)(4 \times 10^{-3})}(10^3) \approx 2.8 \times 10^8 \text{ s} \approx 9 \text{ years}$$

$$\Delta p = \frac{QL}{kA} = \frac{(4 \times 10^{-3})(10^3)}{(2 \times 10^{-9})(2 \times 10^3)} = 10^6 \text{ N m}^{-2} = 10 \text{ bar}$$

2.11.2 Heat extraction from hot dry rock

In a hot dry rock geothermal system, water is pumped at high pressure through natural fissures in the rock. In both aquifer and hot dry rock extraction it is usual to fracture the rocks using controlled explosions in the vicinity of the inlet and outlet boreholes, in order to reduce the overall pressure drop across the system. The water absorbs heat from the surface of the adjacent rock and the heat is transported by the fluid to the outlet borehole (Fig. 2.14). Heat is lost from the surrounding rock by conduction. The heat flux to the water decreases with time as the layer of cooled rock thickens. As in the case of an aquifer, a 'cold front' moves from the inlet to the outlet, but the temperature behind the front is more diffuse because some heat continues to be supplied from the rock behind the cold front.

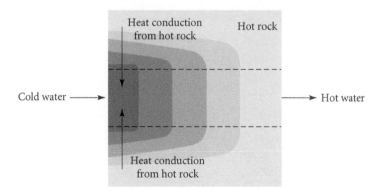

Fig. 2.14 Heat extraction from hot dry rock.

In 2006 a group at MIT estimated that there would be the potential of the order of 100 GWe of enhanced geothermal systems (EGS) within the USA by 2050 (MIT2006). There are demonstration EGS schemes in Australia, France, and Germany. The cost of drilling boreholes to depths of several kilometres is currently very high and the costs associated with exploration and drilling can account for ~40% of the total cost. A breakthrough in drilling technology (from exponential $ per m to linear $ per m) would make EGS more cost-competitive. One possibility being developed is hydrothermal spallation, which utilizes water at high temperature and pressure to penetrate through hard rock, and might increase drilling rates by ~50%. Also the nature of the rock formation and rock temperature can be unknown in advance. A hot dry rock project at Camborne in Cornwall (UK) was eventually abandoned in 1989 as a result of unforeseen problems in the rock formation. However, there has been progress in the development of techniques for geothermal exploitation and in the understanding of the geothermal sources.

2.11.3 Potential of geothermal power

Geothermal energy is a largely untapped source of energy, with only a small fraction of the estimated available resource (approximately 1–2 TWe) currently being exploited. The cost of geothermal power plants is generally higher than that of conventional fossil fuel plants but the cost of electricity can be competitive with natural gas plants, where the cost of fuel is approximately two-thirds of the cost. It can also compete with wind and nuclear generation, with the California Energy Commission (2007) estimating levelized costs at ~9¢ per kWh for geothermal-binary, wind, and nuclear.

Table 2.2 shows the top ten countries exploiting geothermal energy for electricity production and for direct use in 2005.

Table 2.2 Geothermal energy: electricity and direct use in 2005

Country	Electricity production (GWh y^{-1})	Country	Direct use (GWh y^{-1})
USA	17 917	China	12 605
Philippines	9253	Sweden	10 000
Mexico	6282	USA	8678
Indonesia	6085	Turkey	6900
Italy	5340	Iceland	6806
Japan	3467	Japan	2862
New Zealand	2774	Hungary	2206
Iceland	1483	Italy	2098
Costa Rica	1145	New Zealand	1968
Kenya	1088	Brazil	1840
*World**	57 000	*World**	76 000

*end of 2004
Source: Bertani2006.

Geothermal electricity is produced in 24 countries and in five of these the percentage is 15–22% of their total demand. In 72 countries there is direct application of geothermal energy, e.g. for heating. The estimated economic geothermal capacity by 2050 is ~70 GWe using conventional systems and ~140 GWe including EGS, from its present amount of 10 GWe (2004) with a capacity factor of ~70%. The accessible potential by 2050 is therefore ~100 GWe of continuous power. Its low environmental impact and base-load capability (being always available) make it an attractive low-carbon source of heat and electricity.

The amount of thermal power in direct applications of geothermal energy in 2004 was ~9 GW$_{th}$, of which ~3 GW$_{th}$ was in heat pump applications (see Section 2.10). The future of geothermal energy technology depends on the economics of fossil fuels and on future policy towards tackling global warming. The development of low-temperature binary cycle plants has increased the potential resource and if EGS proves cost-effective then large geothermal resources of 1–2 TWe would be available globally. Geothermal energy provides energy security and may soon be competitive with fossil fuel generation, and with sufficient investment could make a very useful contribution to overall energy production.

SUMMARY

- For *steady* heat conduction, Fourier's law states that the heat flow is proportional to the temperature gradient, i.e. $Q = kA \dfrac{(T_1 - T_2)}{d}$.

- For *unsteady* heat conduction, isotherms move a characteristic distance of order $x \approx (\kappa t)^{\frac{1}{2}}$ in a time t.

- In forced convection the rate of heat transfer per unit area is of the form $\dfrac{Q}{A} = Nu \dfrac{k(T_s - T_\infty)}{L}$ where Nu is a dimensionless parameter known as the Nusselt number.

- Improving the thermal insulation of houses is essential for reducing the demand for energy: 30–40% of energy is used in buildings in developed countries.

- In radiative heat transfer the rate at which energy is radiated from a surface of unit area is given by the Stefan–Boltzmann law, $P_e = \varepsilon \sigma T^4$.

- The absorption of infrared radiation by the atmosphere gives rise to the greenhouse effect, which raises the Earth's surface temperature by about 35 °C.

- The first law of thermodynamics, $Q - W = \Delta U$, expresses energy conservation in which thermal energy is taken into account.

- According to the second law of thermodynamics, a system operating in a closed cycle cannot convert all the heat absorbed from a heat reservoir into the same amount of work.

- The maximum possible efficiency of a closed cycle operating between two heat reservoirs T_1 and T_2 (absolute temperatures) is the Carnot efficiency, $\eta_C = 1 - \dfrac{T_2}{T_1}$.

- The concept of entropy arises from the second law of thermodynamics and is a measure of the degree of disorder of a system. There is no change in entropy in a reversible adiabatic process.

- Enthalpy $h = u + pv$ is a useful thermodynamic quantity for describing: (a) heat transfer at constant pressure (e.g. in boilers and condensers), in which the change in enthalpy $h_2 - h_1$ is equal to the heat input Q; (b) adiabatic compression or expansion (e.g. in compressors and turbines), in which the net work done on the system is equal to the change in enthalpy.

- The Gibbs free energy is important for determining the conditions under which a chemical reaction takes place spontaneously and the maximum amount of work available, as from a battery.

- Ground- and air-source heat pumps are a very effective way of heating and cooling buildings and could make a significant contribution to reducing carbon emissions.

- Geothermal energy is a potentially large source of low-carbon electricity (1–2 TWe), which would be available in all countries using EGS technology. It can provide energy security, and the estimated accessible potential by 2050 for electricity generation is ~100 GWe.

FURTHER READING

Andrews, J.G., Richardson, S.W., and White, A.A.L. (1981). Flushing geothermal heat from moderately permeable sediments. *J. Geophysical Research* **86**(B10), 9439–50. Mathematical model of two-dimensional heat flow in an aquifer.

Bertani, R. 2009. Long-term projections of geothermal-electric development in the world, GeoTHERM conference, Offenburg 2009.

Blundell, S. and Blundell, K. (2006), *Concepts in thermal physics*. Oxford University Press, Oxford. A very good textbook on thermal physics.

Carslaw, H.S. and Jaeger, J.C. (1959). *Conduction of heat in solids*, 2nd edn. Clarendon Press, Oxford. Comprehensive treatment of mathematics of heat conduction.

Rogers, G. and Mayhew, Y. (1992). *Engineering thermodynamics*, 4th edn. Longman, Harlow. Engineering approach to thermodynamics with many practical examples.

Taylor, F.W. (2005). *Elementary climate physics*. Oxford University Press, Oxford. Good introduction.

Zemansky, M. and Dittman, R. (1997). *Heat and thermodynamics*. McGraw-Hill, New York. Standard physics textbook on thermodynamics.

WEB LINKS

www.chem1.com/acad/webtext/thermeq/index.html Useful introduction to the thermodynamics of chemical equilibrium.

www.communities.gov.uk/documents/planningandbuilding UK data source giving permitted U-values of building elements.

www.geothermal-energy.org/ Very informative source on geothermal energy.

geothermal.inel.gov/publications/future_of_geothermal_energy.pdf MIT report on the future of geothermal energy (MIT2006).

www.worldenergy.org Neutral source of data and current developments.

www.geothermal-energy.org/pdf/IGAstandard/ISS/2009Slovakia/I.1.Bertani.pdf (Bertani2006).

LIST OF MAIN SYMBOLS

A	area	t	time
c	specific heat	T	temperature
C_p	heat capacity at constant pressure	U	internal energy, thermal conductance
C_v	heat capacity at constant volume	H	enthalpy
h	specific enthalpy, heat transfer coefficient	G	Gibbs free energy
k	thermal conductivity, permeability	S	entropy
Nu	Nusselt number	W	work
P	power	ε	emissivity
p	pressure	η	efficiency
Pr	Prandtl number	κ	thermal diffusivity
Q	heat or volume flow rate (i.e. in an aquifer)	μ	coefficient of viscosity
R	universal gas constant, R-value	ρ	density
Re	Reynolds number	σ	Stefan–Boltzmann constant
		φ	porosity

EXERCISES

2.1 Consider a proposal to extract heat directly from the Earth's core by drilling a cylindrical shaft of radius 1 m and depth 100 km through the Earth's mantle and filling it with copper. Estimate the power output assuming the temperature difference is 1000 °C and the thermal conductivity is $3.5 \times 10^2\,\mathrm{W\,m^{-1}\,K^{-1}}$.

2.2 Derive the form of eqn (2.5) using dimensional analysis. Estimate the characteristic timescale for heat to conduct through a heat shield of thickness 1 cm. ($\rho = 5 \times 10^3\,\mathrm{kg\,m^{-3}}$, $k \approx 10^{-1}\,\mathrm{W\,m^{-1}\,°C^{-1}}$, $c \approx 10^3\,\mathrm{J\,kg^{-1}\,°C^{-1}}$.)

2.3 A composite wall is made of three layers of material with thicknesses and conductivities of $d_1, k_1, d_2, k_2, d_3, k_3$, respectively. Using eqn (2.4), show that the thermal resistivity of the wall R is given by $R = d_1/k_1 + d_2/k_2 + d_3/k_3$.

2.4 Calculate the U-value ($\mathrm{W\,m^{-2}\,K^{-1}}$) of a thermal block wall of thickness 100 mm and thermal conductivity $0.132\,\mathrm{W\,m^{-1}\,K^{-1}}$, assuming that the thermal resistance of the air layers on the inner and outer surfaces are $0.12\,\mathrm{m^2\,K\,W^{-1}}$ and $0.06\,\mathrm{m^2\,K\,W^{-1}}$, respectively.

2.5 An older domestic property loses 25% of its heat through the roof and 10% through the windows. Estimate the percentage decrease in the energy used for space heating by (a) replacing the existing boiler (75% efficient) by a condensing boiler (95% efficient), (b) decreasing the U-value of the roof insulation by 60% and the windows by 70% (by installing double-glazing), (c) reducing the average temperature difference between inside and outside from 12 °C to 10 °C (by turning down the air thermostat).

2.6 The heat loss through air movement is given by $Q = V\rho c_\mathrm{p}(T_\mathrm{i} - T_\mathrm{o})$, where V is the air volume exchanged per hour, ρc_p is the heat capacity of air ($0.018\,\mathrm{Btu\,°F^{-1}\,ft^{-3}}$), and T_i and T_o are the air temperatures inside and outside the house. Calculate the heat loss for a house with 2000 square feet of floor space and 8 ft ceilings for $T_\mathrm{i} = 65\,°\mathrm{F}$ and $T_\mathrm{o} = 10\,°\mathrm{F}$. Assume the number of air changes per hour is 0.5.

2.7 A useful measure of the space heating requirement of buildings in any given location is the number of degree-days. It is assumed that buildings are heated only when the ambient temperature falls below a nominal base temperature, typically 18 °C (60 °F). The number of degree-days in a given month is then defined as the difference between the base temperature and the average ambient temperature over a whole month, multiplied by the number of days in the month; if the difference is negative then there are no degree-days in that month. (The unit of temperature is normally chosen to be appropriate to the country concerned.) For example, for a location in the USA in January with a maximum temperature of 42 °F and minimum temperature of 34 °F the number of degree-days is $\sim[60 - (42 + 34)/2] \times 31 = 682$. Using this definition, calculate the total number of degree-days over a whole year for (a) Long Island and (b) San Francisco, using the data in the tables below.

Long Island	Jan	Feb	Mar	Apr	May	Jun	Jul	Aug	Sep	Oct	Nov	Dec
High °F	37	40	50	61	71	80	85	84	76	65	54	42
Low °F	25	26	34	44	54	64	68	67	60	48	41	30

San Francisco	Jan	Feb	Mar	Apr	May	Jun	Jul	Aug	Sep	Oct	Nov	Dec
High °F	55	58	60	64	66	70	71	72	74	70	62	56
Low °F	41	45	45	47	48	52	54	55	55	51	47	42

2.8 A house in Pittsburgh designed for an outdoor temperature of 4 °F (the temperature is only lower that this ~1% of the time) has a heating load of $36\,\mathrm{MJ\,h^{-1}}$ for an inside temperature of 65 °F. Calculate the heating required (in kWh) for a year with 5500 degree-days (for a nominal temperature of 65 °F).

2.9 A solar concentrator produces a heat flux of $2500\,\mathrm{W\,m^{-2}}$ on the outside of a tube of diameter 50 mm. Water flows through the tube at a rate of $0.015\,\mathrm{kg\,s^{-1}}$. If the water temperature at the inlet is 15 °C, what length of pipe is required to produce water at a temperature of 85°C? (Water at 50°C has: $\rho=990\,\text{kg}\,\text{m}^{-3}$, $k=0.64\,\text{W}\,\text{m}^{-1}\,\text{K}^{-1}$, $c=4180\,\text{J}\,\text{kg}^{-1}\,\text{K}^{-1}$.)

2.10 Redo the calculation in Example 2.3 for a tube with a diameter of 10 mm, with the rest of the data unchanged.

2.11 Assess the relative merits of the materials given in the following table from the point of view of heat storage in building construction.

	Melting point °C	Latent heat of fusion kJ kg^{-1}
Potassium fluoride tetrahydrate	18.5	231
Zinc nitrate hexahydrate	36.4	147
Butyl stearate	19	140
1-dodecanol	26	200

Source: Feldman et al. (1991). *Solar Energy Materials* **22**, 231–42.

2.12* Consider the following simple model to describe solar radiation incident on the Earth, in which the atmosphere is included. A fraction f of the incident solar flux is absorbed by the atmosphere and a fraction A is reflected, the rest being absorbed by the Earth. Assume that the only heat transfer to the atmosphere from the Earth is by radiation and that $f=0.25$ and $A=0.3$. Find the temperature T_E of the surface of the Earth for radiative equilibrium.

2.13 Using the empirical correlation for turbulent flow in a pipe, $Nu=\dfrac{f}{2}\dfrac{RePr}{1+2Re^{-1/8}(Pr-1)}$, where $f\approx0.08Re^{-1/4}$, calculate the Reynolds number Re required to give a Nusselt number of $Nu=100$ for a fluid with a Prandtl number $Pr=3.5$.

2.14 Estimate the power radiated from a black body with a surface area of 1 cm^2 at a temperature of 1000 °C.

2.15 Discuss whether the caloric theory is consistent with (a) Fourier's law of heat conduction and (b) the first law of thermodynamics.

2.16* Two systems are in thermal contact. Energy (heat) will flow from one to the other until they are in thermal equilibrium, i.e. at the same temperature. Show that the condition that the entropy S of the combined system is a maximum is given by $\partial S_1/\partial U_1=\partial S_2/\partial U_2\equiv1/T$, where T is the equilibrium temperature and $S=S_1+S_2$, $U=U_1+U_2$. Hence show that $\Delta Q=T\Delta S$ and deduce that in a Carnot cycle $Q_1/T_1=Q_2/T_2$.

2.17 The temperature in an endothermic reaction is increased. In what direction will this change shift the reaction?

2.18 The water shift reaction $CO(g)+H_2O(g)\leftrightarrow CO_2(g)+H_2(g)$ is in equilibrium. Will the concentration of H_2 be increased if (a) a catalyst is added? (b) the concentration of CO is increased? (c) the temperature is increased?

2.19 Consider an aquifer at an initial temperature of 100 °C. The aquifer data are: thickness 50 m, width 100 m, density $3\times10^3\,\text{kg}\,\text{m}^{-3}$, specific heat $1500\,\text{J}\,\text{kg}^{-1}\,°\text{C}^{-1}$, porosity 0.01, permeability $5\times10^{-9}\,\text{m}^3\,\text{kg}^{-1}$. Estimate the required separation of the boreholes and the pressure drop in order to produce an output of 5 MW of heat for 10 years. Assume the water at inlet is at 5 °C (ρ_w=1 kg m^{-3}, c_w=4000 J kg^{-1}°C^{-1}).

2.20 Derive an expression for the cost of drilling a borehole as a function of depth. Assume that the cost of drilling is independent of depth but the cost of lifting the rock material from the borehole increases linearly with depth.

2.21* Consider a two-dimensional aquifer of thickness d, with an inlet borehole at A ($x=-a$, $y=0$) and outlet borehole at B ($x=+a$, $y=0$). The velocity field of the water in the aquifer is given by

$$u(x,y)=\frac{Q}{2\pi d}\left(-\frac{1}{r_1}+\frac{1}{r_2}\right)$$

where

$$r_1=\sqrt{(x-a)^2+y^2} \quad r_2=\sqrt{(x+a)^2+y^2}$$

Derive an expression for the time taken for cold water from the inlet borehole to reach the outlet borehole.

2.22 An EGS 1 GWe power station extracts heat from a volume of 100 km^3 of granite that initially is at a temperature of 200 °C. Estimate how long the power station could operate before the rock temperature has fallen to 175 °C (Density of granite=2750 kg m^{-3}; specific heat of granite=0.79 kJ kg^{-1}°C^{-1}.)

3 Energy from fossil fuels

Introduction

We now describe the different forms of fossil fuel, their availability, their combustion, and the thermo-dynamic cycles used in various types of fossil fuel power stations. We also discuss technologies under development for mitigating the impact of greenhouse gases on global warming due to emissions from fossil fuel stations.

3.1 Coal

Coal is a carbon-rich solid which originated about 300 million years ago in vast swamps containing large trees and leafy plants. The remains of this vegetable matter accumulated on the bottom of the swamps and turned into peat. This peat was then covered by layers of soil and sandy material over many millions of years, subjecting the peat to intense heat and pressure, thereby driving out the water and transforming it into a solid rock-like material: coal. Coal is a combustible material, and is graded according to its carbon content, varying from 60% to 75% for lignite, to 75% to 90% for sub-bituminous and bituminous coals, to over 90% for anthracite. Coal also contains other combustible elements, including hydrogen and sulfur, and various incombustible elements, e.g. nitrogen, water, and ash-forming minerals.

Before it can be burned in power stations, coal has to be pulverized into a fine powder (which is a potential explosion risk). Also, for lignite and bituminous coals, the carbon content is often enriched prior to combustion by removing moisture and some pollutants, which adds to the cost of electricity production.

There are major environmental issues associated with coal-fired power stations, notably global warming, due to the release of carbon dioxide into the atmosphere, and acid rain, from coals with a high sulfur content, which causes damage to buildings. There are also serious health issues, including lung cancer, heavy metal poisoning, and radiation exposure from fly ash. The World Health Organization has estimated that there are 1.3 million premature deaths per year from urban outdoor pollution.

3.2 Crude oil and natural gas

Crude oil originated about 300 million years ago, from micro-organisms in the sea that used sunlight to create energy-rich hydrocarbons. The remains of these organisms settled on the seabed, forming a dense sludge. This sludge became covered by layers of sand and mud, which were transformed into sedimentary rock over many millions of years. The intense pressure due to this rock and the heat from the interior of the Earth eventually converted the sludge into a highly viscous oil. Sweet crude oil is sulfur-free and is the most economic, whereas sour crude oil contains sulfur, which needs to be removed. Oil is refined into petrol (gasoline), jet fuel, diesel oil, fuel oil, asphalt, lubricants, and kerosene, and is used in the manufacture of plastics, fertilizers, cosmetics, foodstuffs, and many other products used in the modern world. However, only a comparatively small fraction of oil is used for electricity production.

In those geological formations where the temperature was high enough to crack the oil, pockets of natural gas were formed, trapped between layers of impervious rock. The largest proven reserves of natural gas are in Russia and the Middle East. Natural gas is a mixture of methane (CH_4) and smaller proportions of ethane (C_2H_6) and other hydrocarbons, together with water vapour, carbon dioxide, and various impurities. In the oilfields of the nineteenth and early twentieth century, natural gas was generally regarded as a waste product, and was burned off ('flaring'). It is now regarded as a valuable resource, and is used for electricity generation, space heating of buildings, transportation, and industrial processes. Natural gas is burned in gas turbines to generate electricity, in particular for meeting peak-load demand, and gas turbines are also used in combination with steam turbines in combined cycle power stations. Gas turbines operate at much higher temperatures than steam turbines. Also, natural gas produces about 45% less carbon dioxide than the equivalent amount of coal.

3.3 Unconventional oil and unconventional gas

Oils and gases are described as conventional or unconventional, according to how they can be extracted from the ground and how much treatment they require to bring them up to refinery specification. Thus, crude oil and natural gas are conventional, in that they are easy to extract and do not require much treatment, whereas unconventional oils, notably shale oil, heavy oils, and bitumen, are more difficult to extract and require considerable extra treatment, which significantly adds to the cost. In the past, apart from periods of high oil prices, e.g. during the oil price shocks of the 1970s, unconventional oils were regarded as uneconomic.

As conventional oilfields become exhausted in the coming decades, unconventional oils will become more economically viable.

Shale oil is the oil derived from oil shale—a sedimentary rock containing a high proportion of an oily material called kerogen. In the most common process, the oil shale is brought to the surface, where it is crushed in order to increase the surface area, and then heated for a long period in an oxygen-free environment to separate the condensable gases, which are condensed to form shale oil, and the non-condensable shale gas, which is combustible. Shale gas can also be extracted *in situ* by hydraulic fracturing of the oil shale formations (fracking), but this is a controversial technology since it is believed by some experts to be a potential cause of earthquakes. There are serious environmental issues associated with unconventional oils and gases, notably pollution of groundwater by toxic elements in the waste materials, disposal of spoil, enormous water usage, and larger emissions of greenhouse gas than for conventional fossil fuels.

3.4 **Fossil fuel production and reserves**

The proven reserves in different regions of the world of crude oil, natural gas, and coal at the end of 2010 and the reserve to present production per annum ratio R/P are shown in Table 3.1, where the proven reserves are 'generally taken to be those quantities that geological and engineering information indicates with reasonable certainty can be recovered in the future from known reservoirs under existing economic and operating conditions' (BP2011). The R/P ratio therefore gives a rough estimate of the number of years of supply that are left and shows that Africa in particular has relatively small fossil fuel reserves.

From the R/P ratios, the global production in 2010 of oil, gas, and coal was 4.09 Gt, 2.19 Tm^3, and 7.30 Gt, respectively, with the main producers' percentage shares of global production and reserve shown in Table 3.2. The ratio (%R)/(%P) is proportional to the R/P ratio. For oil, both the USA and China have ~10 years of reserves left at present production levels; for gas, both the USA and Canada have ~12 years; and for coal, China has 35 years, and Indonesia 18 years.

According to Hubbert peak theory, the rate of production of any particular fossil fuel follows a bell-shaped curve with time, the peak rate of production being known as the Hubbert

Table 3.1 Proven reserves and R/P values for oil, gas, and coal (189 Gt of oil is equivalent to 1380 Gb)

Region	Oil (%)	Oil R/P (y)	Gas (%)	Gas R/P (y)	Coal (%)	Coal R/P (y)
North America	5.4	148	5.3	13.0	28.5	231
South & Central America	17.3	93.9	40	45.9	1.5	148
Europe & Eurasia	10.1	21.7	33.7	60.5	35.4	257
Middle East	54.4	81.9	40.5	>100	0.1	>500
Africa	9.5	35.8	7.9	70.5	3.7	127
Asia Pacific	3.3	14.8	8.7	32.8	30.9	57
World	189 Gt	46.2	187 Tm^3	58.6	861 Gt	118

Source: BP2011.

Table 3.2 The main producers of oil, gas, and coal, and the share of global production (%P) and reserve (%R)

Oil	%P	%R	Gas	%P	%R	Coal	%P	%R
Russia	13	6	USA	19	4	China	48	13
Saudi Arabia	12	19	Russia	18	24	USA	15	28
USA	9	2	Canada	5	0.9	Australia	6	9
Iran	5	10	Iran	4	16	India	6	7
China	5	1.1	Qatar	4	14	Indonesia	5	0.6

Source: BP2011.

peak. Prior to the peak, the production rate increases as new fields are discovered and the infrastructure to exploit the fuel is developed. After the peak the production rate declines as the resource becomes scarcer and more expensive to extract.

Hubbert was an American geologist who in 1956 argued that the amount of oil extracted would follow an approximately sigmoid (logistic) curve and accurately predicted that the maximum production of oil in the USA would peak between 1965 and 1970.

In the case of oil, discoveries of conventional oil have been declining for the last 40 years, as shown in Fig. 3.1. This figure gives the number of so-called giant (>0.5 Gb or >10^5 b d^{-1} for >1 y) oilfields that accounted for more than 60% of global production in 2005, and their volume found within the time periods indicated. These are conventional oilfields, which means that the oil can largely be recovered from the reservoir as a free-flowing liquid.

By 2030 the output from existing fields could be down by over 50% and the world will eventually become reliant on unconventional sources of oil at ever increasing costs of production. The reserve of a resource expands as technology improves, and also depends on accessibility, e.g. whether the deposit is under a National Park. It is also very much a function of cost, as is illustrated in Fig. 3.2 which shows IEA estimates of the economic price versus cumulative production of oil. The estimated amount of unconventional oil can be seen to increase the global reserves of oil by approximately a factor of 2. From an assessment of 32 countries the

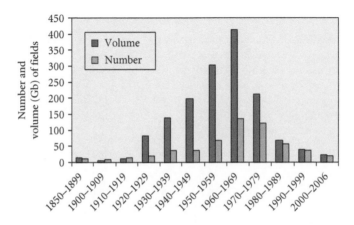

Fig. 3.1 Number and volume of giant oilfields discovered between 1850 and 2006. *Source:* Oilfields.

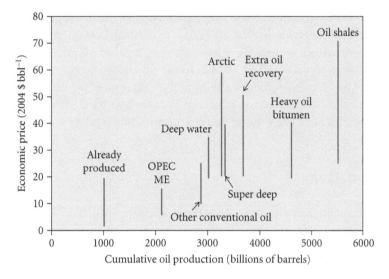

Fig. 3.2 Economic price and cumulative production of conventional and unconventional oils. *Source:* IEA2005.

IEA has estimated that shale gas reserves are similar to the proven reserves of conventional sources of natural gas, i.e. ~190 Tm^3.

3.5 Oil and fossil fuel prices

The cost of fossil fuels is subject not only to the economic laws of supply and demand but also to political pressures. The oil price crisis of 1973 arose after the members of OPEC declared an oil embargo as a result of the USA's support of Israel in the Yom Kippur war (which started in October 1973), and ended in March 1974 when a peace settlement was in sight. The oil price after the embargo was higher than before, as shown in Fig. 3.3.

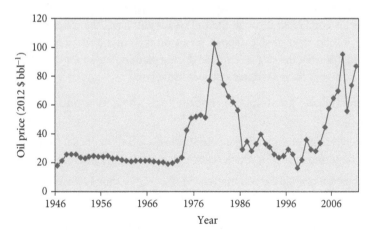

Fig. 3.3 Oil price variation during the period 1946–2011. *Source:* Oilprice.

In 1979 and 1980 events in Iran and Iraq caused another rise in the price of oil, and during the decade and a half following the 1973 crisis, research and development into alternative energy sources flourished around the world. However, by 1986 oil prices were almost back to their previous levels and funding and interest in alternative energy diminished. Greater appreciation of the risks from global warming and the increase in the oil price has rekindled efforts to develop low-carbon sources of energy. For the other fossil fuels, the cost of gas has tended to follow that of oil while that of coal has historically been more stable.

3.6 **Combustion**

The combustion of methane in natural gas is the exothermic reaction

$$CH_4 + 2O_2 \rightarrow CO_2 + 2H_2O \tag{3.1}$$

which produces 55 MJ of heat per kilogram of methane. The atomic masses of H, C, and O are in the proportion $1:12:16$, so the burning of 16 kg of methane releases 44 kg of carbon dioxide.

The combustion of the hydrocarbons in oil follows similar exothermic reactions. For example, the combustion of 72 kg of pentane

$$C_5H_{12} + 8O_2 \rightarrow 5CO_2 + 6H_2O \tag{3.2}$$

releases 220 kg of carbon dioxide.

A typical 500 MW coal-fired plant consumes around 250 tonnes of coal an hour. Coal needs to be in the form of a fine powder before it can be burned. The coal lumps are ground in large coal mills and the coal powder is injected through nozzles into a combustion chamber (Fig. 3.4), where it burns in a huge fireball. The combustible material undergoes exothermic reactions with oxygen, and the suspended particles of carbon and ash emit and absorb radiation. The flames are optically thick, i.e. most photons produced in the interior of the flame are reabsorbed in the flame before reaching the walls of the furnace. The outer surface of the flame is a close approximation to a black body and radiation is the dominant form of heat transfer to the boiler tubes lining the walls of the combustion chamber. The radiant heat incident on the boiler tubes is conducted through the tube walls and heats the water flowing inside. Over a long period, a solid layer of slag deposit forms on the outer surface of the boiler tubes and reduces the heat transfer; the slag is removed during plant outages for general maintenance.

The combustion of coal is a complex process, involving

- the evaporation of water trapped inside the coal (which uses some of the energy content of the fuel);
- the production of combustible gases from the dissociation of coal (notably methane CH_4 and carbon monoxide CO), which react with oxygen and release heat;
- the combustion of solid carbon matter, $C + O_2 \rightarrow CO_2$; thus 12 kg of carbon releases 44 kg of CO_2.

Incomplete combustion of the carbon results in the formation of carbon monoxide:

$$2C + O_2 \rightarrow 2CO \tag{3.3}$$

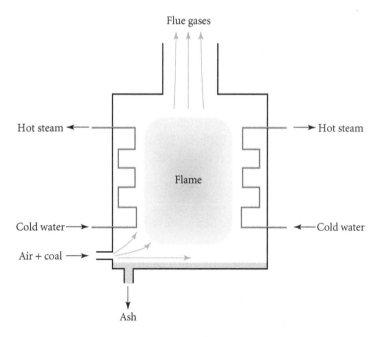

Fig. 3.4 Coal-fired combustion chamber.

Since carbon monoxide is poisonous, sufficient air must be injected into the furnace to ensure that the production of carbon monoxide is minimized, which also improves the fuel efficiency. The design of efficient coal-fired furnaces that minimize the production of environmentally harmful gases is an active area of coal technology.

Controlling the emission of sulfur dioxide from fossil fuel power stations is also an important issue since it is a major contributor to acid rain. There are various ways of tackling the problem, including

- removing the sulfur prior to combustion by coal scrubbing and oil desulfurization;
- gasification of coal under pressure with air and steam to form gases which can be burned to produce electricity;
- flue gas desulfurization, in which the waste gases are scrubbed with a chemical absorbent (e.g. limestone).

Though these processes are beneficial to the environment, they typically add about 10% to the overall cost of electricity, so power companies need to be given incentives to incorporate such measures.

3.7 Carbon capture and storage

The main scenario of the IEA's World Energy Outlook 2010 (IEA2010a) forecasts a 21% increase in the global emission of carbon dioxide from energy-related sources between 2008 and 2035. In order to offset the impact on global temperature, various carbon capture and storage (CCS) technologies are being actively pursued. There is also the possibility of carbon capture and utilization (CCU), where CO_2 is used in the production of building materials and fuels, for enhancing plant (e.g. algae) growth, or in the chemical or beverage industry. However, to implement CCS or CCU (sometimes collectively called CCUS) on a global scale presents enormous technical, economic, and political challenges.

The main sources of carbon dioxide are power stations and other industrial plants which burn fossil fuels. Since the concentration of carbon dioxide in the atmosphere decreases rapidly with distance from a point source, the best place to capture the gas is at the source itself. The three main methods of capture under consideration are

- post-combustion capture, where the CO_2 is scrubbed from the flue gases exiting the combustion chamber;

- pre-combustion capture, where the fuel is converted into a mixture of CO_2-rich gas and H_2, prior to combustion;

- oxyfuel capture, where the fuel is burned in oxygen rather than air, resulting in a stream of CO_2-rich gas.

Technologies exist for the first two processes, but not on the scale to deal with the emissions from a large power station. One process proposed for post-combustion is to pass the flue gases through an amine solution in which the CO_2 dissolves, with the nitrogen and other gases passing through. When the amine solution is saturated the flue gas is switched to another tank of amine while the first tank is heated to release the dissolved CO_2, which is then compressed and liquefied. The liquid CO_2 is then pumped to where it will be stored, e.g.in an underground aquifer. A plant at Monstag in Norway employing this technology opened in 2012.

A technology for pre-combustion capture is the integrated gasification combined cycle (IGCC) plant. In this process a pulverized coal–water slurry is burned in oxygen from a cryogenic (liquid oxygen) air separator unit to produce synthesis gas or 'syngas', a mixture of carbon monoxide and hydrogen, which is then passed through a shift converter. There the syngas is reacted with superheated steam over a catalyst to produce hydrogen and carbon dioxide. The principal reactions are

$$3C+O_2+H_2O \rightarrow 3CO+H_2 \quad \text{Syngas production}$$

$$CO+H_2O \rightarrow H_2+CO_2 \quad \text{Shift reaction}$$

In a conventional IGCC plant the syngas would be sent to a combined cycle gas turbine (CCGT; see Section 3.12). Adding a shift converter produces hydrogen and a relatively pure CO_2 stream that can be captured, but it does make the plant quite complex and hence expensive. The efficiency when using hydrogen as a fuel is also reduced since ~20% of the heat produced is taken away by the steam generated in the combustion process. The final efficiency of IGCC-CCS is about 35%, similar to that of a good post-combustion CCS plant.

The oxyfuel route is still at the research and development stage. The plant is like that for an air-fired pulverized coal boiler, but with O_2, and involves recycling a portion of the resultant flue gases. The remainder is rich in CO_2 which then has to be captured. All three processes require significant energy input; this 'energy penalty' adds 15–40% to the cost of electricity generated. After capture, the CO_2 is treated (desulfurization, dehydration and compression) and then sent by pipe or by ship to a storage location.

The long-term storage of CO_2 is known as carbon sequestration. The main lines of attack are summarized below.

Geological storage (geo-sequestration) by direct injection of liquid CO_2 into underground geological formations, including

- crude oil fields;
- heavy oil fields (enhanced oil recovery);
- unmineable coal seams, displacing the methane trapped in the coal;
- saline reservoirs.

The IPCC reported in 2009 that suitable geological storage sites had been identified to store all the CO_2 produced in the USA for the next 500 years.

Ocean storage, by

- injecting gaseous CO_2 into oceans at depths of more than 3000 m, where it liquefies and forms 'lakes' of liquid CO_2 on the ocean floor;
- injecting CO_2 into oceans at depths of between 1000 m and 3000 m; the CO_2 then rises towards the surface and dissolves, forming carbonic acid;
- reacting the CO_2 with suitable minerals to produce bicarbonates, which are then dumped on the ocean floor.

Biological storage, by

- burying trees when they die;
- the pyrolysis of biomass, which produces a charcoal called biochar that sequesters the carbon in a chemically inert state;
- changing farming practices so as to reduce CO_2 emissions from soil, e.g. by not tilling the land so often.

It is clear from the above that CCS technologies are expensive and will probably require internationally binding agreements on the carbon price before they are widely employed. Furthermore, there are serious environmental issues associated with certain technologies, e.g. leakage of carbon dioxide from underground storage sites and acidification of the sea. As of 2012, there are eight full-scale CCS projects in operation. The timescale for proving the technical viability of such projects and for international agreements on the incentives to ensure their commercial success is fairly short if CCS is to make a significant impact on mitigating the effects of increased carbon dioxide emissions over the next two decades. (NB Policies that affect global warming and the implementation of measures such as CCS capture are discussed in Chapter 12.)

Support for CCS has been somewhat undermined by shortage of funds, regulatory issues, scepticism that the technology can be developed to a large enough scale to have much impact on global warming, and public opposition due to fears about the possibility of leakage, notably in sites under land in Germany and Holland. In Europe, only 4 of the original 12 CCS demonstration plants are expected to be operating by 2015, 2 of which are briefly described in

Case study 3.1 The Sleipner carbon capture and storage project (Norway)

The Sleipner gas field was discovered in 1974 and is situated in the North Sea about 250 km from the coast of Norway. It produces natural gas and light oil condensates from sandstone structures about 2500 m below sea level. About 10% of the gas is carbon dioxide, which needs to be separated from the natural gas before it can be sold. After separation, the carbon dioxide is injected (in liquid form) into a porous and permeable layer of rock (the Utsira Sand), about 3000 m below sea level, rather than simply being vented to the atmosphere and thereby contributing to global warming. The CCS project was started in 1996 and sequesters about a million tonnes of carbon dioxide a year. The distribution of liquid carbon dioxide in the Utsira Sand is regularly monitored by seismic reflection analysis and has been found to be a distinct plume trapped between layers of shale.

The project provides powerful support for the technical feasibility of using suitable geological formations for the long-term storage of carbon dioxide. It has been estimated that even if only 1% of the Utsira Sand were used, it could nonetheless absorb 50 years' worth of carbon dioxide emissions from 20 coal-fired, or 50 gas-fired, 500 MW power stations.

Case study 3.2 The Lacq carbon capture and storage project (France)

Lacq, an industrial complex in south-west France, is the site of a demonstration CCS project where a gas-fired 30 MW boiler has been converted to the oxyfuel process. The captured carbon dioxide is piped 27 km to Rousse (in the Pyrenees), where it is injected as a liquid at 80 bars pressure into a depleted natural gas reservoir at a depth of 4500 m. By June 2012 the project had captured 120 000 tonnes of CO_2, equivalent to that emitted by 40 000 cars in 2 years. Existing capture costs are around $70 per tonne of CO_2, about two-thirds of the total cost. The aims are to make significant reductions in the capture cost by optimizing the oxyfuel technology and to verify that the reservoir is leak-proof. If successful, it is planned to use the technology to sequester all the CO_2 emissions from the Lacq site. The oxyfuel process is estimated to cost about half as much as existing post-combustion technologies, so proof of its success would significantly enhance its commercial potential for large-scale carbon sequestration.

the following case studies. On a global scale, there are probably enough large-scale projects to provide a realistic appraisal of the various CCS technologies by 2020.

In the aggressive IEA BLUE map scenario (IEA2010b), it is projected that capture from power generation and in industry would each reduce emissions in 2050 by $\sim 4\,GtCO_2\,y^{-1}$ corresponding to $\sim 1000\,GWe$ continuous generation (assuming an equivalence of $\sim 0.5\,kgCO_2/kWh$). But at the current rate of progress, this looks like it is a very optimistic projection.

3.8 Thermodynamics of steam power plants

In a conventional thermal power plant cycle the working fluid is water and, at various stages in the cycle, the working fluid changes phase from water, to a two-phase mixture of water and steam, to dry steam, and finally back to water. Some knowledge of the thermal properties of water and steam is essential in order to understand the operation of such a plant. We now apply thermodynamic considerations which are specific to the physical properties of water and steam in a thermal power plant.

The most convenient thermodynamic variables for describing thermal power plants are temperature T and entropy s. The T–s diagram for water and steam is shown in Fig. 3.5.

There are three distinct regions of interest:

I water;

II two-phase mixture of water and steam;

III dry steam.

The bell-shaped curve represents the phase boundary. The solid blue lines are isobars (constant pressure) and the dashed lines in region II are lines of constant steam quality x, i.e. the fraction by mass of steam in the two-phase mixture, defined by eqn (3.5).

To illustrate how to interpret the T–s diagram, consider the process of boiling water in a kettle. Since the fluid remains at atmospheric pressure throughout the heating process, we follow the isobar ABCD (at a pressure of 1 bar). Along AB water is heated from cold to the boiling point of 100 °C. Water starts to boil at point B but the temperature remains at 100 °C along BC as the fluid absorbs the latent heat of evaporation. At point C all the water has converted to dry steam. (This is an idealization of what happens in practice, since some water droplets may still exist for a while beyond point C.) The line CD represents superheated fluid (i.e. the temperature of the steam is above the boiling point), and the temperature of the dry steam rises at constant pressure as more heat is supplied, but the kettle should have switched off before getting this far!

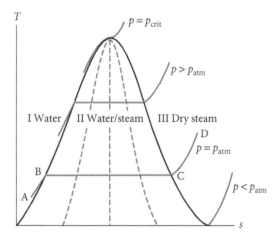

Fig. 3.5 T–s diagram for water and steam (not to scale).

The properties of water (Region I) and dry steam (Region III) can be obtained directly from steam tables. However, in order to use the steam tables in the two-phase region (Region II), some further explanation is necessary. Consider a mass of water m at B that converts into an equal mass of dry steam at C. At any point along the isobar BC the mixture of water and steam contains a mass m_f of water and m_g of steam (where the subscripts refer to fluid and gas, respectively). The total mass of the mixture is $m=m_f+m_g$. If v_f is the specific volume of liquid water at B and v_g the specific volume of dry steam at C, then the total volume of the mixture is $V=V_f+V_g=m_f v_f+m_g v_g$. Hence the specific volume v of the water–steam mixture is given by

$$v=\frac{V}{m}=\frac{m_f v_f+m_g v_g}{m}=\frac{(m-m_g)v_f}{m}+\frac{m_g v_g}{m}=\left(1-\frac{m_g}{m}\right)v_f+\frac{m_g}{m}v_g \tag{3.4}$$

The ratio

$$x=\frac{m_g}{m} \tag{3.5}$$

represents the proportion by mass of steam in the mixture and is called the steam quality. Likewise, $(1-x)$ represents the proportion by mass of water in the mixture. Thus the steam quality at B is $x=0$ (all water) and at C is $x=1$ (all steam). Substituting for x from eqn (3.5) in eqn (3.4) gives the specific volume of the mixture as

$$v=(1-x)v_f+xv_g \tag{3.6}$$

The numerical values of the coefficients v_f and v_g are obtained from steam tables. Equation (3.6) can then be used to determine the specific volume of the mixture for any particular value of the steam quality x.

Likewise the corresponding values of u, h, and s in the mixture are of the form

$$u=(1-x)u_f+xu_g \tag{3.7}$$

$$h=(1-x)h_f+xh_g \tag{3.8}$$

$$s=(1-x)s_f+xs_g \tag{3.9}$$

where the numerical values of the coefficients u_f, u_g, h_f, h_g, s_f, s_g are also given in steam tables.

To illustrate how to use the T–s diagram to solve practical problems, we consider a steam power plant operating in a Carnot cycle (Example 3.1).

EXAMPLE 3.1 Steam power plant operating in a Carnot cycle

A steam power plant operates in the Carnot cycle shown in Fig. 3.6. The boiler is at $T_1=352\,°C$, $p=170\,bar$ and the condenser is at $T_2=30\,°C$, $p=0.04\,bar$. Calculate (a) the efficiency of the cycle, (b) the heat input to the boiler, (c) the work done on the turbine, (d) the heat output in the condenser, and (e) the fraction of heat used in a complete cycle, using the steam table data below.

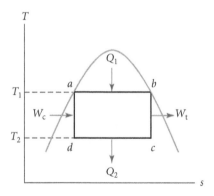

Fig. 3.6 *T-s* diagram for a steam plant operating in a Carnot cycle.

Steam table data

T (°C)	p (bar)	h_f	h_g	s_f	s_g
		h (kJ kg^{-1})		s (kJ kg^{-1} K^{-1})	
30	0.04	126	2556	0.436	8.452
352	170	1690	2548	3.808	5.181

(a) From eqn (2.17) the efficiency of the cycle is

$$\eta_C = 1 - \frac{T_2}{T_1} = 1 - \left(\frac{273+30}{273+352} \right) \approx 0.52$$

(b) The boiler operates at constant pressure, so we can use eqn (2.21) to calculate the heat input Q_1 from the change in enthalpy along ab, i.e.

$$Q_1 = h_b - h_a = 2548 - 1690 = 858 \text{ kJ kg}^{-1}$$

(c) Since bc is an adiabatic in a Carnot cycle, we can use eqn (2.22) to calculate the work W_t done on the turbine, i.e.

$$W_t = h_b - h_c = 2548 - h_c$$

To evaluate h_c we use eqn (3.8), i.e. $h_c = (1-x_c)h_f + x_c h_g$, where x_c is the steam quality at c. In order to determine x_c we use the fact that the expansion in the turbine is adiabatic. Equating the entropy at b and c gives $s_c = s_b = 5.181$ kJ kg^{-1} K^{-1}. Using eqn (3.9), we then have the following equation for x_c:

$$s_c = (1-x_c)s_f + x_c s_g = 5.181$$

From the steam table, $s_f = 0.436$ kJ kg^{-1} K^{-1} and $s_g = 8.452$ kJ kg^{-1} K^{-1} on the isobar $p = 0.04$ bar. Hence

$$0.436 (1-x_c) + 8.452 x_c = 5.181$$

which yields $x_c \approx 0.59$.

We can now evaluate the enthalpy at c as

$$h_c = (1-x_c)h_f + x_c h_g \approx (1-0.59)(126) + (0.59)(2556) \approx 1560 \text{ kJ kg}^{-1}$$

The work done on the turbine is then given by

$$W_t = h_b - h_c = 2548 - 1560 = 988 \text{ kJ kg}^{-1}$$

(d) Point a lies on the phase boundary, where the specific entropy is $s_a = 3.808$ kJ kg^{-1} K^{-1}. Since there is no change in entropy in the compressor, the entropy at d and a must be identical. Hence $s_d = 3.808$ kJ kg^{-1} K^{-1}. From eqn (3.9), the steam quality at d is then given by

$$(1-x_d)s_f + x_d s_g = 3.808$$

Using the steam table, $s_f = 0.436$ kJ kg^{-1}K^{-1} and $s_g = 8.452$ kJ kg^{-1}K^{-1} on the isobar at $p = 0.04$ bar. Hence

$$0.436(1-x_d) + 8.452x_d = 3.808, \text{ yielding } x_d \approx 0.42$$

Also $h_d = (1-x_d)h_f + x_d h_g = (1-0.42)(126) + (0.42)(2556) = 1147$ kJ kg^{-1}.

Since the condenser operates at constant pressure, the heat lost to the environment is minus the change in enthalpy, so that

$$Q_2 = h_c - h_d = 1560 - 1147 = 413 \text{ kJ kg}^{-1}$$

(e) The fraction of heat used in the complete cycle is given by

$$1 - \frac{Q_2}{Q_1} = 1 - \frac{413}{858} \approx 0.52, \text{ i.e. the Carnot efficiency derived in part (a)}$$

3.9 Disadvantages of a Carnot cycle for a steam power plant

Despite the fact that a Carnot cycle yields the maximum possible efficiency for a thermal power plant operating in a closed cycle, it suffers from a number of disadvantages that make it impractical for a real working fluid such as water.

To begin with, a Carnot cycle requires the temperature T_1 of the upper reservoir to be constant. However, from the T–s diagram for water/steam (Fig. 3.5) we see that this can only be achieved by operating the boiler along an isobar in the two-phase region II. It is not possible to operate the boiler in the dry steam region since the temperature rises along any given isobar in Region III. Hence the upper temperature of the cycle, T_1, is constrained by the maximum temperature of the two-phase boundary.

Another problem arises in the turbine. In a Carnot cycle the turbine operates with a two-phase mixture of water and steam (Region II in Fig. 3.5). The momentum of fast-moving water droplets in the mixture damages the turbine blades and shortens the life of the turbine. Similarly, since the compressor is required to compress a mixture of water and steam into high pressure water, water droplets in the mixture damage the blades of the compressor.

Finally, the volume of steam in the mixture is very large, which means that the compressor needs to be very large and therefore very expensive. The combined effect of all these factors makes a Carnot cycle impractical for a steam power plant.

3.10 Rankine cycle for steam power plants

Fortunately, there is a thermodynamic cycle that does not suffer from the problems of the Carnot cycle. It is the Rankine cycle, named in honour of Thomas Rankine, one of the founders of thermodynamics. We begin by considering the simplest case of a Rankine cycle without reheat.

3.10.1 Rankine cycle without reheat

The Rankine cycle without reheat is shown in Fig. 3.7. Unlike in a Carnot cycle, all the steam in a Rankine cycle is converted into water in the condenser (*de*) before entering the compressor. The compressor increases the pressure of the water (*ef*) adiabatically before the water enters the boiler. In modern steam power plants, boilers are normally constructed in three distinct sections, each made with a different grade of steel, using cheaper steels in the lower temperature sections and more expensive steels in the higher temperature sections. In the economizer section (*fa*) water is heated at high pressure until it starts to boil. In the evaporator section (*ab*) a two-phase mixture of water and steam is heated at constant pressure until all the water has been converted into dry steam. The dry steam is then heated at constant pressure in the superheater section of the boiler (*bc*). Dry steam enters the turbine at high pressure and does work on the turbine blades (*cd*). On leaving the turbine, wet steam enters the condenser (at sub-atmospheric pressure), where it condenses on the cold outer surfaces of a large bank of condenser tubes containing cold water from the external environment. This is a simplification of the situation in a real plant. In practice there are many complicating features,

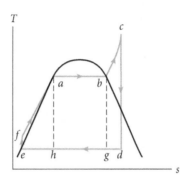

Fig. 3.7 *T–s* diagram for Rankine cycle without reheat (*abcdefa*). *abgha* represents a Carnot cycle with the same lower temperature.

e.g. a pressure drop through the boiler due to frictional losses, the incorporation of a recirculating loop in the economizer section (to take advantage of natural circulation), and instabilities in the position of the two-phase boundaries.

In order to calculate the efficiency of a Rankine cycle without reheat it is not possible to use the Carnot formula (2.17), because the temperature of the upper reservoir of heat is not constant. The average upper temperature in a Rankine cycle lies between the lowest temperature in the economizer and the maximum temperature in the superheater. The method of calculating the heat transfer processes in each stage of the cycle and the overall efficiency is shown in Example 3.2.

EXAMPLE 3.2 Thermal power plant operating in a Rankine cycle without reheat

A steam power plant operates in a Rankine cycle without reheat, as shown in Fig. 3.7. The boiler and the condenser are at $p=170$ bar and $p=0.04$ bar, respectively. The temperature of the evaporator is 352 °C, the maximum temperature in the superheater is 600 °C, and the temperature in the condenser is 30 °C. Calculate (a) the work done by the compressor, (b) the heat input in the boiler, (c) the work done on the turbine, (d) the heat output in the condenser, (e) the efficiency of the Rankine cycle without reheat, and (f) the efficiency of the Carnot cycle operating between a reservoir at 352 °C and a reservoir at 30 °C.

Steam table data

T (°C)	p (bar)	h_f	h_g	s_f	s_g
		h (kJ kg^{-1})		s (kJ kg^{-1}K^{-1})	
30	0.04	126	2556	0.436	8.452
352	170	1690	2548	3.808	5.181
600	170		3564		6.603

(a) Assuming that water is incompressible, the work done by the compressor per unit mass of water is given by $W_c=v_f\,(p_f-p_e)$; see Derivation 2.2(b). Putting $v_f=10^{-3}$ m³ kg^{-1} and $p_f-p_e=(170-0.04)\times10^5$ Nm^{-2}, we have $W_c\approx17$ kJ kg^{-1}.

(b) We first calculate the enthalpy at the entrance to the boiler (*f*). Assuming the compressor is adiabatic, the work done is equal to the change of enthalpy, i.e. $W_c=h_f-h_e$, or $h_f=W_c+h_e\approx17+126=143$ kJ kg^{-1}.

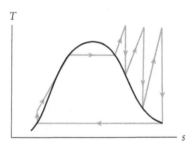

Fig. 3.8 *T–s* diagram for Rankine cycle with reheat.

The boiler operates at constant pressure, so the heat input is given by $Q_{in}=h_c-h_f=3564-143\approx3421$ kJ kg^{-1}.

(c) The work done on the turbine is $W_t=h_c-h_d=3564-h_d$, where h_d is given by $h_d=h_f(1-x_d)+h_gx_d=126(1-x_d)+2556x_d$. To obtain x_d we use the fact that the expansion in the turbine is adiabatic, so that $s_d=s_c=6.603$.

Hence $0.436(1-x_d)+8.452x_d=6.603$, which yields $x_d\approx0.77$.

Thus $h_d=126(1-0.77)+2556(0.77)\approx1995$ kJ kg^{-1},

and $W_t=3564-1995=1569$ kJ kg^{-1}.

(d) The heat output in the condenser is $Q_{out}=h_d-h_e=1995-126\approx1869$ kJ kg^{-1}.

(e) The efficiency of the cycle is $\eta_R=\dfrac{W_t-W_c}{Q_{in}}=\dfrac{1569-17}{3421}\approx0.45$.

(f) The efficiency of the Carnot cycle is $\eta_C=1-\dfrac{T_2}{T_1}=1-\left(\dfrac{273+30}{273+352}\right)\approx0.52$.

3.10.2 Rankine cycle with reheat

The Rankine cycle without reheat does not completely eliminate the production of water droplets with high momentum that damage the turbine blades. To overcome the problem, modern power plants tend to use the Rankine cycle with reheat, in which the steam is reheated several times before leaving the turbine. Figure 3.8 shows the T–s diagram with three reheat stages. Steam is reheated after leaving the high pressure (HP) turbine before entering the intermediate pressure (IP) turbine, and is again reheated between the IP turbine and the low pressure (LP) turbine. The overall efficiency of the cycle is greater than that of a Rankine cycle without reheat and the formation of water droplets is much reduced.

In order to maximize the efficiency of a steam power plant it is desirable to operate at as high a temperature as possible in the superheater. However, above about 650 °C various forms of metal fatigue become significant because of the very high temperatures and pressures that the walls of the boiler tubes have to withstand. Erosion/corrosion of the tubes due to the presence of trace chemicals in the water can also be a lifetime-limiting factor. Replacement of boiler tubing in a power plant is a major operation that puts the plant out of service for a considerable period. The net result is that the maximum operating temperature of a steam

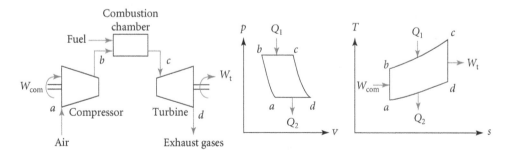

Fig. 3.9 Brayton (or Joule) cycle.

power plant is limited to about 650 °C, above which the benefits from improved efficiency are outweighed by the cost of tube replacement and outage costs. Modern plants using Rankine cycles with reheat achieve overall efficiencies of around 40–45%.

3.11 Gas turbines and the Brayton (or Joule) cycle

As we observed in Section 3.10.2, the maximum temperature in a steam power station must be kept below about 650 °C in order to avoid excessive metallurgical damage. In a gas turbine, however, the gaseous products of combustion are typically at around 1300 °C. The turbine blades are covered by a ceramic coating of low thermal conductivity so that the hot gases do not make direct contact with metal surfaces. In addition, the blade assembly is water-cooled so that the temperature of the blades is maintained below the metallurgical limit.

Gas turbines for electricity generation originally evolved from jet turbine engines. In a jet turbine engine, the thrust arises from the combustion of gaseous fuel and the expansion of the exhaust gases. Since the working fluid does not change phase, a condenser is not involved in the process, so the overall size and cost of a gas turbine plant is less than that of an equivalent steam plant. Gas turbines operate in a Brayton (or Joule) cycle, as shown in Fig. 3.9. It is an open cycle but is equivalent to a closed cycle in the sense that the atmosphere acts as a heat exchanger that cools the air entering the combustion chamber.

Air enters the compressor (a) at atmospheric pressure and is compressed to around 10–20 bar (b). It is then mixed with fuel in the combustion chamber, producing hot combustion gases that do work on the turbine (c). The exhaust gases are then vented to atmosphere (d). Assuming the change in the compressor is an adiabatic process ($Q=0$), then, since work is done on the system, from eqn (2.22) we can equate the net work done per unit mass by the compressor to the increase in enthalpy, i.e.

$$W_{com}=h_b-h_a=c_p\,(T_b-T_a)$$

where c_p is the specific heat at constant pressure. Similarly, assuming the turbine is adiabatic, the work done per unit mass on the turbine is

$$W_t=h_c-h_d=c_p\,(T_c-T_d)$$

The net heat supplied is

$$Q=h_c-h_b=c_p\,(T_c-T_b)$$

Hence the efficiency of the cycle is given by

$$\eta=\frac{W_t-W_{com}}{Q}=\frac{(T_c-T_d)-(T_b-T_a)}{T_c-T_b} \tag{3.10}$$

A more useful expression for the efficiency of a gas turbine is given by the formula

$$\eta=1-r^{-\left(\frac{\gamma-1}{\gamma}\right)} \tag{3.11}$$

where

$$r = \frac{p_b}{p_a} = \frac{p_c}{p_d} \tag{3.12}$$

is called the **pressure ratio** and $\gamma = c_p/c_v$ is the ratio of the specific heats at constant pressure and at constant volume. (See Derivation 3.1.)

Gas turbines are relatively low capital cost devices that can be started up quickly and are employed for satisfying sudden surges in electricity demand. Efficiencies of simple gas turbines are up to around 40%.

Derivation 3.1 Pressure ratio and efficiency of Brayton (or Joule) cycle

Consider the adiabatic expansion of a perfect gas. Putting $đQ=0$, the first law of thermodynamics (eqn (2.15)) becomes $du+pdv=0$, or $c_v dT + pdv = 0$. Differentiating the equation of state for a perfect gas $pv = RT$, we have $pdv + vdp = RdT$. Putting $R = c_p - c_v$ and $v = RT/p$, and eliminating dv, we obtain

$$\frac{dT}{T} = \left(\frac{\gamma-1}{\gamma} \right) \frac{dp}{p}$$

where $\gamma = c_p/c_v$. Integrating, we obtain $\ln T = \left(\frac{\gamma-1}{\gamma} \right) \ln p + \text{const.}$, or $T = A p^{\frac{\gamma-1}{\gamma}}$. Hence T_b and T_c can be expressed in terms of T_a and T_d, respectively, as

$$T_a = T_b \left(\frac{p_b}{p_a} \right)^{\frac{1-\gamma}{\gamma}} = T_b r^{\frac{1-\gamma}{\gamma}} \text{ and } T_d = T_c \left(\frac{p_c}{p_d} \right)^{\frac{1-\gamma}{\gamma}} = T_c r^{\frac{1-\gamma}{\gamma}} \tag{3.13}$$

where $r = \frac{p_b}{p_a} = \frac{p_c}{p_d}$ is the pressure ratio. Substituting for T_a and T_d in eqn (3.10), we obtain

$$\eta = \frac{(T_c - T_d)-(T_b - T_a)}{T_c - T_b} = \frac{T_c - T_c r^{\frac{\gamma-1}{\gamma}} - T_b + T_b r^{\frac{1-\gamma}{\gamma}}}{T_c - T_b} = 1 - r^{\frac{1-\gamma}{\gamma}}$$

EXAMPLE 3.3 Gas turbine

Calculate the exhaust temperature and the efficiency η of an ideal gas turbine operating with a maximum temperature of 1300 K for a pressure ratio $r=8$. (Assume $\gamma=1.4$.)

From eqn (3.13) the exhaust temperature is given by

$$T_d = T_c r^{-\left(\frac{\gamma-1}{\gamma} \right)} = 1300 \times 8^{-\left(\frac{1.4-1}{1.4} \right)} \approx 718 \text{ K}$$

Equation (3.11) yields the efficiency as $\eta = 1 - r^{-\left(\frac{\gamma-1}{\gamma} \right)} = 1 - 8^{-\left(\frac{1.4-1}{1.4} \right)} \approx 0.45$.

3.12 **Combined cycle gas turbine**

The overall efficiency of a gas turbine can be increased by feeding the heat of the exhaust gases into a steam power plant. The combination of a Brayton cycle and a Rankine cycle is called a combined cycle gas turbine (CCGT), and is shown in Fig. 3.10. The net effect is equivalent to that of a single cycle operating between the upper temperature of a Brayton cycle and the lower temperature of a Rankine cycle. Efficiencies of up to 60% are typical in CCGT plants.

Even greater efficiencies can be achieved in a combined heat and power (CHP) cycle. In a CHP plant the condenser in the steam power cycle is operated at a higher temperature than that in a conventional steam power plant, and the waste heat from the condenser is used to provide district heating in the local community. The total efficiency of CHP schemes is typically around 80%. However, the cost involved in installing the pipework and other infrastructure is high, so the application of CHP is limited to industrial complexes or to densely populated urban areas.

3.13 **Fluidized beds**

Fluidized beds (Fig. 3.11) provide an alternative means of burning coal, biomass, or municipal waste that reduces the emission of environmentally harmful gases. The attraction of fluidized beds lies in their ability to cope with a wide range of feedstock and their simplicity. The fuel in the bed is in the form of small solid particles, suspended in an upward jet of air. Turbulent mixing of the particles with the air results in more complete chemical reactions

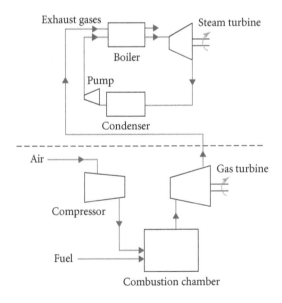

Fig. 3.10 Combined cycle gas turbine (CCGT) generation.

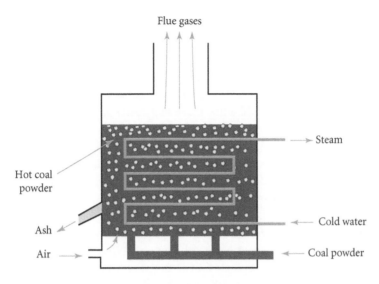

Fig. 3.11 Fluidized bed.

than in a conventional combustion chamber. Most of the sulfur in coal is removed by using limestone or dolomite to precipitate the sulfur. The generation of NO_x is reduced by operating the bed at a lower temperature than in a conventional combustion chamber. Heat is removed by an array of boiler tubes immersed in the fluidized bed, and the resulting steam is used to drive a steam turbine.

The first generation of fluidized beds tended to suffer from erosion/corrosion, in that fine particles and gases caused significant damage to the boiler tubes. Later designs have significantly reduced the problem by using a pressurized system that improves the contact between sorbent and flue gas. Furthermore, by the incorporation of a coal gasifier, a fuel gas is produced which provides significant improvements to the efficiency of combined cycle systems.

3.14 Supercritical and ultrasupercritical plants

Conventional power plants operate at pressures around 170 bar, which is below the critical pressure of 220 bar. Supercritical and ultrasupercritical plants (SCP and USCP) operate above the critical pressure with SCP at ~240 bar and USCP at ~250–300 bar. The higher operating temperature increases the efficiency of the plants: an SCP can achieve ~40–42% at ~566 °C and an USCP 43–46% at 600 °C, compared with ~38% at 538 °C for a conventional plant. There is a corresponding reduction in CO_2 emissions, but CCS is needed for effective reduction. The higher efficiency of SCP and USCP means that the cost penalty of CCS would be less than with conventional plants without CCS. Materials such as NiCrMoV steels that can withstand very high temperatures are required and there is considerable R&D (research and development) into new materials, e.g. nickel-base alloys and high boron steels, which can operate at even higher temperatures.

3.15 **Fossil fuel outlook**

The ratio of the proven reserves of conventional oil and gas to their annual production gives times of 46.2 and 58.6 years, respectively. The exploitation of shale gas in the USA has caused a shift away from coal and reduced the cost of gas. Globally, shale gas has approximately doubled the reserves of gas and will allow more fuel switching away from coal, which will lower CO_2 emissions. Unconventional oil, which includes shale oil, has roughly doubled the reserves of oil. Conventional oil is increasingly inaccessible and deep-water drilling has increased the risk of major oil spills, such as occurred after the Deepwater Horizon platform explosion in the Gulf of Mexico in 2010. Oil flowed for three months before the well was secured and ~5 million barrels of oil were released that caused extensive damage to fish, birds, and their habitat.

The ratio of proven coal reserves to the annual production of coal gives a time period of 118 years. With CO_2 emissions about twice as high per MWh of electricity, it is very important that all coal power plants incorporate CCUS. But at an estimated cost for CCS by the US Department of Energy (DOE) of $100–300 per $t\,CO_2$, the economic penalty is currently very high. More funding for R&D on methods of capture, storage, and utilization is very important.

SUMMARY

- Fossil fuel combustion produces carbon dioxide and other environmentally harmful gases.

- As conventional fossil fuels become depleted, unconventional fossil fuels (e.g. shale oil and shale gas) will eventually become economic to exploit, but there are significant environmental and health issues associated with the use of such fuels.

- It has been estimated that the proven reserves of fossil fuels will last for approximately 46 and 59 years for conventional oil and gas, respectively, and 118 years for coal. The amount of unconventional oil and shale gas approximately doubles the oil and gas reserves.

- Carbon sequestration is a potential means of storing carbon dioxide for long periods. But underground stores need to contain the CO_2 for several thousand years.

- Various CCS demonstration projects are being evaluated around the world, but CCS is expensive and will therefore require widespread international support to become economically viable and make a significant impact on global warming.

- In the aggressive and optimistic IEA BLUE map scenario (IEA2010b), CCS from power generation and in industry (including upstream capture) would each reduce emissions in 2050 by $\sim 4\,GtCO_2\,y^{-1}$ corresponding to ~1000 GWe continuous generation (assuming an equivalence of $\sim 0.5\,kgCO_2/kWh$).

- The T–s (Temperature–entropy) diagram is useful for describing thermodynamic cycles in thermal power plants.

- A Carnot cycle is unsuitable for a real fluid such as water, mainly because the upper temperature is limited and water droplets in the two-phase water/steam mixture damage the blades in the compressor and in the turbine.

- A Rankine cycle overcomes the disadvantages of a Carnot cycle.

- Gas turbines use a Brayton (or Joule) cycle and can operate at higher temperatures than steam turbines because the hot gases are prevented from making direct contact with the metal surfaces of the turbine blades.

- Higher efficiencies can be obtained using combined cycle gas turbines which utilize the waste heat of a Brayton (or Joule) cycle in a Rankine cycle of a steam power plant. Combined heat and power schemes are even more efficient in terms of total energy usage by providing district heating.

- Fluidized beds can cope with a wide range of feedstock, produce smaller quantities of environmentally harmful gases than conventional combustion chambers, and are now commercially viable.

FURTHER READING

Berkowitz, N. (1997). *Fossil hydrocarbons: chemistry and technology.* Academic Press, San Diego. Comprehensive treatment of fossil hydrocarbons.

Howard, J.R. (1989). *Fluidized bed technology: principles and applications.* Institute of Physics Publishing, Bristol. Introduction to fluidized beds with a wide range of applications.

Rackley, S. (2009). *Carbon capture and storage.* Butterworth-Heinemann, Oxford. Comprehensive overview of a wide range of technologies involved in carbon dioxide capture and sequestration.

Rogers, G. and Mayhew, Y. (1992). *Engineering thermodynamics*, 4th edn. Longman, Harlow. Engineering approach to thermodynamics with many practical examples.

Rogers, G. and Mayhew, Y. (1995). *Thermodynamic and transport properties of fluids*, 5th edn. Blackwell, Oxford. Steam table data.

Wilson, E. and Gerard, D. (2007). *Carbon capture and sequestration.* Blackwell, Oxford. Informative book covering everything from CO_2 chemical properties to regulatory issues and public perception.

WEB LINKS

bp.com/statisticalreview (BP2011).

www.iea.org/textbase/npsum/oil_gasSUM.pdf (IEA2005).

www.iea.org/Textbase/npsum/weo2010sum.pdf (IEA2010a).

www.iea.org/publications/freepublications/publication/ccs_roadmap_foldout.pdf IEA CCS roadmap (IEA2010b).

www.akersolutions.com/en/Global-menu/Media/Press-Releases/All/2012/Aker-Solutions-celebrates-opening-of-carbon-capture-plant-at-Mongstad/ Monstag CCS plant.

www.bgs.ac.uk/science/co2/home.html Sleipner CCS project.

dx.doi.org/10.1016/j.enpol.2009.02.020 (Oilfields).

inflationdata.com/inflation/inflation_rate/historical_oil_prices_table.asp (Oilprice).

www.claverton-energy.com/integrated-gasification-combined-cycle-for-carbon-capture-storage.html IGCC-CCS.

www.fas.org/sgp/crs/misc/RL33801.pdf Interesting report on CCS prepared for the US Congress.

enpub.fulton.asu.edu/ece340/pdf/steam_tables.PDF Steam tables.

www.fossil.energy.gov/programs/powersystems/cleancoal/ DOE-CCS.

www.iea-coal.org.uk/site/2010/home IEA-clean coal.

www.worldcoal.org/coal-the-environment/coal-use-the-environment/improving-efficiencies/ Supercritical and ultrasupercritical technology.

www.worldenergy.org Neutral source of data and current developments.

There are many websites that quote steam table data for specific thermal conditions.

LIST OF MAIN SYMBOLS

c_p	specific heat at constant pressure	T	temperature
c_v	specific heat at constant volume	U	internal energy
h	specific enthalpy	u	specific energy
P	power	v	specific volume
p	pressure	V	volume
Q	heat or volume flow rate (i.e. in an aquifer)	W	work
		x	steam quality
r	pressure ratio	γ	ratio of specific heats
R	universal gas constant	η	efficiency
s	specific entropy		

EXERCISES

3.1 Comment *critically* on the concept of *peak oil*. Do your comments apply equally well to gas and coal?

3.2 Estimates of the reserve to production ratio, R/P, for coal, oil, and gas are 118, 46, and 59 years, respectively. What would the duration of these reserves be if production increased linearly by 15% per decade? Estimate the reserves in EJ and in TWy.

3.3 Comment on any correlation between investment in renewable technology and the price of oil (see Fig. 3.3).

3.4 Discuss the challenges facing CCS technologies.

3.5 Investigate what methods are being considered for CCU.

3.6 An adiabatic compressor increases the pressure of water from 0.04 bar to 150 bar. Assuming that water is incompressible, calculate the work done per kg of water.

3.7 Using the steam table data below, estimate the work done on an adiabatic turbine by 1 kg of superheated steam entering the turbine at a pressure of 200 bar and a temperature of 600 °C and leaving the turbine at a pressure of 1 bar and a temperature of 100 °C.

Steam table data

T (°C)	p (bar)	h (kJ kg^{-1})
100	1	2676
600	200	3537

3.8 Show that the internal energy u, enthalpy h, and entropy s of a two-phase mixture can be expressed in the form

$$u = (1-x)u_f + xu_g$$

$$h = (1-x)h_f + xh_g$$

$$s = (1-x)s_f + xs_g$$

3.9 Consider a thermal power station operating in a Carnot cycle between an upper reservoir at $T = 400$ °C and $p = 180$ bar, and a lower reservoir at $T = 20$ °C and $p = 0.02$ bar. Calculate (a) the efficiency of the cycle, (b) the heat input in the boiler, (c) the heat output in the condenser, and (d) the work done on the turbine.

Steam table data

		h (kJ kg^{-1})		s (kJ kg^{-1} K^{-1})	
T (°C)	p (bar)	h_f	h_g	s_f	s_g
20	0.02	84	2538	0.296	8.666
400	180	1732	2510	3.872	5.108

3.10 A power station operates in a Rankine cycle without reheat, consisting of (i) an adiabatic compressor, (ii) a three-stage boiler at 200 bar, (iii) an adiabatic turbine, and (iv) a condenser at $p = 0.02$ bar, $T = 20$ °C. The maximum temperature of the boiler is 700 °C. Using the steam table data in the table below, calculate

(a) the specific work done by the compressor

(b) the heat supplied per unit mass to the boiler

(c) the specific work obtained from the turbine

(d) the efficiency of the cycle

Steam table data

	T (°C)	p (bar)	h_f	h_g	s_f	s_g
			\multicolumn{2}{c}{h (kJ kg^{-1})}	\multicolumn{2}{c}{s (kJ kg^{-1}K)}
Water/steam mixture	20	0.02	84	2454	0.296	8.666
Dry steam	700	200		3806		6.796

3.11 Calculate the exhaust temperature and the efficiency η of an ideal gas turbine operating with a maximum temperature of 1600 K for a pressure ratio $r=9$. (Assume $\gamma=1.4$)

3.12* A heat engine operating between heat reservoirs at T_1 and T_2 is modelled by embedding an ideal heat engine operating between T_3 ($<T_1$) and T_4 ($>T_2$). Heat Q_1 flows in for a fraction β of the cycle time t, and heat Q_2 out for a fraction $(f-\beta)$. This is an example of an *endoreversible* system.

The heat flows are given by

$$Q_1 = k_1'(T_1-T_3)\beta t \equiv g_1(T_1-T_3)t$$
$$Q_2 = k_2'(T_4-T_2)(f-\beta)t \equiv g_2(T_4-T_2)t$$

where k_1' and k_2' are the products of the thermal conductivity and area of the hot and cold heat exchangers, respectively. Show that

$$T_3 = \frac{1}{(g_1+g_2)}\left(g_1T_1 + \frac{g_2T_2}{\tau}\right)$$

where $\tau = T_4/T_3$.

3.13* The power P of the heat engine modelled in Exercise 3.12* is given by $P=(Q_1-Q_2)/t$. Show that P can be expressed in the form

$$P = \frac{g_1g_2}{(g_1+g_2)}(1-\tau)\left(T_1 - \frac{T_2}{\tau}\right)$$

Also show that the maximum power is obtained when $\tau = \sqrt{\dfrac{T_2}{T_1}}$, and the efficiency η is then given by

$$\eta = 1 - \sqrt{\frac{T_2}{T_1}}$$

This is called the Chambadal–Novikov–Curzon–Ahlborn value. This analysis is an example of finite-time thermodynamics.

How much has the energy available for doing work (the exergy) been reduced by the finite temperature differences in the heat exchangers from the value $Q_1(1-T_2/T_1)$ in an ideal heat engine?

Explain why the power is zero when $\tau = T_2/T_1$ and the efficiency equals that of a Carnot cycle.

3.14 Compare the efficiency of actual power plants, including sub-critical, supercritical, and ultrasupercritical, with the Chambadal–Novikov–Curzon–Ahlborn value (see Exercise 3.13*) and comment on the agreement.

4 Essential fluid mechanics for energy conversion

→ **Introduction**

In Chapter 3 we showed that it is possible to describe the energy transfer processes in boilers, condensers, and turbines using basic thermodynamic principles without a detailed knowledge of the fluid flow processes involved in each device. However, in order to understand other areas of energy conversion such as hydropower, wave power, and wind power, a basic knowledge of fluid mechanics is essential.

In this chapter we give a brief summary of the basic physical properties of fluids and derive the conservation laws of mass and energy for a fluid in which viscous effects are ignored (known as an ideal or inviscid fluid). We illustrate how the conservation laws can be applied to situations of practical interest to derive useful information about the flow. Also, we describe the effect of viscosity on the motion of a fluid around a body immersed in a fluid (e.g. an aerofoil) and show how the flow determines the forces acting on the body.

4.1 Basic physical properties of fluids

The bulk physical properties of a fluid are

- Density (ρ). Mass per unit volume of a fluid. Unless otherwise stated, it is assumed throughout the book that the density of a fluid is constant (called incompressible flow); the variations in pressure arising from fluid motion (see Example 4.3) are small in comparison with atmospheric pressure. The unit of density is $kg\,m^{-3}$. ($\rho_{water} \approx 10^3\,kg\,m^{-3}$ and $\rho_{air} \approx 1.3\,kg\,m^{-3}$ at $T = -2\,°C$ and $p = 1\,atm$.)
- Pressure (p). Force per unit area in a fluid. Pressure acts in the direction normal to the surface of a body immersed in a fluid. The unit of pressure is the pascal (Pa; $1\,Pa = 1\,N\,m^{-2}$; $1\,atm \approx 1\,bar = 10^5\,Pa$).

- Viscosity. Force per unit area due to internal friction in a fluid arising from the relative motion between neighbouring elements in a fluid. Viscous forces act in the tangential direction to the surface of a body immersed in a flow (see Section 4.5).

4.2 Streamlines and stream-tubes

A useful concept for visualizing a velocity field is to imagine a set of streamlines parallel to the direction of motion at all points in the fluid. Any element of mass in the fluid flows along a notional stream-tube bounded by neighbouring streamlines (Fig. 4.1). In practice, streamlines can be visualized by injecting small particles into the fluid. For example, smoke can be used in wind tunnels to investigate the flow over wings, turbine blades, cars, buildings, etc.

4.3 Mass continuity

One of the fundamental laws of fluid mechanics is conservation of mass (also known as mass continuity). Consider the flow along a stream-tube in a steady velocity field. Suppose that the speed of the fluid and the cross-sectional area of the stream-tube at any point are u and A, respectively. By definition, the direction of flow is parallel to the boundaries of the stream-tube, so the fluid is confined to the stream-tube and the mass flow per second is constant along the stream-tube. Hence

$$\rho u A = \text{const.} \tag{4.1}$$

Thus the speed of the fluid is inversely proportional to the cross-sectional area of the stream-tube (Example 4.1).

EXAMPLE 4.1 Flow along a stream-tube

An incompressible ideal fluid flows at a speed of $1\,\mathrm{m\,s^{-1}}$ through a pipe of $1\,\mathrm{m}$ diameter in which a constriction of $0.1\,\mathrm{m}$ diameter has been inserted. What is the speed of the fluid inside the constriction?

Putting $\rho_1 = \rho_2$ and using eqn (4.1), we have $u_1 A_1 = u_2 A_2$, or

$$u_2 = u_1 \frac{A_1}{A_2} = (1\,\mathrm{ms^{-1}}) \times \left(\frac{1}{0.1}\right)^2 = 100\,\mathrm{m\,s^{-1}}$$

Fig. 4.1 Stream-tube.

4.4 Energy conservation in an ideal fluid: Bernoulli's equation

In many practical situations, viscous effects are much smaller than those due to gravity and pressure gradients over large regions of the flow field. We can then ignore viscosity to a good approximation and derive an equation known as Bernoulli's equation (or Bernoulli's theorem) for energy conservation in a fluid. For steady flow, Bernoulli's equation is of the form

$$\frac{p}{\rho} + gz + \frac{1}{2}u^2 = \text{const.} \tag{4.2}$$

(For a proof of Bernoulli's equation, see Derivation 4.1.)

For a stationary fluid, $u=0$ everywhere in the fluid, and eqn (4.2) reduces to

$$\frac{p}{\rho} + gz = \text{const.} \tag{4.3}$$

Equation (4.3) is the equation for hydrostatic pressure. It shows that the fluid at a given depth z is all at the same pressure p (see Example 4.2).

EXAMPLE 4.2 Hydrostatic pressure

The atmospheric pressure on the surface of a lake is $10^5\,\mathrm{N\,m^{-2}}$. Assuming the water is stationary, what is the pressure at a depth of 10 m? (Assume $\rho_{\text{water}} = 10^3\,\mathrm{kg\,m^{-3}}$ and $g = 10\,\mathrm{m\,s^{-2}}$.)

From eqn (4.3), we have $\dfrac{p_1}{\rho} + gz_1 = \dfrac{p_2}{\rho} + gz_2$. Putting $p_1 = 10^5\,\mathrm{N\,m^{-2}}$ at $z_1 = 0$, and $z_2 = -10\,\mathrm{m}$, we have $p_2 = p_1 - \rho g(z_2 - z_1) = 10^5 - (10^3)(10)(-10) = 2\times10^5\,\mathrm{N\,m^{-2}}$.

The significance of Bernoulli's equation is that it shows that the pressure in a moving fluid decreases as the speed increases. The practical importance of this effect is illustrated in Examples 4.3, 4.4, and 4.5.

EXAMPLE 4.3 Effect of wind on air pressure

Assuming the pressure of stationary air is $10^5\,\mathrm{N\,m^{-2}}$, calculate the percentage change in pressure due to a wind of $20\,\mathrm{m\,s^{-1}}$. (Assume $\rho_{\text{air}} \approx 1.2\,\mathrm{kg\,m^{-3}}$.)

From eqn (4.2) we have $\dfrac{p_1}{\rho} + \dfrac{1}{2}u_1^2 = \dfrac{p_2}{\rho} + \dfrac{1}{2}u_2^2$. The change in pressure is given by $p_2 - p_1 = \dfrac{1}{2}\rho(u_1^2 - u_2^2) = -\dfrac{1}{2}(1.2)(20)^2 = -2.4\times10^2\,\mathrm{N\,m^{-2}}$. Hence, the percentage change in pressure is $-\dfrac{2.4\times10^2}{10^5}\times100 \approx -0.24\%$.

Derivation 4.1 Bernoulli's equation for steady flow

Consider the steady flow of an ideal fluid in the control volume shown in Fig. 4.2.

The height, cross-sectional area, speed, and pressure at any point are denoted by z, A, u, and p, respectively. The increase in gravitational potential energy of a mass δm of fluid between z_1 and z_2 is $\delta m g (z_2 - z_1)$. In a small time interval δt the mass of fluid entering the control volume at P_1 is $\delta m = \rho\, u_1 A_1 \delta t$ and the mass leaving P_2 is $\delta m = \rho\, u_2 A_2 \delta t$.

In order for the fluid to enter the control volume it has to do work to overcome the pressure p_1 exerted by the fluid. The work done in pushing the elemental mass δm a small distance $\delta s_1 = u_1 \delta t$ at P_1 is $\delta W_1 = p_1 A_1 \delta s_1 = p_1 A_1 u_1 \delta t$. Similarly, the work done in pushing the elemental mass out of the control volume at P_2 is $\delta W_2 = -p_2 A_2 u_2 \delta t$ (note change of sign). The net work done is $\delta W_1 + \delta W_2 = p_1 A_1 u_1 \delta t - p_2 A_2 u_2 \delta t$. By energy conservation, this is equal to the increase in potential energy plus the increase in kinetic energy, so that

$$p_1 A_1 u_1 \delta t - p_2 A_2 u_2 \delta t = \delta m\, g(z_2 - z_1) + \frac{1}{2}\delta m(u_2^2 - u_1^2)$$

Putting $\delta m = \rho u_1 A_1 \delta t = \rho u_2 A_2 \delta t$ and tidying up, we obtain

$$\frac{p_1}{\rho} + gz_1 + \frac{1}{2}u_1^2 = \frac{p_2}{\rho} + gz_2 + \frac{1}{2}u_2^2$$

Finally, since points P_1 and P_2 are arbitrary it follows that

$$\frac{p}{\rho} + gz + \frac{1}{2}u^2 = \text{const.}$$

everywhere along the stream-tube.

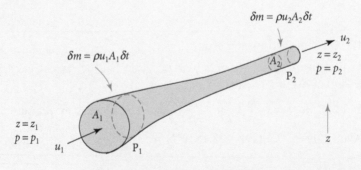

Fig. 4.2 Control volume for Bernoulli's equation.

EXAMPLE 4.4 Pitot tube

A Pitot tube is a device for measuring the velocity of a fluid. Essentially, it consists of two tubes, (a) and (b). Each tube has one end open to the fluid and one end connected to a pressure gauge. Tube (a) has the open end facing the flow and the tube (b) has the open end normal to the flow (Fig. 4.3). Derive an expression for the velocity of the fluid in terms of the difference in pressure between the gauges.

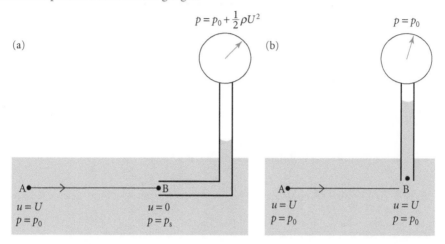

Fig. 4.3 Pitot tube. (a) Open end facing incident flow; (b) open end normal to incident flow.

Consider the fluid moving along the streamline AB. In case (a), the fluid slows down as it approaches the stagnation point B. Putting $u=U$, $p=p_0$ at A and $u=0$, $p=p_s$ at B, eqn (4.2) becomes

$$\frac{p_0}{\rho} + \frac{1}{2}U^2 = \frac{p_s}{\rho} \tag{4.4}$$

Rearranging eqn (4.4) yields the velocity in the undisturbed fluid as

$$U = \left[\frac{2(p_s - p_0)}{\rho} \right]^{\frac{1}{2}} \tag{4.5}$$

p_s is measured by tube (a) and p_0 is measured by tube (b). Note that p_s is larger than p_0 by an amount $p_s - p_0 = \frac{1}{2}\rho U^2$. The quantities $\frac{1}{2}\rho U^2$, p_0, and $p_s = p_0 + \frac{1}{2}\rho U^2$ are called the dynamic pressure, static pressure, and total pressure, respectively.

EXAMPLE 4.5 Venturi meter

In a Venturi meter an ideal fluid flows with a volume flow rate Q and pressure p_1 through a horizontal pipe of cross-sectional area A_1 (Fig. 4.4). A constriction of cross-sectional area A_2 is inserted in the pipe and the pressure is p_2 inside the constriction. Derive an expression for the volume flow rate Q in terms of p_1, p_2, A_1, and A_2.

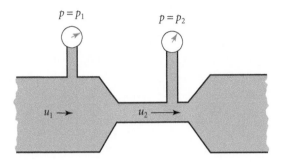

Fig. 4.4 Venturi meter.

From Bernoulli's equation (eqn (4.2)) we have

$$\frac{p_1}{\rho}+\frac{1}{2}u_1^2=\frac{p_2}{\rho}+\frac{1}{2}u_2^2 \tag{4.6}$$

Also, by mass continuity (eqn (4.1)), $\rho u_1 A_1 = \rho u_2 A_2$ or

$$u_2=u_1\frac{A_1}{A_2} \tag{4.7}$$

Eliminating u_2 between eqns (4.6) and (4.7) we obtain the volume flow rate as

$$Q=A_1u_1=A\left[\frac{2(p_1-p_2)}{\rho}\right]^{\frac{1}{2}} \tag{4.8}$$

where $A=A_1A_2(A_1^2-A_2^2)^{-1/2}$.

4.5 Dynamics of a viscous fluid

In general, the motion of a viscous fluid is more complicated than that of an inviscid fluid. A simple case is that of laminar viscous flow between two parallel plates. Consider two parallel plates separated by a small distance d, with one plate moving at constant velocity U and the other plate at rest (Fig. 4.5).

There is no relative velocity between the fluid next to the plates and the plates, owing to strong forces of attraction between the fluid and the surface of the plates. The velocity profile in the fluid is given by

$$u(y)=U\frac{y}{d} \qquad (0\leq y\leq d) \tag{4.9}$$

The viscous shear force per unit area in the fluid is proportional to the velocity gradient, i.e.

$$\frac{F}{A}=-\mu\frac{du}{dy}=-\mu\frac{U}{d} \tag{4.10}$$

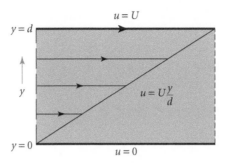

Fig. 4.5 Laminar viscous flow between parallel plates in relative motion.

where the coefficient μ is known as the coefficient of dynamic viscosity.

In the above example, the flow arises from the viscous shear force due to the relative motion of the two plates. For a viscous fluid flow along a pipe, a force needs to be applied in the axial direction (such as gravity or a pressure gradient) to overcome the viscous drag force.

A viscous fluid can exhibit two different kinds of flow regime: laminar flow and turbulent flow. In laminar flow (Fig. 4.6(a)), the fluid slides along distinct stream-tubes and tends to be quite stable, but in turbulent flow the motion is disorderly and unstable (Fig. 4.6(b)).

The particular flow regime that exists in any given situation depends on the ratio of the inertial force to the viscous force. The typical magnitude of this ratio is given by the Reynolds number, defined as

$$Re = \frac{\rho U L}{\mu} = \frac{U L}{\nu} \tag{4.11}$$

where U, L, and $\nu = \dfrac{\mu}{\rho}$ are the characteristic speed, the characteristic length, and the kinematic viscosity. Re is named after Osborne Reynolds, who conducted pioneering experiments on laminar and turbulent flow. He discovered that flows at small Re are predominantly laminar whereas flows at large Re contain regions of turbulence.

(a)   (b)

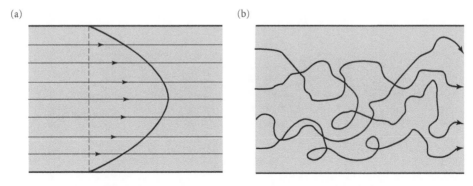

Fig. 4.6 (a) Laminar flow in a pipe. (b) Turbulent flow in a pipe.

EXAMPLE 4.6 Reynolds number

Estimate the Reynolds number for (a) treacle flowing over a plate ($v=10^{-1}\,\mathrm{m^2\,s^{-1}}$, $u=10^{-2}\,\mathrm{m\,s^{-1}}$, $L=0.1\,\mathrm{m}$) and (b) air flowing around a jet aircraft ($v=1.5\times10^{-5}\,\mathrm{m^2\,s^{-1}}$, $u=3\times10^2\,\mathrm{m\,s^{-1}}$, $L=10\,\mathrm{m}$).

The Reynolds numbers in each case are

(a) $Re=\dfrac{UL}{v}\approx\dfrac{10^{-2}\times10^{-1}}{10^{-1}}\ll1$

(b) $Re=\dfrac{UL}{v}\approx\dfrac{3\times10^2\times10^1}{1.5\times10^{-5}}\gg1$

Hence in case (a) the flow is predominantly laminar but in case (b) the flow contains regions of turbulence.

Another important aspect of the Reynolds number is that two different flows with the same Reynolds number, i.e.

$$Re=\frac{\rho_1 U_1 L_1}{\mu_1}=\frac{\rho_2 U_2 L_2}{\mu_2}$$

exhibit geometrically similar behaviour. This is important in engineering because it implies that results obtained from tests on a small scale can be applied to a full-scale model with the same Reynolds number (see Exercise 4.10).

We can derive the above algebraic form of Re from the following dimensional considerations. Consider a fluid flowing with speed U through a cross-sectional area of order L^2, where L is some characteristic length (e.g. the radius in the case of flow around a cylinder). The mass flowing per second is $\sim\rho UL^2$, so the inertial force (i.e. the rate of change of momentum) is $\sim\rho UL^2\times U=\rho U^2 L^2$. Also, from eqn (4.10) the viscous force $\sim\mu AU/L=\mu UL$. Hence the ratio of the inertial force to the viscous force is of order

$$Re=\frac{[\text{inertial force}]}{[\text{viscous force}]}\approx\frac{\rho U^2 L^2}{\mu UL}=\frac{\rho UL}{\mu}=\frac{UL}{v} \tag{4.12}$$

(a)   (b)

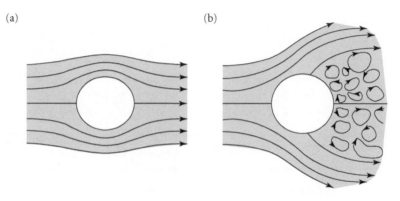

Fig. 4.7 Flow around a cylinder for (a) an inviscid fluid; (b) a viscous fluid.

Figure 4.7 shows the flow around a cylinder for (a) an inviscid fluid and (b) a viscous fluid.

For inviscid flow the velocity fields in the upstream and downstream regions are symmetrical. Hence, the corresponding pressure distribution is symmetrical, and it follows that the net force exerted by the fluid on the cylinder is zero. This startling result is in contradiction to common experience and is an example of d'Alembert's paradox.

For a body immersed in a viscous fluid, the component of velocity tangential to the surface of the fluid is zero at all points on the surface of the body. At large Reynolds numbers ($Re \gtrsim 10^3$), the viscous force is negligible in the bulk of the fluid but is very significant in a viscous boundary layer close to the surface of the body. Rotational components of flow known as vorticity are generated within the boundary layer. At a certain point (known as the separation point) the boundary layer becomes detached from the surface and vorticity is discharged into the body of the fluid. The vorticity is transported downstream of the cylinder in the wake. Thus, the pressure distributions on the upstream side and the downstream side of the cylinder are not symmetrical in the case of a viscous fluid. As a result, the cylinder experiences a net force in the direction of motion, known as the drag force. There is no component of force normal to the direction of the flow past the cylinder, owing to the symmetry of the velocity field above and below the cylinder. However, in the case of a spinning cylinder, a force (called the lift force) does arise at right angles to the direction of flow (see Section 4.6).

We can derive the algebraic form of the lift and drag forces by the following dimensional arguments. By physical intuition, it is reasonable to assume that the lift force $\mathcal{L}$ depends only on the density ρ, the speed U, and the cross-sectional area A projected by the body to the incident flow. Thus we assume the lift force is an algebraic function of the form

$$\mathcal{L} = \frac{1}{2} C_L \rho^a U^b A^c \tag{4.13}$$

where C_L is a dimensionless constant (known as the lift coefficient) and a, b, and c are unknown indices to be determined. Substituting for the physical dimensions of each quantity, we have

$$M^1 L^1 T^{-2} = (ML^{-3})^a (LT^{-1})^b (L^2)^c = M^a L^{-3a+b+2c} T^{-b} \tag{4.14}$$

Equating the indices of like quantities on each side of the equation, we obtain $a = 1$, $b = 2$, and $c = 1$. Hence the lift force is of the form

$$\mathcal{L} = \frac{1}{2} C_L \rho U^2 A \tag{4.15}$$

and likewise the drag force $\mathcal{D}$ is of similar form

$$\mathcal{D} = \frac{1}{2} C_D \rho U^2 A \tag{4.16}$$

where C_D is known as the drag coefficient.

Birds are able to control the lift and drag forces by changing the shape of their wings, the ruffle of their feathers, and the angle of attack of the wings relative to the incident flow. Man has copied nature in designing the shape of an aerofoil for aircraft wings and turbine blades. For small angles of attack, the pressure distribution on the upper surface of an aerofoil is significantly lower than that on the lower surface, resulting in a net lift force on the aerofoil (Fig. 4.8).

Fig. 4.9 shows the variation of the lift and drag coefficients C_L and C_D with angle of attack for a typical aerofoil (NB The drag coefficient has been enlarged by a factor of 5). The lift

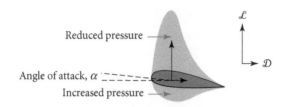

Fig. 4.8 Pressure distribution over aerofoil for small angle of attack.

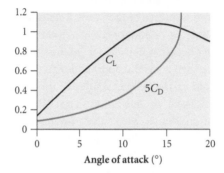

Fig. 4.9 Lift and drag coefficients.

and drag coefficients C_L and C_D cannot be determined by dimensional analysis because they depend on non-dimensional parameters such as the Reynolds number Re, the shape of the body, and the surface roughness. In practice C_L and C_D are obtained from wind tunnel tests on model shapes or from numerical models of the flow.

4.6 Lift and circulation

It is possible to explain lift using inviscid fluid dynamics by introducing the concept of circulation. In order to understand circulation, it is helpful to begin by considering why a spinning ball swerves sideways as it flies through the air, an effect well known to golfers! Spinning creates an imbalance in the pressure on either side of the ball, and generates a net force at right angles to the direction of motion. This is known as the Magnus effect and is illustrated in Fig. 4.10.

We consider an inviscid fluid that is both passing over and rotating around a stationary cylinder, rather than flowing over a rotating cylinder, since it illustrates the same effect and is more like the flow over a stationary aerofoil. Figure 4.10(a) shows the flow of a uniform stream incident on a cylinder. The velocity profile is symmetrical on the upper and lower surfaces, so the resulting pressure distribution is also symmetrical and there is no net sideways force on the cylinder. Figure 4.10(b) shows an inviscid fluid rotating around a cylinder, with a circumferential velocity u_θ that varies inversely with distance r from the centre of the cylinder, i.e.

$$u_\theta = \frac{\Gamma}{2\pi r} \tag{4.17}$$

where Γ is a constant called the circulation. Superposing the velocity profiles shown in Figs 4.10(a) and (b) produces a velocity field in which the fluid moves faster on the upper side

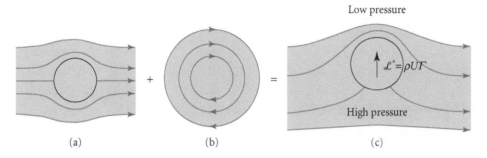

Fig. 4.10 Magnus effect. (a) Flow of a uniform stream around a cylinder; (b) rotating flow around a cylinder; (c) superposition of (a) and (b).

than on the lower side (Fig. 4.10(c)). It follows from Bernoulli's equation (eqn (4.2)) that the pressure is lower on the upper side than on the lower side, so that a net force is exerted on the cylinder at right angles to the incident stream.

The lift force on a body (per unit length in the z-direction) is given by the Kutta–Joukowski lift theorem:

$$\mathcal{L}^* = \rho U \Gamma \tag{4.18}$$

This expression is in fair agreement with experimental observation, despite the fact that the inviscid theory allows the fluid to slip over the surface of the body. However, inviscid theory cannot account for the drag force on a body and it is necessary to consider the effects of viscosity, making the analysis more difficult. A derivation of eqn (4.18) for a cylinder is given in Derivation 4.2. For the equivalent expression for the lift on an aerofoil, see Derivation 4.3.

Derivation 4.2 Magnus effect

Figure 4.10(c) shows the superposition of a uniform stream and a rotational flow around a cylinder. The velocity is enhanced on the upper side of the cylinder and reduced on the lower side. The circumferential component of velocity on the surface of the cylinder is given by (see Exercise 4.13)

$$u_\theta = -2U \sin \theta - \frac{\Gamma}{2\pi a} = -U(2 \sin \theta + B) \tag{4.19}$$

where U is the speed of the uniform stream, Γ is a constant (i.e. the circulation), a is the radius of the cylinder, and B is a non-dimensional parameter given by $B = \dfrac{\Gamma}{2\pi U a}$. For $B < 2$, there are two stagnation points on the surface of the cylinder, obtained by putting $u_\theta = 0$ in eqn (4.19), i.e. $\sin \theta_s = -\frac{1}{2}B$. The fluid is stationary at these points, so from Bernoulli's theorem (eqn (4.2)), the pressure is a maximum. The circulation Γ is given by the angle subtended by the stagnation points, i.e. $\Gamma = -4\pi U a \sin \theta_s$.

From Bernoulli's theorem, the pressure on the surface is given by

$$p = k - \frac{1}{2}\rho u_\theta^2$$

(4.20)

where k is a constant. Substituting for u_θ from eqn (4.19) we have

$$p = k - \frac{1}{2}\rho U^2 (4\sin^2\theta + 4B\sin\theta + B^2) = k - \frac{1}{2}\rho U^2(4\sin^2\theta + B^2) - 2\rho U^2 B\sin\theta$$

(4.21)

The first two terms on the right-hand side of eqn (4.21) are symmetrical on the upper and lower sides of the cylinder. The last term, $-2\rho U^2 B\sin\theta$, is negative for $0 \le \theta \le \pi$ (upper side) and positive for $\pi \le \theta \le 2\pi$ (lower side). Hence there is a net vertical lift force per unit length, given by

$$\mathcal{L}^* = -\int_0^{2\pi} pa\sin\theta \, d\theta = 2\rho U^2 Ba \int_0^{2\pi} \sin^2\theta \, d\theta = \frac{\rho U \Gamma}{2\pi} \int_0^{2\pi} (1 - \cos 2\theta) \, d\theta = \rho U \Gamma$$

(4.22)

Derivation 4.3 Lift and drag of an aerofoil or turbine blade

We begin by assuming that the length of the aerofoil or blade is very long, so that end effects can be ignored. It turns out that it is easier to analyse a cascade of aerofoils than a single aerofoil, because the symmetry of the cascade makes it possible to simplify the calculation.

Consider the control volume $A_1B_1B_2A_2A_1$ enclosing unit length of a single aerofoil, as shown in Fig. 4.11. The streamlines ψ_1 and ψ_2 separate the flow passing over neighbouring aerofoils and are a fixed vertical distance h apart. The vertical planes A_1A_2 and B_1B_2 are chosen to be in regions of roughly uniform flow. Since there is no mass flow across a streamline, it follows that $u = U$. There is also no momentum transfer across the streamlines ψ_1 and ψ_2, and, by symmetry, there is no pressure gradient in the vertical direction on the streamlines ψ_1 and ψ_2. Hence the rate of change in momentum of the fluid in the vertical direction is equal to the vertical component of force f_y on unit length of the aerofoil.

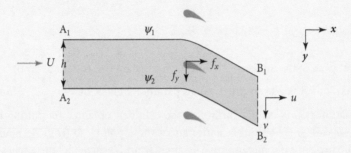

Fig. 4.11 Flow through a cascade of aerofoils.

The mass flow per second crossing B_1B_2 is ρUh, so f_y is given by

$$f_y = -\rho Uhv \tag{4.23}$$

The circulation Γ is given by the closed integral $\Gamma = \oint_C \boldsymbol{u} \cdot d\boldsymbol{s}$, where C is taken to be the contour $C = A_1B_1B_2A_2A_1$. Noting that the contribution along A_1A_2 is zero, and the contributions along the streamlines ψ_1 and ψ_2 cancel one other, the only non-zero contribution is along B_2B_1, given by

$$\Gamma = hv \tag{4.24}$$

Hence

$$f_y = -\rho U\Gamma \tag{4.25}$$

For an aeroplane of mass m, wingspan s, the lift force $F_L = -f_y s = mg$, so the circulation is

$$\Gamma = \frac{mg}{\rho s U} \tag{4.26}$$

Alternatively, putting $v = U \tan \alpha$ (where α is the angle of deflection of the incident stream) in eqn (4.23), then

$$f_y = -\rho h U^2 \tan \alpha \tag{4.27}$$

The kinetic energy of the flow has increased (since the aerofoil imparts a component of velocity v in the y-direction), so power must be supplied, given by $F_D U$, where F_D is called the induced drag force, equal to the increase per unit time in the kinetic energy of the air flow. For a wingspan s, the mass of air flowing per unit time $dm/dt = \rho hsU$, so we have $F_D U = \frac{1}{2}(dm/dt)v^2 = \frac{1}{2}(\rho hsU)U^2 \tan^2 \alpha$ or

$$F_D = \frac{1}{2}\rho hs U^2 \tan^2 \alpha \tag{4.28}$$

The lift force $F_L = -f_y s = mg$, for an aeroplane of mass m, so, using eqn (4.27),

$$mg = \rho hs U^2 \tan \alpha \tag{4.29}$$

Using eqn (4.29) to give $\tan \alpha$, eqn (4.28) becomes

$$F_D = \frac{(mg)^2}{2\rho AU^2} \tag{4.30}$$

where $A = hs$ is the cross-sectional area of air that is deflected by the wing.

We can also deduce the induced drag force F_D from the effect of the shed vorticity (circulation) from a wing of finite length. For such a wing, the circulation Γ decreases towards the wingtip. The shed vorticity produces a downward flow of air, or downwash, at the trailing edge of the wing. Figure 4.12 illustrates the wingtip vortices that are generated, together with the starting vortex and bound vortex that are created when an aircraft takes off.

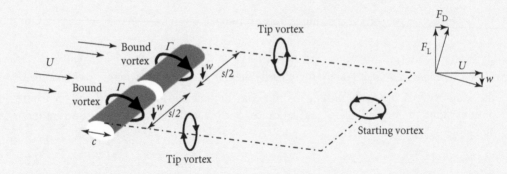

Fig. 4.12 Illustration of the two finite length wings of an aeroplane, showing bound, tip, and starting vortices.

The downwash flow can be estimated by approximating the shed vortices by two tip vortices of strength Γ (the same as that of the bound and starting vortex) and taking the downwash velocity w to be that at the centre of the wing span a distance $s/2$ from one tip vortex. The result is $w = \Gamma/\pi s$, or, using eqn (4.26), $w = mg/(\rho U \pi s^2)$.

The effect of the downwash is to tilt the lift force by an angle $\tan^{-1}(w/U) \sim w/U$, as $w \ll U$ (see Fig. 4.12), and the induced drag force F_D is given approximately by

$$F_D = F_L \frac{w}{U} = \frac{a(mg)^2}{\rho \pi s^2 U^2} \tag{4.31}$$

where a is a correction factor to take account of the shape of the wing and the actual effect of all the shed vorticity. For an elliptical wing shape, $a = 2$, and it follows that the effective cross-sectional area of air deflected downwards in this case is, on comparing eqns (4.30) and (4.31), that of a circle of diameter s.

The total drag force on an aeroplane is the sum of the induced and parasitic drag that arises from the effects of viscosity. From eqn (4.16) the parasitic drag depends on U^2 while the induced drag depends on U^{-2}, so there is an optimum speed when the total drag is a minimum and this determines the speed passenger aeroplanes fly (see Exercise 4.15).

According to inviscid flow theory for flow over an aerofoil with a sharp trailing edge (see, for example, Acheson 1990), there is only one value of the circulation such that the velocity is finite at all points on the surface of the aerofoil, given by

$$\Gamma = \pi U c \sin \alpha \tag{4.32}$$

where c is the width (or chord) of the aerofoil. (Note the mathematical similarity between the expression for the circulation around a spinning cylinder, $\Gamma = -4\pi U a \sin \theta$, and eqn (4.32) for an aerofoil.)

Combining eqns (4.18) and (4.32), we can write the lift force on a single aerofoil of length s in the form

$$\mathcal{L} = \rho U s \Gamma = \pi \rho U^2 s c \sin \alpha \tag{4.33}$$

Equation (4.33) is of the same algebraic form as eqn (4.15), obtained from simple dimensional analysis. The essential difference is that eqn (4.33) gives an explicit expression for the dimensionless lift coefficient, i.e. $C_L = (2\pi sc/A)\sin\alpha = 2\pi\sin\alpha$. This predicts that C_L equals unity at an angle of 9°, which is close to what is observed (see Fig. 4.9).

The physical justification for the concept of circulation arises from the way that vorticity is generated when a body starts to move from rest. In the early stages vorticity is generated around the leading edge, which is swept towards the trailing edge and then shed downstream in the wake, leaving an equal and opposite rotational flow around the aerofoil. This is why aircraft have to wait on the runway to allow time for the shed vortices generated by the previous aircraft to disperse. The effect of viscosity is therefore to produce circulation and the lift force on the aerofoil then follows from inviscid theory.

4.7 Euler's turbine equation

In most types of power generation the kinetic energy of a moving fluid is converted by a turbine into the rotational motion of a shaft. The turbine blades deflect the fluid and the rate of change of angular momentum of the fluid is equal to the net torque on the shaft.

A fluid of density ρ flowing through the turbine with a volume flow rate Q has a mass flow per second given by ρQ. Suppose that the fluid enters at a radius r_1 with a circumferential velocity v_{t1} and exits at a radius r_2 with a circumferential velocity v_{t2} (Fig. 4.13).

The torque exerted on the turbine is equal to the rate of change of angular momentum. Thus

$$T = \rho Q(r_1 v_{t1} - r_2 v_{t2}) \tag{4.34}$$

The power delivered to a turbine rotating with angular velocity ω is given by

$$P = \omega T \tag{4.35}$$

Substituting for T from eqn (4.34) in eqn (4.35) yields the power as

$$P = \omega\rho Q(r_1 v_{t1} - r_2 v_{t2}) \tag{4.36}$$

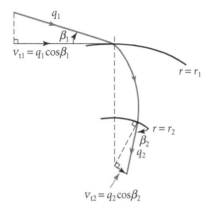

Fig. 4.13 Velocity diagram for Euler's turbine equation.

Writing the tangential velocity in the form $v_t = q\cos\beta$, where q is the total velocity of the fluid and β is the angle between the direction of motion of the fluid and the tangent to the wheel, eqn (4.36) becomes

$$P = \omega\rho Q(r_1 q_1 \cos\beta_1 - r_2 q_2 \cos\beta_2) \tag{4.37}$$

Equation (4.37) is known as Euler's turbine equation. The importance of Euler's turbine equation is that the details of the flow inside the turbine are irrelevant. All that matters is the total change in the angular momentum of the fluid between the inlet and the outlet. The maximum torque is achieved when the fluid flows out in the radial direction, i.e. when $\cos\beta_2 = 0$. Equation (4.37) then reduces to

$$P = \omega\rho Q r_1 q_1 \cos\beta_1 \tag{4.38}$$

EXAMPLE 4.7

Estimate the maximum power output of a water turbine operating at $50\,\text{Hz}$ with $\rho = 10^3\,\text{kg}$ m^{-3}, $Q = 1\,\text{m}^3\,\text{s}^{-1}$, $r_1 = 1\,\text{m}$, $q_1 = 1\,\text{m s}^{-1}$, $\cos\beta_1 = 0.5$.

Substituting in eqn (4.35) yields the power output as $P = (2\pi \times 50)(10^3)(0.5) \approx 160\,\text{kW}$.

 SUMMARY

- The basic physical properties that describe a fluid are density ρ (mass per unit volume), pressure p (normal force per unit area), and viscosity (shear force due to relative motion).

- Mass flow is conserved along a stream-tube, i.e.

  $\rho u A = \text{const.}$

- Conservation of energy for steady fluid flow of an inviscid fluid is given by Bernoulli's equation

  $$\frac{p}{\rho} + gz + \frac{1}{2}u^2 = \text{const.}$$

- The viscous shear force per unit area is proportional to the velocity gradient. For one-dimensional flow

  $$\frac{F}{A} = -\mu\frac{du}{dy}$$

  where μ is the coefficient of dynamic viscosity.

- The Reynolds number

  $$Re = \frac{\rho U L}{\mu}$$

is a useful parameter for distinguishing between laminar flow and turbulent flow. Flows at $Re \ll 10^3$ are essentially laminar; flows at $Re \gg 10^3$ exhibit turbulence.

- Two flows with the same Reynolds number are dynamically similar.

- Dimensional analysis gives an algebraic expression for the forces acting on a body in a moving fluid. The drag force and the lift force have the same form

$$F = \frac{1}{2} C \rho\, u^2 A \qquad (C = C_L \text{ for lift and } C = C_D \text{ for drag})$$

C_L and C_D depend only on non-dimensional parameters (Reynolds number, shape of the body, angle of attack, surface roughness).

- For an aerofoil inclined at a small angle of attack to the incident flow, the pressure on the upper surface is lower than that on the lower surface, creating an upwards lift force.

- The existence of lift can be explained using inviscid theory by introducing the concept of circulation.

- The lift force per unit length is given by the Kutta–Joukowski theorem

$$\mathcal{L}^* = \rho U \Gamma$$

(to a good approximation), where Γ is the circulation. For an aerofoil, the circulation is given by

$$\Gamma = \pi U c \sin \alpha$$

- For an aircraft travelling at a speed U, the induced drag depends on U^{-2} while the parasitic drag depends on U^2, so there is an optimum speed when the total drag is a minimum, and this determines the speed at which passenger aeroplanes fly.

- Euler's turbine equation relates the power output P from a turbine to the rate of change of angular momentum of the fluid between the inlet and outlet of the turbine. The details of the flow inside the turbine are irrelevant.

FURTHER READING

Acheson, D.J. (1990). *Elementary fluid dynamics*. Clarendon Press, Oxford. Mathematical introduction to fluid mechanics.

Douglas, J.F., Gasiorek, J.M., and Swaffield, J.A. (2001). *Fluid mechanics*. Prentice-Hall, Englewood Cliffs, NJ. Textbook on fluid mechanics–good discussion of dimensional analysis and of turbines.

Massey, B.S., Ward-Smith, J., Ward-Smith, A.J. (2005). *Mechanics of fluids*, 8th edn. Taylor & Francis, London. Fluid mechanics for engineers.

LIST OF MAIN SYMBOLS

A	area	C_D	drag coefficient
c	wing chord	C_L	lift coefficient

$\mathcal{D}$	drag force	T	torque
g	acceleration due to gravity	U	characteristic speed
$\mathcal{L}$	lift force	u,v,w	velocity
$\mathcal{L}^*$	lift force per unit length	z	vertical coordinate
L	characteristic length	β	angle
p	pressure	μ	coefficient of viscosity
q	total speed	ρ	density
Q	volume flow rate	v	kinematic viscosity
Re	Reynolds number	ψ	stream function
s	wing span	ω	angular velocity
t	time	Γ	circulation

? EXERCISES

4.1 What happens to the velocity of cars when two slow-moving lanes of cars converge into a single lane, assuming the spacing between cars remains constant?

4.2 Why does an open door swing shut when air blows through the doorway?

4.3 A Pitot tube uses two water manometers to measure pressure. Calculate the speed due to a difference in height of 1 cm between the manometers.

4.4 Verify that the volume flow rate through a Venturi meter is given by

$$Q=A_1u_1=A\left[\frac{2(p_1-p_2)}{\rho}\right]^{\frac{1}{2}}, \text{ where } A=A_1A_2(A_1^2-A_2^2)^{-\frac{1}{2}}$$

4.5 Verify that the Reynolds number is a dimensionless parameter.

4.6 Estimate the Reynolds number Re for a body in an air stream, for $L=10\,\text{mm}$, $u=1\,\text{m s}^{-1}$, $\rho=1.3\,\text{kg m}^{-3}$, and $v=10^{-6}\,\text{m}^2\,\text{s}^{-1}$.

4.7 Verify that all the terms appearing in Bernoulli's equation (eqn (4.2)) have physical dimensions of the form L^2T^{-2}.

4.8 A fountain shoots vertically upwards with speed u_0. Use dimensional analysis to derive an expression for the maximum height h in terms of u_0 and g.

4.9 A jet of water emerges from an orifice in a dam at a depth h below the water surface. Using Bernoulli's equation, and the fact that the surface of the water in the dam and the jet are both at atmospheric pressure, show that the velocity of the jet on leaving the orifice is given by $u=\sqrt{2gh}$.

4.10 It is desired to examine the flow over a model wind turbine using water instead of air. Assuming the kinematic viscosities ($v=\mu/\rho$) of air and water are $1.5\times10^{-5}\,\text{m}^2\text{s}^{-1}$ and $10^{-6}\,\text{m}^2\text{s}^{-1}$, respectively, and that the model is 100 times smaller than the full size turbine, what is the ratio of the speed in the water to that in air, in order for the Reynolds number to be the same in both cases?

4.11* A viscous fluid flows in the x-direction between parallel plates at $y=0$ and $y=b$, under the action of a pressure gradient $\dfrac{dp}{dx}=$ const. Given the momentum equation for the fluid is $\mu\dfrac{d^2u}{dy^2}=-\dfrac{dp}{dx}$, show that the velocity profile is given by $u=-\dfrac{1}{2\mu}\dfrac{dp}{dx}\,y(y-b)$.

4.12 Design an experiment to examine the Magnus effect on a rotating cylinder in a flowing stream of water to investigate how the sideways force varies with the angular velocity of the cylinder and the velocity of the stream.

4.13* A uniform stream of an ideal fluid flows around a cylinder. (a) Given the radial and azimuthal components of velocity in polar coordinates, $u_r=\dfrac{1}{r}\dfrac{\partial\psi}{\partial\theta}$, $u_\theta=-\dfrac{\partial\psi}{\partial r}$, show that u_r and u_θ satisfy the equation

$$\frac{1}{r}\frac{\partial}{\partial r}(ru_r)+\frac{1}{r}\frac{\partial u_\theta}{\partial\theta}=0$$

and also show that ψ satisfies Laplace's equation,

$$\frac{1}{r}\frac{\partial}{\partial r}\left(r\frac{\partial\psi}{\partial r}\right)+\frac{1}{r^2}\frac{\partial^2\psi}{\partial\theta^2}=0$$

(b) Verify that $u_r\equiv0$ and $u_\theta=\dfrac{\Gamma}{2\pi r}$ are valid solutions.

(c) Show that $\psi=Ur\sin\theta\left(1-\dfrac{a^2}{r^2}\right)$ satisfies Laplace's equation, and hence show that

$$u_r=U\left(1-\frac{a^2}{r^2}\right)\cos\theta \text{ and } u_\theta=-U\left(1+\frac{a^2}{r^2}\right)\sin\theta$$

(d) Derive expressions for velocity components due to the superposition of (b) and (c).

4.14 Estimate the lift on an aircraft with wingspan $s=10$ m, chord $c=2$ m, $\alpha=1°$, flying at $900\,\mathrm{m\,s^{-1}}$, (see Fig. 4.12; assume $\rho=1\,\mathrm{kg\,m^{-3}}$.)

4.15 By considering the aerodynamic and induced drag on an aeroplane, show that its optimum speed is given by

$$u_0=\left[\frac{4(mg)^2}{AC_D\pi s^2\rho^2}\right]^{\frac{1}{4}}$$

where m is the mass, s is the wingspan, C_D is the drag coefficient, A is the frontal area of the aeroplane, ρ is the density of air, and g is the acceleration due to gravity.

Calculate the optimum speed for an airliner of mass 325 tonnes travelling at an altitude of 10 km, where the air density is $0.41\,\mathrm{kg\,m^{-3}}$. The airliner has a drag coefficient $C_D=0.06$, a frontal area $A=110\,\mathrm{m^2}$, and a wingspan $s=70$ m.

4.16 Use Euler's turbine equation to estimate the maximum power output of a water turbine operating at 50 Hz such that

$$\rho=10^3\,\mathrm{kg\,m^{-3}},\,Q=10\,\mathrm{m^3\,s^{-1}},\,r_1=10\,\mathrm{m},\,q_1=5\,\mathrm{m\,s^{-1}},\,\cos\beta_1=0.4$$

4.17 Write an article of about 100 words for a popular science magazine on the proposition that 'without viscosity, birds could not fly and fish could not swim'. (NB Assume the readers have no knowledge of mathematics or fluid mechanics.)

5 Hydropower, tidal power, and wave power

Introduction

In this chapter we investigate three different forms of power generation that exploit the abundance of water on Earth: hydropower, tidal power, and wave power. Hydropower taps into the natural cycle of

$$\text{solar heat} \rightarrow \text{seawater evaporation} \rightarrow \text{rainfall} \rightarrow \text{rivers} \rightarrow \text{sea}$$

Hydropower is an established technology, which accounted for ~16% of global electricity production in 2010, making it by far the largest source of renewable energy. The energy of the water is either in the form of potential energy (reservoirs) or kinetic energy (e.g. rivers). In both cases electricity is generated by passing the water through large water turbines.

Tidal power is a special form of hydropower that exploits the bulk motion of the tides. Tidal barrage systems trap seawater in a large basin and the water is subsequently drained through low-head water turbines. In recent years, rotors have been developed that can extract the kinetic energy of underwater currents.

Wave power is a huge resource that is largely untapped. The need for wave power devices to be able to withstand violent sea conditions has been a major problem in the development of wave power technology. The energy in a surface wave is proportional to the square of the amplitude and typical ocean waves transport about 30–70 kW of power per metre width of wave-front. Large amplitude waves generated by tropical storms can travel vast distances across oceans with little attenuation before reaching distant coastlines. Most of the best sites are on the western coastlines of continents between the 40° and 60° latitudes, above and below the equator.

5.1 **Hydropower**

The power of water was exploited in the ancient world for irrigation, grinding corn, metal forging, and mining. Waterwheels were common in western Europe by the end of the first millennium; over 5000 waterwheels were recorded in the Domesday Book of 1086 shortly after the Norman conquest of England. The early waterwheels were of the undershot design (Fig. 5.1(a)) and very inefficient. The development of overshot waterwheels (Fig. 5.1(b)) and improvements in the shape of the blades to capture more of the incident kinetic energy of the stream led to higher efficiencies (~66%).

A breakthrough occurred in 1832 with the invention of the Fourneyron turbine, a fully submerged vertical-axis device that achieved efficiencies of over 80%. Fourneyron's novel idea was to employ fixed guide vanes which directed water radially outwards into the gaps between moving runner blades as shown in Fig. 5.2. Moreover, the head was not limited to the diameter of the water wheel (as in overshot wheels) since the water was contained in a pipe. Many designs of water turbines incorporating fixed guide vanes and runners have been developed since. Modern water turbines are typically over 90% efficient.

There are two main types of hydropower plant systems: dams and run-of-river (ROR) schemes. RORs use a very small reservoir (pondage) or none at all, while dams have a large

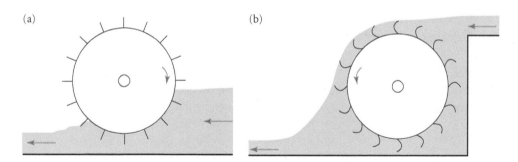

Fig. 5.1 (a) Undershot and (b) overshot waterwheels.

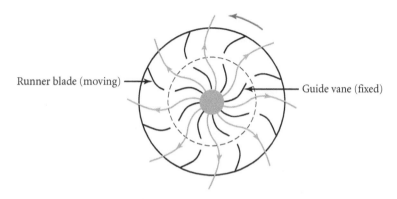

Runner blade (moving) — Guide vane (fixed)

Fig. 5.2 Fourneyron water turbine.

reservoir. ROR plants without pondage are very sensitive to water flow variations, while those with pondage are less so. As ROR systems do not impound a large volume of water their environmental impact is generally small.

The main economic advantages of hydropower are low operating costs, minimal impact on the atmosphere, quick response to sudden changes in electricity demand, and long plant life—typically 40 years or more before major refurbishment. However, the capital cost of construction of dams is high and the payback period is very long. There are also serious social and environmental issues to be considered when deciding about a new hydroelectric scheme, including the displacement of population, sedimentation, changes in water quality, impact on fish, and flooding. A notable example is the Three Gorges dam on the Yangtze River in China, originally envisaged in 1919 with the first plans in the 1930s. These plans were revived in the 1980s as a way of providing electric power as well as reducing the risk of flooding: 18 million people were displaced and 33 000 killed following a flood in 1954. However, while the dam has a maximum output of 22.5 GWe, 1.3 million people were displaced by the project and there are concerns over the increase in landslides around the dam.

Mountainous countries like Norway and Iceland are virtually self-sufficient in hydropower but, in countries where the resource is less abundant, hydropower is mainly used to satisfy peak-load demand. The hydroelectric capacity by country and the largest sites are shown in Tables 5.1. and 5.2, respectively.

5.2 Power output from a dam

Consider a turbine situated at a vertical distance h (called the head) below the surface of the water in a reservoir (Fig. 5.3). The power output P is the product of the efficiency η, the potential energy per unit volume ρgh, and the volume of water flowing per second Q:

$$P = \eta \rho g h Q \tag{5.1}$$

Table 5.1 Installed hydropower capacity in 2010

Country	Capacity (GW)
China	210
Brazil	84
USA	79
Canada	74
Russia	50
India	38
Norway	30
Japan	28
France	21
World total	936

Source: IRENA 2012.

Table 5.2 Largest sites for hydropower

Country	Site	Capacity (GW)
China	Three Gorges	22.5
Brazil/Paraguay	Itaipu	14
Venezuela	Guri Dam	10.2
Brazil	Tucurui	8.5
USA	Grand Coulee Dam	7.0
Russia	Sayano Shushenskaya	6.4

Source: Hydro2011a, Hydro2011b.

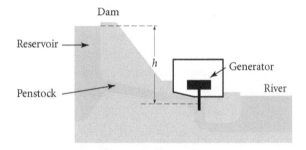

Fig. 5.3 Hydroelectric plant.

Note that the power output depends on the product hQ. Thus a high dam with a large h and a small Q can have the same power output as a ROR installation with a small h and large Q. Which design of water turbine is suitable for a particular location depends on the relative magnitude of h and Q (see Section 5.7).

EXAMPLE 5.1

Estimate the power output of a dam with a head of 50 m and volume flow rate of $20\,\mathrm{m^3\,s^{-1}}$. (Assume $\eta = 1$, $\rho = 10^{-3}\,\mathrm{kg\,m^{-3}}$, $g = 10\,\mathrm{m\,s^{-2}}$.)

From eqn (5.1) we have $P = \eta \rho g h Q \approx 1 \times 10^3 \times 10 \times 50 \times 20 \approx 10\ \mathrm{MW}$.

Case study 5.1 Aswan High Dam (Egypt)

Completed in 1970, the Aswan High Dam lies just north of the border between Egypt and Sudan, and is fed by water from the upper reaches of the River Nile. It contains a vast reservoir known as Lake Nasser (Saad el Aali in Arabic), of surface area $5250\,\mathrm{km^2}$. The Aswan Low Dam, 7 km downriver, came close to being overtopped in 1946, emphasizing the need for urgent action. Prior to the construction of the two dams, the Nile flooded in the rainy seasons, providing fresh water for irrigation and valuable nutrients which sustained the population for many thousands of years—about 95% of the population of Egypt lives within 20 km of the Nile. However, in some years no flooding occurred, leading to drought and famine, whereas in other years there was excess flooding and major crop damage. The vast storage capacity of the reservoir (around 132 km³), together with the ability to control the flow of water through the dam, means that Egypt is now safe from droughts and flooding and, in addition, the Aswan High Dam provides about 2 GWe of hydroelectric power, about 15% of the national generating capacity, from 12 Francis-type turbines (see Section 5.4). Also, the Nile is now navigable throughout the year, boosting shipping and tourism, and the area of land for growing crops along the Nile and in the delta region has been increased significantly. On the negative side, over 100 000 people were displaced by the construction of the Aswan High Dam, the quality of the soil has been degraded so that Egyptian farmers now need to use artificial fertilizers, and the reservoir is a breeding ground for parasites. Sediment retention has also caused increased erosion in the Nile delta and a decline in the local fishing industry. Nonetheless, the consensus of opinion is that the benefits which have accrued from the building of the Aswan High Dam have outweighed the disadvantages.

5.3 Measurement of volume flow rate using a weir

For power extraction from a stream it is important to be able to measure the volume flow rate of water. One particular method diverts the stream through a straight-sided channel containing an artificial barrier called a weir (Fig. 5.4). The presence of the weir forces the level of the fluid upstream of the weir to rise. The volume flow rate per unit width is related to the height of the undisturbed level of water y_{min} above the top of the weir by the formula (see Derivation 5.1)

$$Q = g^{\frac{1}{2}} \left(\frac{2}{3} y_{min} \right)^{\frac{3}{2}}$$ (5.2)

(NB Throughout Chapter 5 the y-axis denotes the vertical direction.)

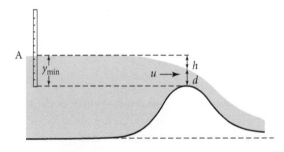

Fig. 5.4 Flow over broad-crested weir.

Derivation 5.1 Flow over a broad-crested weir

Consider a point A on the surface of the water upstream of the weir where the level is roughly horizontal (i.e. $h=0$ in Fig. 5.4) and the velocity u_A. Towards the weir, the level drops and the speed increases. For a broad-crested weir we can ignore the vertical component of velocity and express the volume rate of water per unit width in the vicinity of the crest in the form

$$Q \approx ud \tag{5.3}$$

where d is the depth of the water near the crest. Using Bernoulli's equation (eqn (4.2)), noting that the pressure on the surface is constant (atmospheric pressure), we have $\frac{1}{2}u^2 - gh \approx \frac{1}{2}u_A^2$. Hence, if the depth of the water upstream of the weir is much greater than the minimum depth over the crest of the weir, then $u_A^2 \ll u^2$ and $u \approx (2gh)^{1/2}$. Substituting for u in eqn (5.3) we obtain

$$d \approx \frac{Q}{(2gh)^{\frac{1}{2}}}$$

The vertical distance from the undisturbed level to the top of the weir is $y=d+h$. Substituting for d we have

$$y = \frac{Q}{(2gh)^{\frac{1}{2}}} + h \tag{5.4}$$

The first term on the right-hand side of eqn (5.4) decreases with h but the second term increases with h. y is a minimum when $\dfrac{dy}{dh}=0$, i.e. $-Q/(8gh^3)^{1/2}+1=0$, or

$$h = \left(\frac{Q^2}{8g} \right)^{\frac{1}{3}} \tag{5.5}$$

Finally, substituting for h from eqn (5.5) in eqn (5.4), yields $y_{min} = \dfrac{3}{2}\left(\dfrac{Q^2}{g} \right)^{\frac{1}{3}}$, so that

$$Q = g^{\frac{1}{2}} \left(\frac{2}{3} y_{min} \right)^{\frac{3}{2}}$$

which is known as the Francis formula.

5.4 Water turbines

When water flows through a waterwheel the water between the blades is almost stationary. Hence the force exerted on a blade is essentially due to the difference in pressure across the blade. In a water turbine, however, the water is fast moving and the turbine extracts kinetic energy from the water. There are two basic designs of water turbines: impulse turbines and reaction turbines. In an impulse turbine, the blades are fixed to a rotating wheel and each blade rotates in air, apart from when the blade is in line with a high speed jet of water. In a

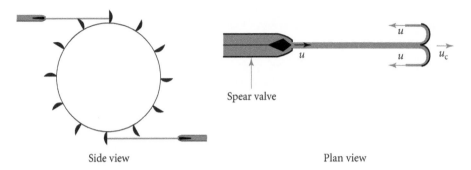

Fig. 5.5 Impulse turbine (Pelton wheel).

reaction turbine, however, the blades are fully immersed in water and the thrust on the moving blades is due to a combination of reaction and impulse forces.

An impulse turbine called a Pelton wheel is shown in Fig. 5.5. In this example there are two symmetrical jets, and each jet imparts an impulse to the blade equal to the rate of change of momentum of the jet. The speed of the jet is controlled by varying the area of the nozzle using a spear valve. Thomas Pelton went to seek his fortune in the Californian Gold Rush during the nineteenth century. By the time he arrived on the scene the easy pickings had already been taken and the remaining gold had to be extracted from rocks that needed to be crushed. Impulse turbines were being used to drive the mills to grind the rocks into small lumps. Pelton observed the motion of the turbine blades and deduced that not all the momentum of the jets was being utilized. He realized that some momentum was being lost because the water splashed in all directions on striking the blades. He redesigned the cups so that the direction of the splash was opposite to that of the incident jet. This produced a marked improvement in efficiency and Pelton thereby made his fortune.

To calculate the maximum power output from a Pelton wheel, we consider a jet moving with velocity u and the cup moving with velocity u_c. Relative to the cup, the velocity of the incident jet is $(u - u_c)$ and the velocity of the reflected jet is $-(u - u_c)$. Hence the total change in the velocity of the jet relative to the cup is $-2(u - u_c)$. The mass of water striking the cup per second is ρQ, so the force on the cup is given by

$$F = 2\rho Q(u - u_c) \tag{5.6}$$

The power output P of the turbine is the rate at which the force F does work on the cup in the direction of motion of the cup, i.e.

$$P = Fu_c = 2\rho Q(u - u_c)u_c \tag{5.7}$$

To derive the maximum power output we put $\dfrac{dP}{du_c} = 0$, yielding $u_c = \frac{1}{2}u$. Substituting in eqn (5.7) then yields the maximum power as

$$P_{max} = \tfrac{1}{2}\rho Q u^2 \tag{5.8}$$

Thus the maximum power output is equal to the kinetic energy incident per second.

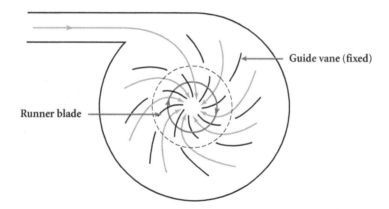

Fig. 5.6 Reaction turbine (plan view).

As in the Fourneyron turbine (see Section 5.1), modern reaction turbines use fixed guide vanes to direct water into the channels between the blades of a runner mounted on a rotating wheel (see Fig. 5.6). However, the direction of radial flow is inward. (In the Fourneyron turbine the outward flow caused problems when the flow rate was either increased or decreased.)

The most common designs of reaction turbines are the Francis turbine and the Kaplan turbine. In a Francis turbine the runner is a spiral annulus whereas in the Kaplan turbine it is propeller-shaped. In both designs the kinetic energy of the water leaving the runner is small compared with the incident kinetic energy.

The term 'reaction turbine' is somewhat misleading in that it does not completely describe the nature of the thrust on the runner. The magnitude of the reaction can be quantified by applying Bernoulli's equation (eqn (4.2)) to the water entering (subscript 1) and leaving (subscript 2) the runner, i.e.

$$\frac{p_1}{\rho} + \frac{1}{2}q_1^2 = \frac{p_2}{\rho} + \frac{1}{2}q_2^2 + E \tag{5.9}$$

where E is the energy per unit mass of water transferred to the runner. Consider two cases: (a) $q_1 = q_2$, and (b) $p_1 = p_2$. In case (a), eqn (5.9) reduces to

$$E = \frac{p_1 - p_2}{\rho} \tag{5.10}$$

i.e. the energy transferred arises from the difference in pressure between inlet and outlet. In case (b), E is given by

$$E = \frac{1}{2}\left(q_1^2 - q_2^2\right) \tag{5.11}$$

i.e. the energy transferred is equal to the difference in the kinetic energy between inlet and outlet. In general, we define the degree of reaction R as

$$R = \frac{p_1 - p_2}{\rho E} = 1 - \frac{(q_1^2 - q_2^2)}{2E} \tag{5.12}$$

(see Example 5.2).

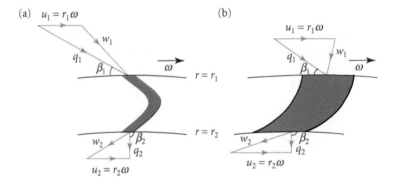

Fig. 5.7 Velocity diagrams for: (a) an impulse turbine; (b) a reaction turbine.

The velocity diagrams in the laboratory frame of reference for an impulse turbine and a reaction turbine are shown in Figs 5.7(a) and (b), respectively. The symbols **u**, **q**, and **w** denote the velocity of the runner blade, the absolute velocity of the fluid, and the velocity of the fluid relative to the blade. Figure 5.7 shows the velocity triangles on the outer radius of the runner, $r=r_1$, and the inner radius, $r=r_2$. The runner rotates with angular velocity ω, so the velocity of the blade is $u_1=r_1\omega$ on the outer radius and $u_2=r_2\omega$ on the inner radius.

The torque on the blade is

$$T=\rho Q(r_1q_1\cos\beta_1-r_2u_2\cos\beta_2)$$

Putting $r_1=u_1/\omega$ and $r_2=u_2/\omega$, the work done per second is given by

$$P=T\omega=\rho Q(u_1q_1\cos\beta_1-u_2q_2\cos\beta_2)$$

The term in brackets represents the energy per unit mass

$$E=u_1q_1\cos\beta_1-u_2q_2\cos\beta_2 \tag{5.13}$$

Equating the incident power due to the head of water h from eqn (5.1) to the power output of the turbine, given by Euler's turbine equation (eqn (4.37)), we have

$$\eta\rho ghQ=\rho Q(u_1q_1\cos\beta_1-u_2q_2\cos\beta_2)$$

The term $\rho Q u_2 q_2 \cos\beta_2$ represents the rate at which kinetic energy is removed by the water leaving the runner.

We define the hydraulic efficiency as

$$\eta=\frac{u_1q_1\cos\beta_1-u_2q_2\cos\beta_2}{gh} \tag{5.14}$$

The maximum efficiency is achieved when the fluid leaves the runner at right angles to the direction of motion of the blades, i.e. when $\beta_2=\dfrac{\pi}{2}$ so that $\cos\beta_2=0$. Equation (5.14) then reduces to

$$\eta_{max}=\frac{u_1q_1\cos\beta_1}{gh} \tag{5.15}$$

EXAMPLE 5.2

Consider a particular reaction turbine in which the areas of the entrance to the stator (the stationary part of the turbine), the entrance to the runner, and the exit to the runner are all equal. Water enters the stator radially with velocity $q_0 = 2$ m s^{-1} and leaves the stationary vanes of the stator at an angle $\beta_1 = 10°$ with an absolute velocity $q_1 = 10$ m s^{-1}. The velocity of the runner at the entry radius $r = r_1$ is u_1 in the tangential direction, and is such that the velocity of the water w_1 relative to the runner is in the radial direction. On leaving the runner, the total velocity is q_2 in the radial direction. Given that the head is $h = 11$ m, calculate the degree of reaction and the hydraulic efficiency.

The volume flow rate is $q_0 A_0$ into the stator, $w_1 A_1$ into the runner, and $q_2 A_2$ out of the runner. Since $A_0 = A_1 = A_2$ it follows by mass conservation that $q_0 = w_1 = q_2$. The energy transfer per unit mass E is given by eqn (5.13). Since the total velocity q_2 leaving the runner is in the radial direction, we have $\beta_2 = \pi/2$. Putting $q_1 \cos \beta_1 = u_1$, then $E = u_1^2$. Also, the square of the total velocity is $q_1^2 = u_1^2 + w_1^2 = u_1^2 + q_2^2$, since w_1 and q_2 are equal and radial. Hence the degree of reaction is $R = 1 - \dfrac{(q_1^2 - q_2^2)}{2E} = 1 - \dfrac{u_1^2}{2u_1^2} = \dfrac{1}{2}$. The hydraulic efficiency is $\eta = u_1 q_1 \cos \beta_1 / gh \approx 0.90$.

5.4.1 **Choice of water turbine**

The choice of water turbine depends on the site conditions, notably the head of water h and the volume flow rate Q. Figure 5.8 indicates which turbine is most suitable for any particular combination of head and volume flow rate. Impulse turbines are suited for large h and low Q, e.g. fast-moving mountain streams. Kaplan turbines are suited for low h and large Q (e.g. ROR sites), and Francis turbines are usually preferred for large Q and large h, e.g. dams. A useful parameter for choosing the most suitable turbine is the shape (or type) number S, described in Derivation 5.2.

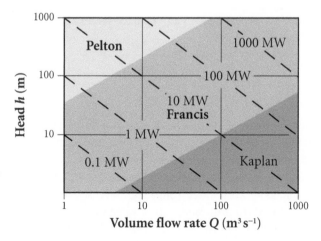

Fig. 5.8 Choice of turbine in terms of head h and volume flow Q.
Source: Boyle2004.

Derivation 5.2 Shape or type number

Dimensional analysis is a useful means for choosing the appropriate type of turbine for a particular combination of h and Q. The power output P from a turbine depends on the head h, the angular velocity ω, the diameter D of the turbine, and the density ρ of water. Various dimensionless parameters can be formed from these physical quantities, the power coefficient $K_p = P/(\omega^3 D^5 \rho)$ and the head coefficient $K_h = gh/(\omega^2 D^2)$ being particularly useful ones. When a turbine of a particular design is operating at its maximum efficiency, K_p and K_h will have particular values which can be used to predict the power and head in terms of the diameter D and the angular velocity ω. We can eliminate the dependence on D (which determines the size of the turbine) by forming the dimensionless ratio

$$S = \frac{K_p^{\frac{1}{2}}}{K_h^{\frac{5}{4}}} = \frac{\omega P^{\frac{1}{2}}}{\rho^{\frac{1}{2}}(gh)^{\frac{5}{4}}} \tag{5.16}$$

called the shape or type number. Substituting $P = \eta \rho gh Q$ from eqn (5.1) and assuming $\eta = 1$, eqn (5.16) becomes

$$S = \frac{\omega Q^{\frac{1}{2}}}{(gh)^{\frac{3}{4}}} \tag{5.17}$$

Putting $Q = v_w A$, $gh = \frac{1}{2}v_w^2$ (where $A = \pi r^2$ is the inlet area and v_w is the speed of the water) and $\omega = 2\pi f = 2\pi \left(\dfrac{v_b}{2\pi R}\right) = \dfrac{v_b}{R}$, where v_b is the speed of the blade tip of radius R, we have

$$S = \frac{\frac{v_b}{R}(v_w \pi r^2)^{\frac{1}{2}}}{\left(\frac{1}{2}v_w^2\right)^{\frac{3}{4}}} = 2^{\frac{3}{4}}\pi^{\frac{1}{2}}\frac{r}{R}\frac{v_b}{v_w} \approx 5\frac{r}{R}\frac{v_b}{v_w} \tag{5.18}$$

For a Pelton turbine $r/R \sim 0.1$, $v_b/v_w \approx 0.5$ and $S \sim 0.25$; for a Kaplan turbine $r \sim R$ and $v_b \sim v_w$, so $S \sim 5$, while for a Francis turbine $S \sim 1$.

5.5 **Micro hydro**

For small or isolated communities with a local supply of flowing water, micro-hydropower installations (typically 5–100 kW) can provide an economical source of electricity. The choice of turbine depends on the volume flow rate and head of water (as explained in Section 5.4), and also on the environmental impact, cost, and reliability. In a typical micro-hydropower installation in a mountainous area, water from a high-level stream could be diverted through a pipe (the penstock) to a Pelton turbine, with a valve to control the rate of flow of water through the turbine. Alternatively, in the case of water flowing over a weir or a waterfall, a simple turbine could be used to convert the kinetic energy into electricity. The electricity can be stored in batteries as a reserve for periods of increased demand. Micro-hydropower schemes avoid the need for dams or connection to transmission networks, they have virtually zero carbon footprint, and they can enable small rural communities to be self-sustaining. One such example is Dyffryn Crawnon, in the Brecon Beacons National Park in Wales, where 15 kW is generated, using a turgo wheel, from a flow of 17 litre s^{-1} and a head of 130 m, enough to sustain a community of about 25 homes.

In the developing world there are numerous micro-hydro installations, for example, in remote communities in the Andes and Himalayas, and in hilly parts of the Philippines, Sri Lanka, and China. When not directly near a community, a local grid can be used to transmit the power, with the voltage stepped up to reduce losses—an example is in Peru where a 10 kV grid is used. In response to fluctuations in demand, the supply can be adjusted by using a load controller or by varying the water supply. Such systems can enable communities to have electricity for the first time. Electricity enables lighting, communications, and the web to be available for schools and homes. Micro-hydro schemes can also complement solar power, since river flows are typically highest in the winter when solar insolation is lowest.

5.6 Impact, economics, and prospects of hydropower

Hydropower sites tend to have a large impact on the local population. Over 1 million people were displaced by the Three Gorges dam in China, and it has been estimated that 30–60 million people worldwide have had to be relocated because of hydropower. Proposed hydropower plants often provoke controversy, and in some countries public opposition to hydropower has stopped all construction except on small-scale projects. Also, dams sometimes collapse for various reasons, e.g. overspilling of water, inadequate spillways, foundation defects, settlement, slope instability, cracks, erosion, or freak waves from landslides in steep-sided valleys around the reservoir. As with nuclear plants, the risk of major accidents is small but the consequences can be catastrophic. Given the long lifetime of dams, even a typical failure rate as low as one per 6000 dam years means that any given dam has a probability of about 1% that it will collapse at some time in its life. In order to reduce the environmental impact and the consequences of dam failure, the question arises as to whether it is better to build a small number of large reservoirs or a large number of small ones. Although small reservoirs tend to be more acceptable to the public than large ones, they need a much larger total reservoir area than a single large reservoir providing the same volume of stored water.

An argument in favour of hydropower is that it does not produce greenhouse gases or acid rain gases. However, water quality may be affected both upstream and downstream of a dam as a result of increases in the concentration of dissolved gases and heavy metals. These effects can be mitigated by inducing mixing at different levels and oxygenating the water by auto-venting turbines. The installation of a hydropower plant can also have a major impact on fish, owing to changes in the habitat, water temperature, and flow regime, and the loss of life around the turbines.

The capital cost of construction of hydropower plants is typically larger than that for fossil fuel plants. Another cost arises at the end of the effective life of a dam, when it needs to be decommissioned. The issue of who should pay for the cost involved in decommissioning is similar to that for nuclear plants: the plant owners, the electricity consumers, or the general public? On the positive side, production costs for hydropower are low because the resource (rainfall) is free. Also, operation and maintenance costs are minimal and lifetimes are long: typically 40–100 years. The efficiency of a hydroelectric plant tends to decrease with age because of the build-up of sedimentation trapped in the reservoir. This can be a lifetime-limiting factor because the cost of flushing and dredging is usually prohibitive.

The economic case for any hydropower scheme depends critically on how future costs are discounted (see Chapter 12). Discounting reduces the benefit of long-term income, disadvantaging hydropower compared with quick payback schemes such as CCGT generation (see Chapter 3). Hydropower schemes therefore tend to be funded by governmental bodies seeking to improve the long-term economic infrastructure of a region rather than by private capital. Typical costs for large, medium, and small hydropower plants are shown in Table 5.3.

The capacity factor of hydropower plants varies but a typical value is 40%. The availability will be affected by the rainfall and so have a seasonal dependence. The amount of rainfall needed to maintain the level in a dam's reservoir can be estimated from the size of the catchment area, the head of the dam, and the power output. For the Three Gorges dam, the catchment area A is $\sim 10^6$ km^2, the head h is ~ 100 m, and the continuous output power P is ~ 10 GWe (22.5 GWe at $\sim 40\%$ capacity factor). The power P is given by $P = \eta \rho g h Q$ (eqn (4.1)). Taking the efficiency $\eta = 90\%$, then $Q = 10^{10}/(0.9 \times 1000 \times 9.81 \times 100) \sim 10^4$ m^3 s^{-1}. Since $Q = A dz/dt$, where dz/dt is the rainfall, then $dz/dt \sim 10^{-8}$ m s^{-1} or ~ 0.3 metres per year, which is about the amount of rainfall required in a region for a large hydropower plant.

5.7 Outlook and potential for hydropower

Hydropower is the largest renewable source of power, producing 3190 TWh of electricity in 2010, equivalent to 364 GWe continuous. This corresponds to 16% of global electricity production and 88% of renewable generation. Concern over climate change from CO_2 from fossil fuel emissions is one of the drivers for developing hydropower and, though climate change could affect water resources, modelling suggests that there would be little overall effect on the global resource.

Table 5.3 The investment costs and levelized cost of electricity (LCE) of large (>10 MW), medium (1–10 MW), and small hydropower plants (0.1–1 MW). The lifetime is assumed to be 50 years, the capacity factor 50%, the discount rate 5%, and the annual O&M 2% of the investment cost

	LHP	MHP	SHP
Investment cost (€/kW)	1200–4600	1400–5500	1800–7300
Typical value	3000	3300	3700
LCE (€/kWh)	0.029–0.08	0.033–0.088	0.04–0.135
Typical LCE	0.051	0.056	0.063

Source: SETIS2011.

Besides being a renewable low-carbon source of energy with a typical carbon footprint of 2–10 gCO_2eq/kWh (mainly related to concrete and steel used in construction), hydropower is relatively cheap, particularly when the dam is also built for flood control or for water storage. It is also a well-established and adaptable source of power: it can turn on quickly, and can therefore meet fluctuations in demand as well providing base load. Many plants operate for more than 50 years. Some hydropower stations are dedicated to supply the power for very energy intensive operations such as the production of aluminium, and hydropower is also used in the production of silicon for solar panels.

Dams can also provide energy storage, enabling the better use of more variable sources of energy, and pumped hydro energy storage (PHES) plants are the most cost-effective form of large energy storage available: ~95 GW of generating capacity is provided by pumped storage, which is ~3% of global generating capacity, and considerably more capacity could be provided by adapting some hydropower plants. A 30 MW seawater PHES plant with a head of 136 m has been built in Okinawa, Japan. Seawater PHES has great potential because of the large number of sites, but prevention of corrosion is important. Another possibility for PHES is to use underground reservoirs, e.g. old mines.

The installed hydropower capacity in 2010 was ~936 GW, making the average capacity factor ~40%. The global technical potential is estimated by the IEA as ~16 400 TWh (~2000 GWe continuous), with over 8000 TWh in China, Russia, United States, Russia, Brazil, and Canada, and over 2500 TWh in India, Indonesia, Peru, and Tajikistan. The potential in the EU is quite well exploited, but considerable expansion is expected over the next decade in China, India, Turkey, Canada, and Latin America, helped in some of these regions by low labour costs. In the developed world the competitive power market has tilted the balance away from capital-intensive projects towards plants with rapid payback of capital.

The capacity factor (currently ~40%) of many systems could be raised by up to 20%, which may be more cost-effective and socially acceptable than large new projects. Research on minimizing the environmental impact of hydropower is particularly important. Whole life-cycle analyses on the impact of projects are necessary to ensure, for example, that water availability for agriculture is not significantly affected. About 19% of the global potential has been developed and the IEA estimates that hydropower could produce ~5750 TWh per year by 2050, equivalent to ~650 GWe continuous, i.e. an *accessible* potential by 2050 of ~650 GWe.

5.8 Tides

There are two high tides and two low tides around the Earth at any instant. One high tide is on the longitude closest to the Moon and the other on the longitude furthest from the Moon. The low tides are on the longitudes at 90° to the longitudes where the high tides are situated. On any given longitude the interval between high tides is approximately 12 hours 25 minutes (see Exercise 5.13). The difference in height between a high tide and a low tide is called the tidal range. The mid-ocean tidal range is typically about 0.5–1.0 m but is somewhat larger on the continental shelves. In the restricted passages between islands and straits the tidal range can be significantly enhanced, by as much as 12 m in the Bristol Channel

(UK) and 13 m in the Bay of Fundy (Nova Scotia). Tidal power has the advantage over other forms of alternative energy of being predictable. For conventional tidal power generation it is necessary to construct huge tidal basins in order to generate useful amounts of electricity. However, in recent years an alternative technology for exploiting strong tidal currents is under development using underwater rotors. The most substantial tidal power plants are at La Rance (France; 240 MW) and Sihwa Lake (South Korea; 254 MW), commissioned in 1966 and 2011, respectively. A 1320 MW plant at Uldolmok in South Korea is due to be completed in 2017.

5.8.1 Physical cause of tides

The main cause of tides is the Moon; the effect of the Sun is about half that of the Moon but increases or decreases the size of the lunar tide according to the positions of the Sun and the Moon relative to the Earth. The daily rotation of the Earth about its own axis only affects the location of the high tides. In the following explanation we ignore the effect of the Sun (see Exercise 5.14).

For simplicity we assume that the Earth is covered by water. Consider unit mass of water situated at some point P, as shown in Fig. 5.9. The gravitational potential due to the Moon is given by $-Gm/s$, where G is the gravitational constant, m is the mass of the Moon, and s is the distance from P to point N the centre of the Moon. For $d \gg r$ we can expand $1/s$ as follows:

$$\frac{1}{s} = \frac{1}{[d^2 + r^2 - 2dr\cos\theta]^{1/2}} = \frac{1}{d}\left[1 + \left(-\frac{2r}{d}\cos\theta + \frac{r^2}{d^2}\right)\right]^{-1/2}$$

$$= \frac{1}{d}\left[1 + \frac{r}{d}\cos\theta + \frac{r^2}{d^2}(\tfrac{3}{2}\cos^2\theta - \tfrac{1}{2}) + \cdots\right]$$

The first term in the expansion does not yield a force and can be ignored. The second term corresponds to a constant force Gm/d^2 directed towards N, which acts on the Earth as a whole and is balanced by the centrifugal force due to the rotation of the Earth–Moon system. The third term describes the variation of the Moon's potential around the Earth. The surface profile of the water is an equipotential surface due to the combined effects of the Moon and the Earth. The potential of unit mass of water due to the Earth's gravitation is gh, where h is the

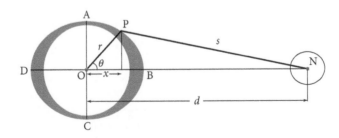

Fig. 5.9 Tidal effects due to the Moon (not to scale).

height of the water above its equilibrium level and $g=GM/r^2$ is the acceleration due to gravity at the Earth's surface, where M is the mass of the Earth. Hence, the height $h(\theta)$ of the tide is given by

$$gh(\theta)-\frac{Gmr^2}{d^3}\left(\frac{3}{2}\cos^2\theta-\frac{1}{2}\right)=0$$

or

$$h(\theta)=h_{\text{max}}\left(\frac{3}{2}\cos^2\theta-\frac{1}{2}\right) \tag{5.19}$$

where

$$h_{\text{max}}=\frac{mr^4}{Md^3} \tag{5.20}$$

is the maximum height of the tide, which occurs at points B and D ($\theta=0$ and $\theta=\pi$). Putting $\frac{m}{M}=0.0123$, $d=384\,400\,\text{km}$, and $r=6378\,\text{km}$, we obtain $h_{\text{max}}\approx0.36$ m, which is roughly in line with the observed mean tidal height.

5.8.2 Tidal waves

There are two tidal bulges around the Earth at any instant. A formula for the speed of a tidal wave in a sea of uniform depth h_0 is obtained from shallow wave theory (see Derivation 5.3) as

$$c=\sqrt{gh_0} \tag{5.21}$$

The tidal bulges cannot keep up with the rotation of the Earth (see Exercise 5.19), so the tides lag behind the position of the Moon, the amount dependent on latitude. The presence of continents and bays significantly disturbs the tides and can enhance their range (see Section 5.11).

5.9 Tidal power

The global resource from tidal energy can be estimated from the power dissipated by tides in the Earth's shelf seas of ~2.5 TW out of a total dissipation of ~3.5 TW. The energy is carried by tidal waves in the oceans into coastal waters where tides and tidal currents, which are potential and kinetic energy stores of tidal energy, respectively, are generated.

The earliest exploitation of tidal power was the tidal mill, created by building a barrage across the mouth of a river estuary. Seawater was trapped in a tidal basin on the rising tide and released at low tide through a waterwheel, providing power to turn a stone mill to grind

Derivation 5.3 Shallow water theory

We consider a wave such that the wavelength λ is much greater than the mean depth of the sea h_0. We also assume that the amplitude of the wave is small compared with the depth, in which case the vertical acceleration is small compared with the acceleration due to gravity, g. Hence the pressure below the surface is roughly hydrostatic and given by

$$p = p_0 + \rho g(h - y) \tag{5.22}$$

where p_0 is atmospheric pressure and $h(x,t)$ is the wave profile on the free surface (Fig. 5.10). The accelerating force on the section of water of unit width between x and $x + \delta x$ is $-\delta x h\, \partial p/\partial x = \rho \delta x h\, \partial u/\partial t$. Hence from differentiating eqn (5.22)

$$\partial u/\partial t = -g\, \partial h/\partial x \tag{5.23}$$

Since $h(x,t)$ is independent of y, it follows that u is also independent of y. This allows us to derive an equation of mass conservation in terms of u and h. Consider a slice of fluid between the planes x and $x + \delta x$. The volume flowing per second is uh across x and $(u + \delta u)(h + \delta h)$ across $x + \delta x$. By mass conservation, the difference in the volume per second flowing from x to $x + \delta x$ is equal to the volume displaced per second $v\delta x$ in the vertical direction. Hence

$$uh = (u + \delta u)(h + \delta h) + v\delta x$$

Putting $v = \dfrac{\partial h}{\partial t}$, $\delta u \approx \dfrac{\partial u}{\partial x}\delta x$, $\delta h \approx \dfrac{\partial h}{\partial x}\delta x$, and noting that $h\partial u/\partial x \gg u\partial h/\partial x$ and $h \approx h_0$ yields the mass continuity equation as

$$-\frac{\partial u}{\partial x} = \frac{1}{h_0}\frac{\partial h}{\partial t} \tag{5.24}$$

Eliminating u between eqn (5.23) and eqn (5.24), we obtain the wave equation

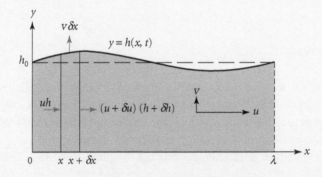

Fig. 5.10 Shallow water wave.

$$\frac{\partial^2 h}{\partial x^2} = \frac{1}{c^2}\frac{\partial^2 h}{\partial t^2} \tag{5.25}$$

for the height profile $h(x,t)$ of the wave, where $c=\sqrt{gh_0}$ is the wave speed or phase velocity. The travelling wave solution to eqn (5.25) is $h=h_0+a\cos(x\pm ct)$, where a is the amplitude of the wave, with $\cos(x-ct)$ representing a wave travelling to the right and $\cos(x+ct)$ a wave to the left. The speed of the water flow u is related to c through eqn (5.24), so $h_0\partial u/\partial x = -ac\sin(x-ct)$ and $h_0 u = ac\cos(x-ct)$. Therefore the speed of water flow u is given by $u = u_0\cos(x-ct)$ where the amplitude of the velocity flow u_0 and of the travelling wave are related by

$$h_0 u_0 = ac \tag{5.26}$$

The power P transmitted by a tidal wave is given by the product of the energy E per unit width per unit length and the group velocity of the wave. Since the speed c is independent of wavelength, the wave is non-dispersive and the group velocity equals the phase velocity, $c=\sqrt{gh_0}$. The energy E is given by $E=\frac{1}{2}\rho g a^2$ (see Derivation 5.6), so the power per unit width $P=\frac{1}{2}\rho g^{3/2}h_0^{1/2}a^2$ or, using eqn (5.26),

$$P=\frac{1}{2}\rho g^{1/2}h_0^{3/2}u_0^2 \tag{5.27}$$

corn. Tidal barrages for electricity generation use large low-head turbines and can operate for a greater fraction of the day. An important issue is whether it is better to use conventional turbines that are efficient but operate only when the water is flowing in one particular direction, or less efficient turbines that can operate in both directions (i.e. for the incoming tide and the outgoing tide).

The first large-scale tidal power plant in the world was built in 1966 at La Rance in France. It generates 240 MW using 24 low-head Kaplan turbines. A number of small tidal power plants have also been built more recently in order to gain operational experience and to investigate the long-term ecological and environmental effects of particular locations. Various proposals during the last century to build a large-scale tidal barrage scheme for the River Severn in the UK have been turned down because of the large cost of construction, public opposition, and the availability of cheaper alternatives—the last time being in 2010 when the estimated cost of £30 billion was deemed too expensive to justify.

5.10 Power from a tidal barrage

A rough estimate of the average power output from a tidal barrage can be obtained from a simple energy balance model by considering the average change of potential energy during the draining process. One mode of operation is for the basin to fill during the last ~3 hours

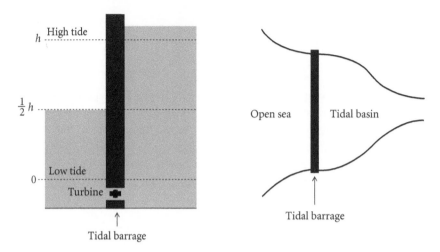

Fig. 5.11 Tidal barrage.

of the flood tide (from mid to high tide), then to raise the height in the basin by pumping for the next ~3 hours (see Exercise 5.15). The situation is then as shown in Fig. 5.11. The basin is drained through the turbines for the next ~6 hours: during the first ~3 hours the difference in level between the water in the basin and outside the barrage is somewhat larger than $h/2$ and then it decreases to zero over the next ~3 hours. The result is that a total mass of water in the tidal basin of area A, $m \approx \rho A h/2$, where h is the tidal range, falls a distance $\approx h/2$. The loss in potential energy is $\sim \rho g A h^2 / 4$. Hence the average power output is

$$P_{\text{ave}} = \frac{\rho g A h^2}{4T} \tag{5.28}$$

where T is the interval between successive tides, i.e. the tidal period. In practice, the power varies with time according to the difference in water levels across the barrage and the volume of water allowed to flow through the turbines. Also, the operating company would seek to optimize revenue by generating electricity during periods of peak-load demand when electricity prices are highest.

EXAMPLE 5.3

Estimate the average power output of a tidal basin with a tidal range of 12 m and a tidal basin area of 520 km² (i.e. Severn barrage).

Substituting in eqn (5.26), noting that the tidal period is $T \approx 4.5 \times 10^4$ s ($T \approx 12.5$ h), the average power output is

$$P_{\text{ave.}} = \frac{\rho g A h^2}{4T} \approx \frac{10^3 \times 10 \times 520 \times 10^6 \times 12^2}{4 \times 4.5 \times 10^4} \approx 4.2 \text{ GW}$$

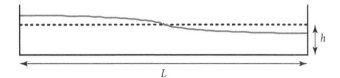

Fig. 5.12 Water of depth h oscillating in a shallow basin of length L.

5.11 **Tidal resonance**

The tidal range varies in different oceans of the world in part because of an effect known as tidal resonance. Figure 5.12 shows water of depth h oscillating in a shallow basin of length L. The surface of the water is horizontal at both ends, since there is no flow there, and the displacement in the middle is zero (node). The water goes up and down at the ends with a frequency f_b given by $f_b = c/\lambda$, where $c = \sqrt{gh_0}$ is the speed of shallow water waves and $\lambda = 2L$ is the wavelength of the oscillation. When the tidal frequency is close to f_b the natural oscillation in the basin builds up and the range—the difference between the maximum and minimum displacements at the ends of the basin—increases. For example, the Atlantic Ocean has a width of about 4000 km and an average depth of about 4000 m, so the speed of a shallow water wave, eqn (5.21), is about $c = \sqrt{gh_0} \approx \sqrt{10 \times 4000} \approx 200\ \mathrm{m\,s^{-1}}$ and $f_b \sim 2.5 \times 10^{-5}$. The tidal frequency is about $2 \times 10^{-5}\ \mathrm{s^{-1}}$, which is close to f_b, so the time taken for the tidal bulge to make the round trip, reflecting off both shores, is about the same as the tidal period, so the amplitude builds up; along the Atlantic shore the tidal range is amplified from ~0.5 m to ~3 m.

River estuaries can also exhibit large tidal resonance if the length and depth of the estuary are favourable. An oscillation in an estuary is represented by the oscillation from the middle to one end of the basin shown in Fig. 5.12: there is no flow at the head but water flows in and out of the mouth of the estuary. The length of estuary for tidal resonance to occur is therefore a quarter of the wavelength of shallow water waves in the depth of the estuary. (See Derivation 5.4.)

Derivation 5.4 Tidal resonance in a uniform channel

For simplicity, consider a uniform channel of length L such that the end at $x=0$ is open to the sea and the other end of the channel at $x=L$ is a vertical wall. The tides at the entrance of the channel will cause the water to flow in and out of the channel. Suppose that the height of the water in the channel is given by

$$h(x,t) = h_0 + a \sin kx \sin \omega t$$

(NB This standing wave solution corresponds to the superposition of two travelling waves going in opposite directions, each with amplitude $a/2$ and where $k = 2\pi/\lambda$.) Then, from the mass continuity equation (5.24), we have

$$-\frac{\partial u}{\partial x} = \frac{1}{h_0}\frac{\partial h}{\partial t} = \frac{\omega a}{h_0} \sin kx \cos \omega t$$

Integrating with respect to x yields the velocity in the horizontal direction as

$$u(x,t) = \frac{\omega a}{h_0 k} \cos kx \cos \omega t$$

At $x=L$, we put $u=0$ (since there cannot be any flow across the barrier); hence $kL=(2n+1)\pi/2$. The shortest channel that satisfies this condition is of length $L=\lambda/4$. When this condition is satisfied, the time taken for the tidal bulge to travel up, be reflected at the head, and travel back to the entrance of the channel matches the time between the tide flowing in and out of the channel. The result is that the tidal range at the head of the channel is amplified—in the River Severn estuary between England and Wales a range of 10–14 m is observed.

EXAMPLE 5.4

The Bristol Channel (UK) has an average depth of ~80 m and the length of the continental shelf in the channel is ~300 km. Show that a tidal resonance enhancement of the tides at the head of the channel is expected.

The frequency of the natural oscillations f_b in the channel is $c/4L$. The speed of shallow waves in the channel is $c=\sqrt{gh}$, so $c=28\,\text{m s}^{-1}$ and $f_b=28/(4\times3\times10^5)=2.3\times10^{-5}\,\text{s}^{-1}$. Since this is close to the tidal frequency of $2\times10^{-5}\,\text{s}^{-1}$, tidal resonance is expected. The actual situation is more complicated since the variable width and depth of the channel also has an effect on the tides.

5.12 Kinetic energy of tidal currents

In particular locations (e.g. between islands) there may be strong tidal currents that transport large amounts of kinetic energy. In recent years, various devices for extracting this energy have been proposed. These devices are essentially underwater versions of wind turbines. For isolated underwater turbines far below the surface, the power generated can be estimated as for a wind turbine (see Chapter 6), with the maximum fraction of the kinetic energy in the flow extracted being given by the Betz limit of 16/27, ~59%, of the amount of kinetic energy flowing through the cross-sectional area of the turbines. However, for turbines in a channel, where the water flow is driven by a tidal oscillation out of phase at either end generating a head of water, the reactive force of the turbines will cause the flow through the channel to decrease. On equating the forces acting on the mass of water in the channel to its acceleration (see Derivation 5.5), the maximum average power P_{max} that can be extracted is given by

$$P_{max} = \gamma \rho g a Q_{max} \tag{5.29}$$

Derivation 5.5 Tidal current potential of channels

For a channel that is short compared with the wavelength of the tide, typically hundreds of kilometres, connecting two basins with different elevations, the equation of motion for the mass of water in a channel of cross-sectional area A and length L is

$$\rho AL \frac{\partial u}{\partial t} = \rho A g a \cos \omega t - F_{\text{turb}} - k'u^2$$

(c.f. eqn (5.23)), where a is the amplitude of the sinusoidal head driving the flow with angular frequency ω, u is the speed of the water flow, $k'u^2$ is the natural friction in the channel, and F_{turb} is the additional frictional force caused by the presence of turbines. When the natural friction parameter k' is large then the acceleration term in the above equation is small compared with the head and can be neglected to a first approximation. Also, since the volume of water flow per second $Q = uA$, the above equation can then be written in the form

$$\rho A g a \cos \omega t \cong F_{\text{turb}} + kQ^2$$

When no turbines are present the maximum flow Q_{max} is given by $Q_{\text{max}}^2 = \rho A g a / k$. The power extracted by the turbines $P = F_{\text{turb}} u = F_{\text{turb}} Q/A$ so

$$P = (\rho A g a \cos \omega t - kQ^2) Q/A$$

This has a maximum, P_{max}, when $Q^2 = \frac{1}{3}(\rho A g a / k) \cos \omega t = \frac{1}{3} Q_{\text{max}}^2 \cos \omega t$, so the maximum average power P_{max} is given by

$$P_{\text{max}} = \frac{2}{3^{3/2}} \rho g a Q_{\text{max}} \langle (\cos \omega t)^{3/2} \rangle = \gamma \rho g a Q_{\text{max}}$$

which is eqn (5.29). The parameter $\gamma = 0.22$ since $\langle (\cos \omega t)^{3/2} \rangle = 0.56$ over half a tidal period (where $\langle X \rangle$ is the average value of X). Consideration of the magnitude and power law dependence of the frictional drag on the flow rate Q gives a 10% uncertainty in γ. Also, the increased flow resistance caused by the turbines could increase the head slightly and thus the power extracted (Garrett and Cummins 2005).

where $\gamma = 0.22 \pm 0.02$, a is the amplitude of the sinusoidal variation with angular frequency ω in the head driving the flow, and Q_{max} is the maximum flow rate through the channel when no turbines are deployed. Note that this expression is algebraically similar to that for the power output from a dam, $P = \eta \rho g h Q$ (eqn (5.1)), where the head h is fixed and the efficiency η is typically ~0.9 (see Example 5.2).

In the majority of designs the axis of rotation of the turbine is vertical and the device is mounted on the seabed or suspended from a floating platform. Before installation, the tidal currents for any particular location need to be measured to depths of 20 m or more in order to determine the suitability of the site. The first generation of prototype kinetic energy absorbers have been operated in shallow water (i.e. 20–30 m) using conventional engineering components. Later generations are likely to be larger and more efficient, and use specially designed low speed electrical generators and hydraulic transmission systems.

5.13 Ecological and environmental impact of tidal power

Tidal stream devices, such as SeaGen in Strangford Lough in Ireland, can be deployed with minimal environmental and ecological effect. However, the installation of a tidal barrage has a major impact on both the environment and ecology of the estuary and the surrounding area:

- The barrage acts as a major blockage to navigation and requires the installation of locks to allow navigation to pass through.
- Fish are killed in the turbines and impeded from migrating to their spawning areas.
- The intertidal wet/dry habitat is altered, forcing plant and animal life to adapt or move elsewhere.
- The tidal regime may be affected downstream of a tidal barrage; for example it has been claimed that a proposed barrage for the Bay of Fundy in Canada could increase the tidal range by 0.25 m in Boston, 1300 km away.
- The water quality in the basin is altered since the natural flushing of silt and pollution is impeded, affecting fish and bird life. In the case of the River Severn in the UK, a barrage would cause the river flow to slow, which would alter the habitat and tend to *increase* biodiversity.

On the positive side, there are the benefits arising from improved flood protection, new road crossings, marinas, and tourism. To reduce the environmental impact, tidal lagoons (see section 5.14) have been suggested that only partially block an estuary.

5.14 Economics and prospects for tidal power

The global tidal resource has been estimated to be ~2.5 TW but with only 3%, i.e. ~75 GW, in regions with sufficient tidal range ($\gtrsim 5$ m) to be economically exploitable by tidal barrages. Large tidal barrages have the economic disadvantages of large capital cost, long construction times, and intermittent operation. On the other hand, they have long plant lives (over 100 years for the barrage structure and 40 years for the equipment) and low operating costs. An alternative idea is to create a closed basin in the estuary known as a tidal lagoon. The wall of a tidal lagoon does not extend across the whole channel so the environmental effects are lessened and the impact on fish and navigation is reduced. Also, by restricting the tidal lagoons to shallow water, the retaining wall can be low and cheap to build. So far, large barrage schemes have not been pursued.

The tidal energy resource can also be exploited by tidal stream devices. The UK has a particularly high fraction of the global resource. A recent (2011) DTI report (based on the method described in Derivation 5.5) estimated that the UK tidal stream technical resource was 2–4 GWe, with the Pentland Firth, a channel connecting the Atlantic to the North Sea with currents of ~3 m s^{-1} accounting for ~35% of the UK resource. The amount of tidal energy dissipated on the European continental shelf is ~200 GW, ~10% of the global dissipation, with about half the European tidal stream resource estimated to be around the UK. Hence, scaling these estimates would give a global tidal stream resource of order 100 GWe, similar to that from tidal barrages. We estimate the accessible potential by 2050 of all tidal energy to be ~50 GWe.

The economics of small tidal current devices (kinetic energy absorbers) has the attraction that they can be installed on a piecemeal basis, thereby reducing the initial capital outlay. They also have a more predictable output than wind turbines and there is no visual impact. The danger to fish is minimal because the blades rotate fairly slowly (typically about 20 revolutions per minute). The long-term economic potential and environmental impact of such devices will become clearer after trials on various designs, notably in the UK, Canada, Japan, Russia, Australia, and China. Estimates (MOTT2011) for the levelized cost of electricity in the UK from tidal stream generation are ~29p per kWh compared with onshore wind at ~9p per kWh and offshore wind at ~17p per kWh. The first large-scale commercial tidal stream generator, SeaGen, was installed in Strangford Lough in Ireland in 2008. It has a generating capacity of 1.2 MWe with a capacity factor of 75–80%, operating on both the flood and ebb tide.

The engineering challenge is to design reliable and durable equipment capable of operating for many years in a harsh marine environment with low operational and maintenance costs.

5.15 **Wave energy**

The waves on the surface of the sea are caused mainly by the effects of wind. The streamlines of air are closer together over a crest, so the air moves faster. It follows from Bernoulli's theorem (see Section 4.4) that the air pressure is reduced, so the amplitude increases and waves are generated. As a wave crest collapses the neighbouring elements of fluid are displaced and forced to rise above the equilibrium level (Fig. 5.13).

The motion of the fluid beneath the surface decays exponentially with depth. About 80% of the energy in a surface wave is contained within a quarter of a wavelength of the surface. Thus, for a typical ocean wavelength of 100 m this layer is about 25 m deep. We now derive an expression for the speed of a surface wave using intuitive physical reasoning. When the depth of water is much greater than the wavelength of the wave, the water particles follow circular trajectories, as shown in Fig. 5.13 (for a mathematical derivation, see Derivation 5.6).

Consider a surface wave on deep water and choose a frame of reference that moves at the wave velocity c, so that the wave profile remains unchanged with time. Noting that the pressure on the free surface is constant (i.e. atmospheric pressure), Bernoulli's equation (eqn (4.2)) yields

$$u_t^2 = u_c^2 + 2gh \tag{5.30}$$

where u_c is the velocity of a particle at a wave crest, u_t is the velocity of a particle at a wave trough, and h is the difference in height between a crest and a trough. If r is the radius of a circular orbit and τ is the wave period, then we can put

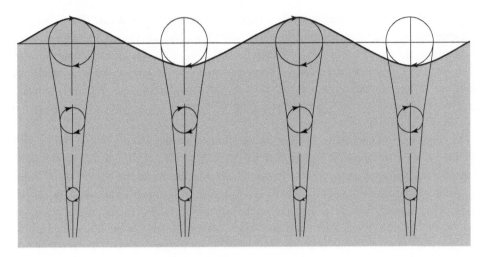

Fig. 5.13 Surface wave on deep water.

$$u_c = \frac{2\pi r}{\tau} - c, \quad u_t = -\frac{2\pi r}{\tau} - c, \quad h = 2r \tag{5.31}$$

Substituting for u_c, u_t, and h from eqn (5.31) in eqn (5.30), and putting $\lambda = c\tau$, we obtain the wave speed as

$$c = \sqrt{\frac{g\lambda}{2\pi}} \tag{5.32}$$

It follows from eqn (5.32) that the wave speed increases with wavelength, so that the surface waves are dispersive. In practice, the wave profile on the surface of the sea is a superposition of waves of various amplitudes, speeds, and wavelengths moving in different directions. The

Derivation 5.6 Potential theory of a surface wave

In potential theory, the viscosity of the fluid is ignored and the velocity field is assumed to be irrotational ($\nabla \times u = 0$). The velocity of the fluid can then be expressed as the gradient of a velocity potential, i.e. $u = \nabla\phi$, so that $u = \dfrac{\partial\phi}{\partial x}$ and $v = \dfrac{\partial\phi}{\partial y}$, where ϕ satisfies Laplace's equation,

$$\nabla^2\phi = \frac{\partial^2\phi}{\partial x^2} + \frac{\partial^2\phi}{\partial y^2} = 0$$

because $\nabla \cdot u = 0$, since the total mass flow out of any volume is zero. Ignoring non-linear terms, the equation of motion of a particle of fluid in the y direction is given by

$$\rho\frac{\partial v}{\partial t} = -\frac{\partial p}{\partial y} - \rho g$$

Putting $v=\partial\phi/\partial y$ and integrating with respect to y, we obtain

$$\rho\frac{\partial\phi}{\partial t}=-p-\rho gy+\text{const.}$$

On the free surface $y=h(x,\ t)$, $p=p_0$ (atmospheric pressure) and we can choose const. $=p_0+\rho gh_0$, yielding the boundary condition

$$\frac{\partial\phi}{\partial t}=-g\left(h-h_0\right) \tag{5.33}$$

A second boundary condition on the free surface is obtained by equating the velocity of a fluid particle on the surface in the vertical direction to $\partial h/\partial t$, i.e.

$$\partial h/\partial t=\partial\phi/\partial y \tag{5.34}$$

We now consider a surface wave of amplitude a and wavelength λ, where $a\ll\lambda$, on a sea of finite depth h_0. We look for a velocity potential of the form

$$\phi=A\cosh ky\ \sin k(x-ct)$$

where $k=2\pi/\lambda$. Note that this expression satisfies Laplace's equation and the boundary condition $v=\partial\phi/\partial y=0$ on the (fixed) bottom surface of the sea, $y=0$.

Applying the boundary condition, eqn (5.34), and putting $y\approx h_0$,

$$\partial h/\partial t=Ak\sinh kh_0\ \sin k(x-ct)$$

Integrating with respect to t yields

$$h=(A/c)\sinh kh_0\cos k(x-ct)+h_0=a\cos k(x-ct)+h_0$$

where $a=(A/c)\sinh kh_0$ is the wave amplitude.

Next, applying the boundary condition, eqn (5.33), we obtain the dispersion relation

$$c^2=\frac{g}{k}\tanh kh_0=\frac{g\lambda}{2\pi}\tanh\left(\frac{2\pi h_0}{\lambda}\right)$$

If $\lambda\gg h_0$, then $\tanh\left(\dfrac{2\pi h_0}{\lambda}\right)\approx\dfrac{2\pi h_0}{\lambda}$ and $c=\sqrt{gh_0}$, i.e. the speed of a tidal wave, eqn (5.21). At the other extreme, $h_0\gg\lambda$, we have $\tanh\left(\dfrac{2\pi h_0}{\lambda}\right)\approx 1$, so that $c\approx\sqrt{\dfrac{g\lambda}{2\pi}}$, the speed of surface waves in deep water, eqn (5.32).

The x- and y-components of velocity (u,v) of a particle of water whose displacement from its mean position at (x,y) is (X,Y) are given by

$$u=\frac{\partial X}{\partial t}=\frac{\partial\phi}{\partial x}=kA\cosh ky\cos k(x-ct),\quad v=\frac{\partial Y}{\partial t}=\frac{\partial\phi}{\partial y}=kA\sinh ky\sin k(x-ct)$$

Integrating with respect to t, we obtain the coordinates of the fluid particle as $X = -\dfrac{A}{c}\cosh ky \sin k(x-ct)$ and $Y = \dfrac{A}{c}\sinh ky \cos k(x-ct)$. Hence, in terms of the wave amplitude $a = (A/c)\sinh kh_0$,

$$\frac{X^2}{\cosh^2 ky} + \frac{Y^2}{\sinh^2 ky} = \frac{a^2}{\sinh^2 kh_0}$$

For $h_0 \gg \lambda$, we have $\sinh kh_0 \approx \cosh kh_0 \approx \tfrac{1}{2}\exp(kh_0)$ and $X^2 + Y^2 = R^2$, where $R = a\exp[-k(h_0-y)]$, and we deduce that a fluid particle at a depth $(h_0 - y)$ rotates in a circular orbit of radius $R = a\exp[-k(h_0-y)]$.

Noting also that then $u^2 + v^2 = k^2 a^2 c^2 \exp[-2k(h_0-y)]$, the total kinetic energy over a complete wavelength, per unit width of wave-front, is given by

$$T = \int_0^\lambda \left[\int_0^{h_0} \frac{1}{2}\rho(u^2+v^2)\,dy \right] dx = \frac{1}{2}\rho k^2 a^2 c^2 \left(\frac{1}{2k}\right)\lambda = \frac{1}{4}\rho k a^2 c^2 \lambda$$

Putting $c^2 = \dfrac{g}{k}$ yields

$$T = \frac{1}{4}\rho g a^2 \lambda$$

net displacement of the surface is therefore more irregular than that of a simple sine wave. Hence, in order for a wave power device to be an efficient absorber of wave energy in real sea conditions, it needs to be able to respond to random fluctuations in the wave profile.

The total energy E of a surface wave per unit width of wave-front per unit length in the direction of motion is given by

$$E = \frac{1}{2}\rho g a^2 \tag{5.35}$$

(see Derivation 5.7). The dependence of wave energy on the square of the amplitude has mixed benefits. Doubling the wave amplitude produces a fourfold increase in wave energy. However, too much wave energy poses a threat to wave power devices and measures need to be taken to ensure they are protected in severe sea conditions.

The power P per unit width in a surface wave is the product of E and the group velocity c_g, given by

$$c_g = \frac{1}{2}\sqrt{\frac{g\lambda}{2\pi}} \tag{5.36}$$

(see Exercise 5.22). Hence the incident power per unit width of wave-front is

$$P = \frac{1}{4}\rho g a^2 \sqrt{\frac{g\lambda}{2\pi}} \tag{5.37}$$

In mid-ocean conditions the typical power per metre width of wave-front is 30–70 kW m^{-1}.

EXAMPLE 5.5

Estimate the power per unit width of wave-front for a wave amplitude $a = 1\,\text{m}$ and wavelength of $100\,\text{m}$.

From eqn (5.37), the power per unit width of wave-front is

$$P = \frac{1}{4}\rho g a^2 \sqrt{\frac{g\lambda}{2\pi}} \approx \frac{1}{4}\times 10^3 \times 10 \times 1^2 \times \sqrt{\frac{10\times 10^2}{2\times 3.14}} \approx 32\,\text{kWm}^{-1}$$

Derivation 5.7 Energy in a surface wave

Consider a wave with a surface profile of the form

$$y = a\sin\left(\frac{2\pi x}{\lambda}\right)$$

as shown in Fig. 5.14. (NB The time dependence is irrelevant for this derivation.)

The gain in potential energy of an elemental mass $\delta m = \rho g\,\delta x\delta y$ of fluid in moving from $-y$ to $+y$ is $\delta V = \delta m\,g(2y) = 2\rho g y\,\delta x\delta y$. Hence the total potential energy of the elevated section of fluid is

$$V = 2\rho g \int_{x=0}^{x=\lambda/2} \int_{y=0}^{y=a\sin(2\pi x/\lambda)} y\,\mathrm{d}y\,\mathrm{d}x = \rho g a^2 \int_{x=0}^{x=\lambda/2} \sin^2(2\pi x/\lambda)\,\mathrm{d}x = \tfrac{1}{4}\rho g a^2 \lambda$$

Assuming equipartition of energy, the average kinetic energy is equal to the average potential energy (see Derivation 5.6), so that the total energy over a whole wavelength is $E = \tfrac{1}{2}\rho g a^2 \lambda$, or

$$E = \tfrac{1}{2}\rho g a^2$$

per unit length in the x-direction per unit width of wave-front.

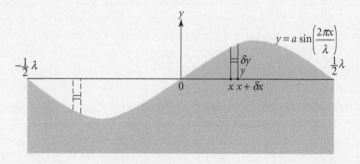

Fig. 5.14 Energy of surface wave.

5.16 Wave power devices

Though the first patent for a wave power device was filed as early as 1799, wave power was effectively a dormant technology until the early 1970s, when the global economy was hit by a series of large jumps in oil prices. Wave power was identified as one of a number of sources of alternative energy that could potentially reduce dependence on oil. It received financial support for assessment of its technical potential and commercial feasibility, resulting in hundreds of inventions for wave power devices, but most of these were dismissed as either impractical or uneconomic. The main concerns were whether wave power devices could survive storms and their capital cost. Through the 1980s, publicly funded research for wave power virtually disappeared as global energy markets became more competitive. However, in the late 1990s interest in wave power technology was revived as a result of increasing evidence of global climate change and the volatility of oil and gas prices. A second generation of wave power devices emerged which were better designed and had greater commercial potential.

In general, the key issues affecting wave power devices are

- survivability in violent storms;
- vulnerability of moving parts to seawater;
- capital cost of construction;
- operational costs of maintenance and repair;
- cost of connection to the electricity grid.

We now describe different types of wave power device and examine how they operate and how they address the above challenges. Good absorbers of wave energy are also good generators of waves; the reason why this is so and how much energy per unit length of wave-front can be extracted is explained in Derivation 5.8.

Derivation 5.8 Wave energy extraction

Linear absorber

A wave incident on a wave power device causes it to oscillate and extract energy and also to generate waves. Figure 5.15 illustrates schematically a wave power device oriented perpendicular to the direction of the wave, i.e. parallel to the wave crests; such a device is called a terminator.

Figure 5.15(a) shows a plane wave travelling in the positive x-direction without the device; Figures 5.15(b) and (c) show the waves generated when the device oscillates forward and backward (pitch) and up and down (surge), respectively; Figure 5.15(d) shows the result when an incident plane wave causes the device to oscillate both ways, each generating waves with *half* the amplitude of the incident wave. The resulting superposition cancels the ongoing part of the incident wave and allows all the incident energy to be extracted. This example illustrates the principle that a good absorber is also a good generator of waves.

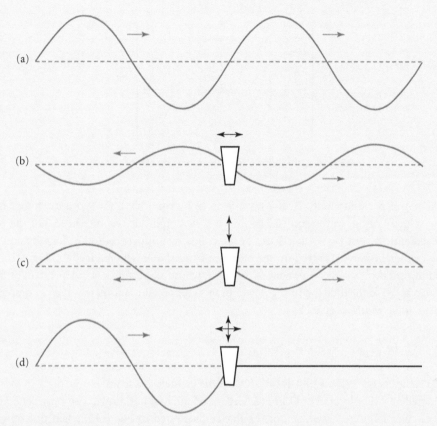

Fig. 5.15 Waves incident on and generated by a wave power device oriented perpendicular to the direction of the wave.

Point absorber

For an absorber whose size is small compared with the wavelength of the wave, the absorber will generate circular waves as it oscillates up and down. To calculate the maximum amount of energy that can be extracted from an incident plane wave, we can express the plane wave in terms of cylindrical waves.

A plane wave $A\cos(kx - \omega t)$ travelling in the positive x-direction can be represented by the real part of the complex exponential function $A\exp[i(kx - \omega t)]$, whose spatial part can be expanded into a sum of ingoing and outgoing cylindrical waves (see Fig. 5.16). At large distances, $kr \gg 1$, from the origin at O,

$$A\exp(ikx) = A\exp(ikr\cos\theta) = A\sum_{n=-\infty}^{\infty}\exp\left[in\left(\theta + \tfrac{1}{2}\pi\right)\right]J_n(kr)$$

$$\rightarrow A\sum_{n=-\infty}^{\infty}\exp\left[in\left(\theta + \tfrac{1}{2}\pi\right)\right](2\pi kr)^{-1/2}\left[\exp(iR_n) + \exp(-iR_n)\right] \tag{5.38}$$

where $k = 2\pi/\lambda$, $x = r\cos\theta$, $R_n = kr - \pi/4 - n\pi/2$, and $J_n(kr)$ is the Bessel function of order n.

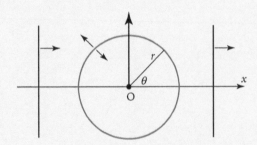

Fig. 5.16 Aerial view of plane and circular waves about origin at O.

The power per unit width, P, of a wave is given by eqn (5.37), $P = \frac{1}{4}\rho g a^2 c$, where c is the phase velocity of the wave (eqn (5.32)) and a is its amplitude. From eqn (5.38), the plane wave extending from $x = -\infty$ to $+\infty$ can be described by a superposition of ingoing and out-going cylindrical waves centred on the origin at O, as shown schematically in Fig. 5.16. The incoming circular ($n=0$) component of the wave, whose amplitude a is independent of θ, has $a = A/(2\pi kr)^{1/2}$, so the total incoming power in this component, noting that the circumference $2\pi r$ is its total 'width', is given by

$$P_0 = \frac{1}{4}\rho g(A^2/2\pi kr)c \times (2\pi r) = \frac{1}{4}\rho g A^2 c/k \tag{5.39}$$

which is the energy in the width $1/k = \lambda/(2\pi)$ of the incident plane wave.

A small absorber located at O in Fig. 5.16, oscillating up and down, can only generate an $n=0$ circular outgoing wave. As for the device illustrated in Fig. 5.15, when this outgoing wave cancels the outgoing circular ($n=0$) wave part of the plane wave, it is possible for the ingoing $n=0$ circular part of the plane wave to be totally absorbed, i.e. the maximum energy that can be absorbed by a small absorber is the energy in the width $1/k = \lambda/2\pi$ of the incident plane wave.

A device of sufficient length whose orientation is parallel to the wave direction, i.e. along the x-axis, called an attenuator, can generate both the $n=\pm1$ as well as the $n=0$ parts of the plane wave, with the result that the maximum energy that can be absorbed is that contained within the width $3/k$ of wave-front of the incident plane wave. An example of such a device is the Pelamis, which is illustrated in Fig. 5.22.

5.17 Spill-over devices

TAPCHAN (TAPered CHANnel) is a Norwegian system in which sea waves are focused in a tapered channel on the shoreline. Tapering increases the amplitude of the waves as they propagate through the channel. The water is forced to rise up a ramp and spill over a wall into a reservoir about 3–5 m above sea level (Fig. 5.17). The potential energy of the water trapped in the reservoir is then extracted by draining the water back to sea through a low-head Kaplan turbine. Apart from the turbine, there are no moving parts and there is

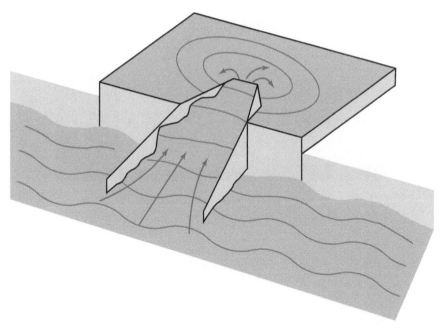

Fig. 5.17 TAPCHAN.

easy access for repairs and connections to the electricity grid. Unfortunately, shore-based TAPCHAN schemes have a relatively low power output and are only suitable for coastal sites where there is a deep-water shoreline and a low tidal range of less than about a metre. To overcome these limitations, a floating offshore version of TAPCHAN called Wave Dragon is under development. It consists of two large wave reflectors which focus sea waves towards a ramp; water overtopping the ramp is collected in a large reservoir and discharged through low-head turbines. A one-third scale device rated at 1.5 MW was successfully tested offshore in Denmark in 2003–2010, and the next stage is to install a pre-commercial device at Milford Haven in Wales.

5.18 Oscillating water columns

The oscillating water column (OWC) uses an air turbine housed in a duct well above the water surface (Fig. 5.18). The base of the device is open to the sea, so that incident waves force the water inside the column to oscillate in the vertical direction. As a result the air above the surface of the water in the column moves in phase with the free surface of the water inside the column and drives the air turbine housed in a duct. The speed of air in the duct is enhanced by making the cross-sectional area of the duct much less than that of the column.

A key feature of the OWC is the design of the air turbine, known as the Wells turbine. It has the remarkable property of spinning in the same direction irrespective of the direction of air flow in the column! Unlike conventional turbine blades, the blades in a Wells turbine are symmetrical about the direction of their motion (Fig. 5.19). Relative to a blade, the direction

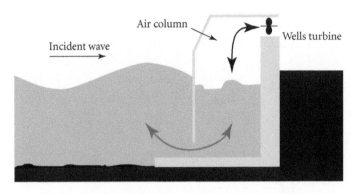

Fig. 5.18 Oscillating water column (OWC).

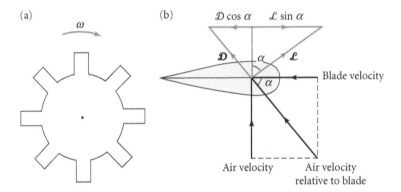

Fig. 5.19 Wells turbine. (a) Plan view of blades; (b) velocity and force triangles in frame of reference of a blade.

of air flow is at a non-zero angle of attack α. The net force acting on the blade in the direction of motion is then given by

$$F = \mathcal{L}\sin\alpha - \mathcal{D}\cos\alpha \qquad (5.40)$$

where $\mathcal{L}$ and $\mathcal{D}$ are the lift and drag forces acting on the blade. It is clear from the force diagram in Fig. 5.19(b) (not to scale: $\mathcal{D} \ll \mathcal{L}$) that the direction of the net force is the same, irrespective of whether the air is flowing upwards or downwards inside the air column.

The shape of the blade is designed to maximize the net force on the blade, and the operational efficiency of a Wells turbine is around 80%. At low air velocities the turbine absorbs power from the generator in order to maintain a steady speed of rotation, while for large air velocities the air flow around the blades is so turbulent that the net force in the direction of motion of the blade becomes erratic and the efficiency is reduced.

Various OWC devices have been developed, with mixed success. The Limpet 500 (Land Installed Marine Pneumatic Energy Transformer) is a small shore-based OWC on the island of Islay off Scotland, and was commissioned in 2000. A larger OWC, known as Osprey (Ocean Swell Powered Renewable Energy), suffered catastrophic damage during installation and was abandoned. A prototype 0.5 MW Australian OWC scheme is also being developed, which uses a 40 m wide parabolic wave reflector to focus waves onto a 10 m wide shoreline OWC; the capital cost is 30% higher but the output is increased by 300%. A large floating OWC known as the Mighty Whale has been successfully developed in Japan. It measures 50 m by 30 m and generates 110 kW from three air turbines; however, its primary role is as a wave breaker to produce calm water for fisheries and other marine activities.

5.19 Submerged devices

Submerged devices have the advantage of being able to survive despite rough sea conditions on the surface. They exploit the change in pressure below the surface when waves pass overhead: the pressure is increased for a wave crest but is decreased in the case of a wave trough. An example of this type of device is the Archimedes wave swing (AWS, Fig. 5.20). The AWS is a submerged air-filled chamber (the 'floater'), 9.5 m in diameter and 33 m in length, which oscillates in the vertical direction in response to the action of the waves. The motion of the floater energizes a linear generator tethered to the seabed. The AWS has the advantage of being a 'point' absorber, i.e. it absorbs power from waves travelling in all directions, and extracts about 50% of the incident wave power. Also, being submerged at least 6 m below the surface, it can avoid damage from violent sea conditions on the surface. The device also has the advantages of simplicity, no visual impact, quick replacement, and cost effectiveness in terms of the power generated per kg of steel. A full-scale pre-commercial pilot system was tested off the coast of Portugal in 2004 and delivered power to the Portuguese grid. The next stage is a long-term demonstration project, using a modified AWS device. A fully commercial AWS system could involve up to six devices per kilometre and it is estimated that the global potential is around 300 GW, corresponding to ~ 20 000 km of coastline.

5.20 Floating devices

In the early 1970s public interest in wave power was aroused by a novel device known as the Salter duck (Fig. 5.21). The device floated on water and rocked back and forth with the

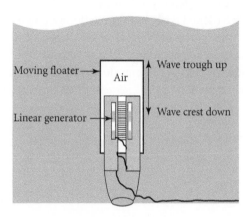

Fig. 5.20 Archimedes wave swing.

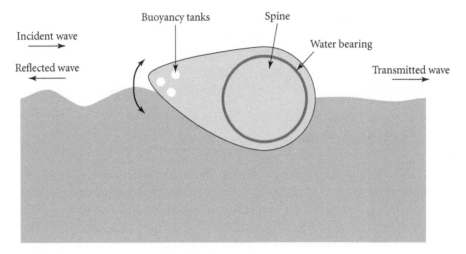

Fig. 5.21 Salter duck.

incident waves and had a cam-shaped cylindrical cross section. Its rear surface was circular so that no waves would be generated behind it by its rocking motion. The front surface profile accommodated the circular trajectories of water particles, so that most of the incident wave energy was absorbed with only minimal reflection and transmission. Efficiencies in excess of 80% were achieved. The complete system envisaged a string of Salter ducks several kilometres in total length, aligned parallel to an incident wave-front. A spinal column, of 14 m diameter, used the relative motion between each duck and the spine to provide the motive force to generate power. The device was designed to be used in the Atlantic Ocean for wavelengths of order 100 m but never got beyond small-scale trials owing to lack of funding in the 1980s when governmental support for wave power in the UK was dropped in favour of wind and nuclear power. Nonetheless, the Salter duck provides a useful benchmark for comparing the efficiencies of all other wave power devices.

Pelamis (Fig. 5.22) is an offshore wave power device designed to operate in depths of 50 m or more. It is of a semi-submerged serpentine construction, consisting of a series of cylindrical hinged segments that are pointed towards the incident waves. As waves move along the device, the segments rock back and forth and the relative motion between adjacent segments activates hydraulic rams that pump high pressure oil through hydraulic motors and drive electrical generators. The second-generation version—Pelamis 2—has five segments, is 180 m long, 4 m in diameter, and weighs 1350 tonnes (including sand ballast). The combination of great length and small cross section to the incident waves provides good protection against

Fig. 5.22 Pelamis.

large amplitude waves. Pelamis 2 has been tested in moderate sea conditions (up to 2.25 m wave height), where it generated an average output of 270 kW over a 30-minute interval, with a capture width of around 20 m. The ultimate aim is to generate 50 MW from an array of 66 Pelamis devices at the European Marine Energy Centre (EMEC), situated in the Orkney Islands north of the Scottish mainland. In order to prevent unwanted interference effects, the devices would be spaced about 60–90 m apart.

5.21 Environmental impact, economics, and prospects for wave power

As with most forms of alternative energy, wave power does not generate harmful greenhouse gases. Opposition to shore-based sites could be an issue in areas of scenic beauty, on account of the visual impact (including the connections to the electricity transmission grid), and the noise generated by air turbines in the case of OWCs. The visual impact is much less significant for offshore devices but providing cables for electricity transmission to the shore is an added cost.

The main challenges for wave power are to reduce the capital costs of construction, to generate electricity at competitive prices, and to be able to withstand extreme conditions at sea. Wave power is generally regarded as a high risk technology. Moving to shore-based and near-shore devices reduces the vulnerability to storms but the power available is less than that further out at sea. Even the largest floating devices are vulnerable in freak storms: every 50 years in the Atlantic Ocean there is a wave with an amplitude about ten times the height of the average wave, so any device must be able to withstand a factor of a hundred times the wave energy; the solution offered by the Pelamis device may be very effective. Measures to combat such conditions, such as submerging the devices, can provide an effective means of defence but adds to the cost of the system. Another factor to consider is that the frequency of incident sea waves is only about 0.2 Hz, much lower than the frequency of 50–60 Hz for electricity transmission. Though this not a difficult electrical engineering problem, the challenge is to find cost-effective solutions.

The global wave energy potential is estimated to be 3 TW when areas affected by sea ice and with very low power ($P < 5 \, \text{kW m}^{-1}$) are excluded. The WEC give ~2000 TWh y^{-1} as the global market potential (~200 GWe) and the DTI (2001) estimated a UK wave power potential of 6 GW. Wave power is beginning to look competitive in certain niche markets, especially in remote locations where electricity supplies are expensive. However, it is likely to take one or two decades to gather sufficient operational experience for wave power to compete with other alternative energy technologies. By 2050, we estimate the accessible potential of wave power at 50 GWe output. In the long term, as fossil fuel reserves become scarce, an estimate of an eventual global potential for wave power to provide about ~5% of total electricity production does not seem unreasonable, as part of a diverse mix of alternative energy sources.

SUMMARY

- The power output P from a dam is $P = \eta \rho g h Q$.

- The dimensionless shape number $S = \omega Q^{1/2} / (gh)^{3/4} \approx 2^{3/4} \pi^{1/2} \dfrac{r}{R} \dfrac{v_b}{v_w}$ is a useful parameter in choosing the most suitable type of turbine for a particular combination of head h and volume flow rate Q.

- The volume flow rate per unit width over a weir is given by $Q = g^{1/2} \left(\frac{2}{3} y_{min} \right)^{3/2}$, where y_{min} is the height of the undisturbed water level above the top of the weir.

- The Fourneyron water turbine employed fixed guide vanes to direct water radially outwards into the gaps between moving runner blades, and was over 80% efficient.

- In an impulse turbine the thrust arises from the momentum imparted by high speed water jets striking the cups. In a Pelton wheel the cups are shaped so that the jets splash in the opposite direction to the incident jet, in order to maximize the transfer of momentum.

- In a reaction turbine the blades are fully immersed in water. Fixed guide vanes direct the water into the gaps between the blades of a runner. The thrust is due to a combination of reaction and impulse forces.

- Combining the formula for the power output of a dam and Euler's turbine equation yields the hydraulic efficiency of a turbine as $\eta = (u_1 q_1 \cos \beta_1 - u_2 q_2 \cos \beta_2)/(gh)$.

- Hydroelectric installations have a high capital cost but low operational costs. Large dams can provide relief from flooding but their environmental impact can be a concern and there can be significant social, safety, and economic issues.

- Hydropower is a large resource of low-carbon energy. The global technical potential from hydro is estimated to be ~2000 GWe and the accessible potential by 2050 ~650 GWe.

- Tidal power is an underdeveloped technology, mainly because of its high capital cost and environmental impact.

- The main cause of tides is the variation of the gravitational attraction of the Moon around the surface of the Earth. There are two tidal bulges, one facing the Moon and the other diametrically opposite.

- The speed of a tidal wave in a sea of uniform depth h_0 is given by $c = \sqrt{gh_0}$.

- The average power output of a tidal barrage is $P_{ave} = \rho g A h^2/(4T)$, tidal range h, period T.

- The height of tides can be increased by tidal resonance.

- Underwater rotors can absorb the energy of water currents and are an alternative means of exploiting tidal power.

- The global tidal energy technical potential is ~2500 GWe with a practical potential of ~75 GWe from tidal barrages and ~100 GWe from tidal stream devices, with ~2–4 GWe around the UK. The accessible potential from tidal power by 2050 is estimated to be 50 GWe.

- The total energy E of a surface wave per unit width of wave-front per unit length in the direction of motion is given by $E = \frac{1}{2} \rho g a^2$. About 80% of the energy is contained within a quarter of a wavelength below the surface.

- The power per unit width of wave-front is $P = \frac{1}{4}\rho g a^2 \sqrt{g\lambda/(2\pi)}$. In mid-ocean conditions the typical power per metre width of wave-front is 30–70 kW m^{-1}.

- Wave power is vast natural resource with a global technical potential of ~3000 GW, with the WEC estimating ~200 GWe for the global economic potential. Significant issues need to be resolved, especially survivability in storms and capital cost. The accessible potential from wave power by 2050 is estimated to be 50 GWe.

- Some shore-based wave power schemes (such as TAPCHAN and OWC) have been shown to be feasible for small-scale operation.

- Large-scale submerged and floating devices (e.g. the Archimedes wave swing and Pelamis) can generate much more power and are undergoing sea trials prior to commercial development.

FURTHER READING

Acheson, D.J. (1990). *Elementary fluid dynamics*. Clarendon Press, Oxford. Good account of shallow water and deep-water waves.

Boyle, G. (ed.) (2004). *Renewable energy*, 2nd edn. Oxford University Press, Oxford. Good qualitative description and case studies.

Douglas, J.F., Gasiorek, J.M., and Swaffield, J.A. (2001). *Fluid mechanics*. Prentice-Hall, Englewood Cliffs, NJ. Textbook on fluid mechanics—good discussion of dimensional analysis and of turbines.

Falnes, J. (2002), *Ocean waves and oscillating systems*. Cambridge University Press, Cambridge. Advanced treatment of energy extraction from waves.

Garrett, C. and Cummins, P. (2005). The power potential of tidal currents in channels. *Proc. R. Soc. A* **461**, 2563–72. Excellent analysis of tidal stream energy extraction.

Kuznetsov, N., Moz'ya, V., and Vainberg, B. (2002). *Linear water waves: a mathematical approach*. Cambridge University Press, Cambridge. Advanced textbook, including modelling of submerged devices.

WEB LINKS

www.ashden.org/micro-hydro Micro-hydro projects in developing countries.

www.bwea.com/ UK wind and marine energy.

www.carbontrust.com/media/77264/ctc799_uk_tidal_current_resource_and_economics.pdf UK Tidal Stream Resource, DTI 2011.

www.ieahydro.org/ Good summary of hydropower generation in the world.

www.oceanor.no/related/59149/paper_OMAW_2010_20473_final.pdf Assessing the global wave energy potential.

www.see.ed.ac.uk/~shs/Tidal%20Stream/April%2007%20I%20MechE%20papers/ Blunden%20and%20Bahaj.pdf Tidal energy resource assessment.

setis.ec.europa.eu/newsroom-items-folder/hydropower-generation Useful source of information on hydropower.

www.thegreenvalleys.org Micro hydro in Wales.

www.worldenergy.org Useful data and overview of current developments.

en.wikipedia.org/wiki/List_of_largest_hydroelectric_power_stations (Hydro2011a).

www.greenworldinvestor.com/2011/03/29/list-of-worlds-largest-hydroelectricity-plants-and-countries-china-leading-in-building-hydroelectric-stations/ (Hydro2011b).

www.irena.org/DocumentDownloads/Publications/RE_Technologies_Cost_Analysis-HYDROPOWER.pdf (IRENA2012).

hmccc.s3.amazonaws.com/Renewables%20Review/MML%20final%20report%20for%20 CCC%209%20may%202011.pdf Marine costs (MOTT2011).

setis.ec.europa.eu/about-setis/technology-map/2011_Technology_Map1.pdf (SETIS2011).

LIST OF MAIN SYMBOLS

a	wave amplitude	R	degree of reaction
c	wave speed	S	shape factor
E	energy	t	time
g	acceleration due to gravity	u, v	velocity components
G	gravitational constant	V	potential energy
h	head, tidal range	x,y,z	coordinates
k	wave number	β	angle
P	power	η	efficiency
p	pressure	λ	wavelength
q	total velocity	ρ	density
Q	volume flow rate	ω	angular velocity

EXERCISES

5.1 Check the units to verify the expression $P = \eta\rho g h Q$ for the power output from a dam.

5.2 Estimate the power output of a dam with a head $h = 100\,\text{m}$ and volume flow rate $Q = 10\,\text{m}^3\text{s}^{-1}$. (Assume efficiency is unity, $\rho = 10^3\,\text{kg m}^{-3}$, $g = 9.81\,\text{m s}^{-2}$.)

5.3 Repeat Example 5.2 with $\beta_1 = 20°$, $q_0 = 6\,\mathrm{m\,s^{-1}}$, $q_1 = 32\,\mathrm{m\,s^{-1}}$, and $h = 100\,\mathrm{m}$.

5.4 Assuming the volume flow rate per unit width over a weir is of the form $Q = g^a y_{min}^b$, use dimensional analysis to determine the numerical values of a and b.

5.5 An abrupt change from h_1, u_1 to h_2, u_2 in the depth and speed of a rapidly flowing stream of water in a horizontal channel (a hydraulic jump) can occur when the initial flow speed u_1 is greater than the speed $(gh_1)^{1/2}$ of shallow water waves. Equate the retarding force arising from the mean pressure difference across the jump to the rate of change of momentum, use the conservation of mass flow, and deduce $h_1 u_1^2 - h_1^2 u_1^2 / h_2 = \frac{1}{2} g \left(h_2^2 - h_1^2 \right)$.

Show $h_2 / h_1 = \frac{1}{2} \left[\sqrt{1 + 8 Fr^2} - 1 \right]$, where the Froude number $Fr = u_1 / \sqrt{gh_1}$.

5.6 The energy flow in a stream of water of depth h and speed u is given by
$$E = \int_0^h \left(p + \tfrac{1}{2} \rho u^2 + \rho g y \right) u \, dy \quad \text{where } p = \rho g (h - y)$$

Show that the energy flow in a region where the depth and speed are h_1 and u_1 is given by $\rho g u_1 h_1^2 + \frac{1}{2} \rho u_1^3 h_1$. Deduce that the energy lost across a hydraulic jump (Exercise 5.5) is given by $\rho g u_1 (h_2 - h_1)^3 / 4 h_2$. Calculate the fractional energy loss when the Froude number for the initial (upstream) flow is 20.

5.7 Draw a sketch of an impulse turbine consisting of four jets.

5.8 Verify that the power output of an impulse turbine is a maximum when $u_c = \frac{1}{2} u$, and that the maximum power delivered to the cup is given by $P_{max} = \frac{1}{2} \rho Q u^2$.

5.9 For a head of water h and a cross-sectional area A of the water jet, show that the maximum power P_m from a Pelton turbine can be written as $P_m \propto A h^{3/2}$. Find the value of the constant of proportionality when P_m is in kW, A is in $\mathrm{m^2}$, and h is in m.

5.10 Explain how a rotary lawn sprinkler works.

5.11 Discuss who should pay for the cost involved in decommissioning dams when they reach the end of their life.

5.12 A turbine is required to rotate at 6 rpm with a volume flow rate of $5\,\mathrm{m^3 s^{-1}}$ and a head of 30 m. What type of turbine would you choose?

5.13 If there are two high tides around the Earth at any instant, explain why the interval between successive high tides is 12 hours 25 minutes rather than 12 hours?

5.14* Compare the magnitude of the effect of the Sun on the tides (a) when the Sun and Moon are both on the same side of the Earth, and (b) when the Sun and the Moon are on opposite sides of the Earth ($m_{Sun} = 2 \times 10^{30}\,\mathrm{kg}$, $m_{Moon} = 7.4 \times 10^{22}\,\mathrm{kg}$, $d_{Sun} = 1.5 \times 10^{11}$ m, $d_{Moon} = 3.8 \times 10^8$ m).

5.15 A tidal barrage is filled during flood tide to a height h above low tide. The level is then raised by pumping to a height y and held until near low tide, when the water within the barrage is discharged quickly through the turbines. Show that the average net power output is given by $P_{ave} = \rho g A [(h + y)^2 - y^2] / (2T)$. Evaluate the fractional increase in output power using a pump that raises the level from $h = 10$ m to 11 m compared with no pumping. Find the optimum value for y when the efficiency for pumping is half that for generating.

5.16* In a tidal basin power plant, seawater is trapped in a basin of area A at high tide and allowed to run out through a turbine at low tide. The difference in sea level between high and low tides, the tidal range R, varies sinusoidally throughout a month, from a maximum R_S for spring tides to a minimum R_N for neap tides. Show that the maximum mean power P_m produced over the month is $P_m = (\rho Ag / 4\tau)R_S^2[(3 + 2\alpha + 3\alpha^2)/8]$ where $R_N = \alpha R_S$, τ is the tidal period, g is the acceleration due to gravity, and ρ is the density of seawater. Calculate the mean power for $A = 12\,\text{km}^2$, $R_S = 6\,\text{m}$, $\alpha = 0.6$, $\rho = 10^3\,\text{kg m}^{-3}$.

5.17   The Pentland Firth is a channel 10.2 km wide with an average depth of 58 m. The maximum flow speed is $3.4\,\text{m s}^{-1}$ and the amplitude of the tidal head driving this flow is 1.2 m. Calculate the maximum average power that could be generated by placing turbines in the channel.

5.18* (a) Show by substitution that the profile $h = a\cos(kx - \omega t) + b\cos(kx + \omega t)$ satisfies the tidal wave equation $\dfrac{\partial^2 h}{\partial x^2} = \dfrac{1}{c^2}\dfrac{\partial^2 h}{\partial t^2}$.
       (b) A uniform channel of length L is bounded at both ends by a vertical wall. Derive the height and velocity profiles of shallow water waves in the channel.

5.19   Show that the speed of a tidal bulge on the equator in the Atlantic Ocean (depth ~ 4000 m) is less than the speed, due to the Earth's rotation, of the seabed.

5.20   Assuming the speed c of surface waves on deep water depends only on the acceleration due to gravity g and the wavelength λ, use dimensional analysis to derive an algebraic expression of the form $c = kg^a\lambda^b$, where k is a dimensionless constant.

5.21   Calculate the speed of a surface wave on deep water of wavelength $\lambda = 100$ m.

5.22   Given that the phase velocity and group velocity of a surface wave are $c = \sqrt{\dfrac{g\lambda}{2\pi}}$ and $c_g = d\omega/dk$, respectively, where ω is the angular velocity and $k = 2\pi/\lambda$, prove that the group velocity is given by $c_g = \dfrac{1}{2}\sqrt{\dfrac{g\lambda}{2\pi}}$.

5.23   Verify that the velocity potential for a surface wave $\phi = A\cosh ky\sin k(x - ct)$ satisfies Laplace's equation and the boundary conditions on the surface and on the bottom of the sea.

5.24   An OWC has an air duct of cross-sectional area $1\,\text{m}^2$ and a water duct of cross-sectional area $10\,\text{m}^2$. If the average vertical speed of the water surface is $1\,\text{m s}^{-1}$ calculate the average speed of the air.

5.25   The maximum width of a wave-front that can be absorbed by a Pelamis wave energy generator is about $3/k$, where $k = 2\pi/\lambda$ and λ is the wavelength. Estimate the average power output of a wave farm containing 25 Pelamis devices moored in an area where the average wave amplitude is 1.0 m and the wave period is 8 s. Take the conversion efficiency of each Pelamis device to be 70%. The density of water is $1000\,\text{kg m}^{-3}$.

5.26   Discuss whether it is better to build a large number of small dams or one large dam.

6 Wind power

Introduction

The international oil crises of the 1970s and, more recently, concern over global warming have renewed interest in wind power. The global installed capacity has grown at an annual rate of 27% since 2000 to 194 GW in 2010, providing ~425 TWh of electricity, 2.1% of the world's demand. The wind is a source of carbon-free and pollution-free energy and wind power could produce ~10–20% of the electrical power used globally by 2050 (IEA2009, GWEC2010). Wind power would therefore save a considerable amount of fuel resources. The modern wind turbine is some 100 times more powerful than the traditional windmills of the seventeenth and eighteenth centuries, and wind farms already generate significant amounts of electricity in some countries—e.g. 16% (42.7 TWh) of Spain's electricity demand in 2010.

We look first at the global wind resources and at the energy available in the wind. How this energy can be extracted using a wind turbine and its design are then described. We then consider the issues associated with siting of turbines, generally as wind farms, which are important for both their output and their environmental impact. We conclude the chapter with a discussion of wind variability and of the economics and potential of wind power.

6.1 **Source of wind energy**

The original source of wind energy is radiation from the Sun, which is primarily absorbed by the land and the sea, which in turn heat the surrounding air. Materials absorb radiation differently, so temperature gradients arise causing convection and pressure changes, which result in winds. A simple example is the offshore night-time wind often found on coasts, caused by the sea retaining the heat from the Sun better than the land. On a global scale, the higher intensity of solar radiation at the equator than elsewhere causes warm air to rise up from the equator and cooler air to flow in from the north and south. The direction of a wind is traditionally described in terms of where it comes from, so in the northern hemisphere the warm air rising up from the equator would give rise to a northerly wind at ground level.

An enormous amount of power resides in the winds as about 0.5% of the incident solar power of $1.37\,\mathrm{kW\,m^{-2}}$ is converted into wind. The radius of the Earth is approximately 6400 km so the cross-sectional area receiving solar radiation is about $2\times10^{14}\,\mathrm{m^2}$ and the power in the winds is $\sim 10^{15}\,\mathrm{W}$. This is some 100 times the total global power usage. However, the wind is a diffuse source and it is only practical to harness a very small fraction of this amount.

Winds are variable both in time and in location, with some parts of the world exposed to frequent high winds and some to almost no wind. Places where high and low winds occur are, in particular, determined by the effect of the rotation of the Earth. Over distances of tens of kilometres, the Earth's rotation has no significant effect on the direction of a wind; however, over hundreds of kilometres the effect is very noticeable. We will now explain how the Earth's rotation affects the global winds.

6.2 **Global wind patterns**

In a simple model, the higher intensity of solar radiation at the equator would set up a north–south convective flow of air if the Earth were not rotating. However, the Earth's rotation causes a point on the equator to have a velocity towards the east that is highest at the equator, decreasing towards the poles. Therefore a wind moving north or south as seen by an observer on the equator will initially have a component of velocity towards the east to an observer in space. As the wind moves away from the equator its distance to the Earth's axis decreases so its component of velocity towards the east increases. (This is a similar effect to ice skaters spinning faster when their arms are pulled in.) Air initially moving north will therefore reach a northern latitude at a point which is east of its origin. For the observer on Earth the wind appears to be accelerating towards the east and the apparent force is called the Coriolis force.

The wind speed would in principle reach large values by high latitudes, but by latitude 30° the flow becomes unstable. As a result, the north or south motion of the wind is dissipated and such winds are thereby restricted to within the 30° latitudes. In the northern hemisphere the sinking air near 30° latitude gives rise to the north-east trade winds and the westerly wind belt, which is the prevailing wind over Europe. A map of the resulting global wind patterns expected in this simple model, which ignores the effects of the underlying configuration of

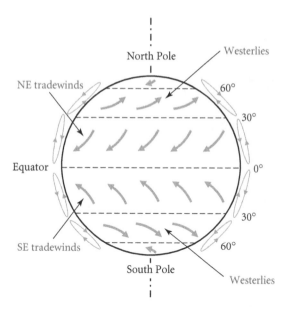

Fig. 6.1 Simplified map of global wind patterns.

oceans and continents, is shown in Fig. 6.1. Notice that there are three regions, called cells, in each hemisphere.

In practice, the effects of surface friction and large-scale eddy motions have a big influence, as do seasonal variations, and only the cell nearest the equator, the Hadley cell, is clearly seen. The mid-latitude Ferrel cell is quite weak and the 'polar' cell is hardly observed. There are many areas where the winds are strong and reliable and it is in these locations that the energy in the wind can be best exploited.

Box 6.1 History of wind power

The first recorded use of wind power was in the tenth century in the Sistan region of Persia, an area where winds can reach speeds of ~45 m s⁻¹, though windmills had probably been in use in the region for several centuries earlier. The windmills had a vertical axis (see Fig. 6.2(a)), and were used to pump water and grind grain. Similar windmills were used in China and may have been first developed there. In these vertical-axis machines the wind pushes the sails around with a force, called a drag force, dependent on the relative speed of the wind and the sail.

Vertical-axis windmills using a drag force are inherently less efficient than horizontal-axis windmills, which are driven by a lift force (see Section 6.6). Horizontal-axis windmills first appeared in Europe in England, France, and Holland around the twelfth century. Their origin is unknown (though it is possible they evolved from the horizontal-axis waterwheel) but such windmills spread rapidly eastward in Europe in the thirteenth

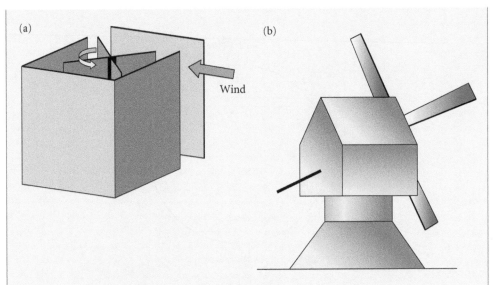

Fig. 6.2 (a) Persian vertical-axis windmill. (b) Tower-mill.

century. They were used for grinding corn, pumping water, and sawing wood. The first mills were post-mills, where the whole mill swivelled on a post so that it could be manually turned to face the oncoming wind. In later (and larger) mills only the top with the sails and windshaft swivelled; these so-called tower-mills (Fig. 6.2(b)) were introduced around the fourteenth century. From experience it was found that more power could be obtained by twisting the sails from the root to the tip of the sail.

The use of windmills peaked around the eighteenth century after which they were displaced by coal-powered steam engines, which were more compact and adaptable, and were continuously available. However, windmills continued in regions where the land was sparsely populated, e.g. in the USA, USSR, Australia, and Argentina. In the nineteenth century small multi-vane windmills were developed in the USA for pumping water, where they became very common. They were eventually displaced with the development of a national electricity grid in the 1930s.

From the late nineteenth century up to the 1960s a number of wind machines were developed for generating electricity and were called wind turbines (to distinguish them from windmills). In the early twentieth century, Poul La Cour built turbines using a four-bladed windmill design that produced about 25 kW. The electricity was used to produce hydrogen, which was then used for lighting. These were subsequently displaced by the introduction of diesel engines, but the production of hydrogen as a fuel is now being seriously considered as hydrogen produces no CO_2 in use (see Section 11.16 on fuel cells). In the late 1930s a massive 1.25 MW two-bladed wind turbine was proposed by Palmer Putman and built in the USA in Vermont by the Morgan Smith

Company. Though it ran successfully for a short while, a blade failure in 1945 caused the project to be terminated.

During the Second World War Denmark used wind energy when oil was not available, though this was only a temporary measure. However, during the international oil crisis in 1973 there was renewed interest in wind power as many countries began looking at sources of alternative energy. In California, concern over high fossil fuel costs led to large-scale investment in wind farms, which was aided by state and federal tax incentives. The technology, though, was not then well developed and several wind farms were unsuccessful. With the cessation of tax incentives and the fall in oil prices in the mid-1980s, investment in wind power in the USA declined. However, in Europe, particularly in Denmark, support was maintained and the more recent alarm about global warming has stimulated considerably more interest and investment in wind turbines worldwide.

6.3 Modern wind turbines

The vast majority of current designs are horizontal-axis wind turbines (HAWTs). The turbine blades are aerofoils. These provide lift forces that drive the turbine. A modern HAWT is illustrated in Fig. 6.3. It consists of a tower on top of which is mounted an enclosure called the nacelle. Inside the nacelle are the bearings for the turbine shaft, the gearbox (if used), and the generator. The wind turbine blades, generally three or two, are mounted to the shaft and

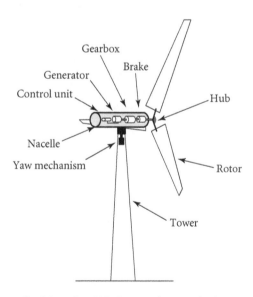

Fig. 6.3 Modern 5 MW horizontal-axis wind turbine.

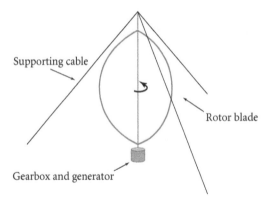

Fig. 6.4 Darrieus vertical-axis wind turbine.

the nacelle is oriented by a drive mechanism, the yaw control, into the wind. The rotor is typically upwind of the tower to avoid the tower shielding the blades from the wind.

There has also been some research and development of vertical-axis wind turbines (VAWTs), in particular the Darrieus design shown in Fig. 6.4. In this design the turbine is driven by lift forces generated by the wind flow over the aerofoil-shaped blades. The torque is a maximum when the blades are moving across the direction of wind flow and zero when the blades are moving parallel to the direction of wind flow (see Exercise 6.9). VAWTs do not require any yaw mechanism and are easier to maintain since the gearbox and generator are situated at ground level. However, VAWTs have proven to be less cost-effective than HAWTs. In the Darrieus design (Fig. 6.4) cables are required to anchor the top of the rotor, and these limit the mean height of the rotors and lose the advantage of the stronger winds that occur at greater heights. Darrieus VAWTs produce larger gearbox torques than those produced by a HAWT of the same power output, and therefore need more robust construction. (Recently, though, a VAWT design for offshore that mimics a spinning sycamore leaf—the 10 MW Aerogenerator X—has been proposed (Aero2010).)

We will therefore concentrate on HAWTs. We first consider how much energy there is in the wind and then how HAWTs extract the energy.

6.4 Kinetic energy of wind

The energy of wind is in the form of kinetic energy. For a wind speed u and air density ρ, the energy density E (i.e. energy per unit volume) is given by

$$E = \tfrac{1}{2}\rho u^2 \tag{6.1}$$

The volume of air flowing per second through a cross-sectional area A normal to the direction of the wind is uA. Hence the kinetic energy per second of the volume of air flowing through this area is given by $P = EuA$ or

$$P = \tfrac{1}{2}A\rho u^3 \tag{6.2}$$

Thus the power of wind P varies as the cube of the wind speed u. Hence much more power is available at higher speeds. However, fluctuations in wind speed can cause the output of a wind turbine to vary significantly.

EXAMPLE 6.1

Calculate the power of the wind moving with speed $u = 5 \, \text{m s}^{-1}$ incident on a wind turbine with blades of 100 m diameter. How does the power change if the wind speed increases to $u = 10 \, \text{m s}^{-1}$? (Assume the density of air is $1.2 \, \text{kg m}^{-3}$.)

Substituting in eqn (6.2) we have

$$P = \tfrac{1}{2} A \rho u^3 = \tfrac{1}{2} \times (\pi \times 50^2) \times 1.2 \times 5^3 \approx 0.6 \, \text{MW}$$

A power of 0.6 MW is sufficient to meet the average electricity usage of about 1000 European households. Doubling the wind speed increases the power by a factor of $2^3 = 8$, so the power would increase to $\sim 8 \times 0.6 = 4.8 \, \text{MW}$.

6.5 Principles of a horizontal-axis wind turbine

Unfortunately, not all of the power in the wind can be extracted by a wind turbine. This is because some of the kinetic energy is carried downstream of the turbine in order to maintain air flow. This effect imposes a theoretical maximum efficiency of 59% for extracting power from the wind, known as the Betz limit and described in detail in Derivation 6.1.

As the wind flows through a turbine it slows down as part of its energy is transferred to the turbine. The air flow looks like that shown in Fig. 6.5. Upstream, the speed of the wind is u_0 and it passes through an area A_0. By the time the wind reaches the turbine it has slowed to u_1 and the area of the stream-tube has increased to A_1, the area swept out by the blades of the turbine. Downstream of the turbine the wind's cross-sectional area is A_2 and its speed is u_2. The drop in speed of the wind before and after the turbine gives rise to a pressure drop across the turbine, through Bernoulli's principle, so there is a thrust on the turbine blades.

The maximum power is generated when downstream of the turbine the wind speed is one-third of the upstream speed u_0 and at the turbine the wind speed is two-thirds of u_0; i.e. $u_2 = \tfrac{1}{3} u_0$ and $u_1 = (2/3) u_0$ (see Derivation 6.1). Under these conditions the power P extracted is given by

$$P = \frac{1}{2} \rho A_1 (16/27) u_0^3 \tag{6.3}$$

The power P_w in the wind passing through an area A_1 with a speed u_0 is given by eqn (6.2) as

$$P_\text{w} = \frac{1}{2} \rho A_1 u_0^3$$

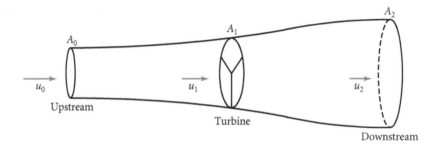

Fig. 6.5 Wind flow through a turbine.

so the fraction of the power that is extracted by the turbine, which is called the power coefficient C_P, given by

$$C_P = P/\left(\tfrac{1}{2}\rho u_0^3 A_1\right) \quad \text{or} \quad P = \tfrac{1}{2}C_P\rho u_0^3 A_1 \tag{6.4}$$

is $16/27 \approx 59\%$ of the power in the incident wind passing freely through an area equal to that of the turbine, A_1. This limit for the power coefficient C_P of $16/27$ of the incident wind power is called the Betz or Lanchester–Betz limit; it was first derived by Lanchester in 1915 and then independently by Betz in 1921.

Derivation 6.1 Maximum extraction efficiency

We can obtain an estimate of the maximum efficiency by modelling the turbine as a thin disc (called an actuator disc) that extracts energy. Consider the stream-tube of air (see Section 4.2) shown in Fig. 6.5, that moves with speed u_1 through a wind turbine of cross-sectional area A_1. Upstream of the turbine the cross-sectional area of the stream-tube is A_0 and the air speed is u_0. Downstream of the turbine the cross-sectional area of the stream-tube is A_2 and the air speed is u_2.

Since the turbine extracts energy from the wind, the air speed decreases as it passes through the turbine and the cross-sectional area of the stream-tube increases, as shown in Fig. 6.5. The thrust T exerted on the turbine by the wind is equal to the rate of change of momentum, given by

$$T = \frac{dm}{dt}(u_0 - u_2) \tag{6.5}$$

where dm/dt is the mass of wind flowing through the stream-tube per second.

The power P extracted is given by the product of the thrust and the air speed at the turbine, u_1, so that

$$P = Tu_1 = \frac{dm}{dt}(u_0 - u_2)u_1 \tag{6.6}$$

We can also express the power extracted as the rate of loss of kinetic energy of the wind, i.e.

$$P = \frac{1}{2}\frac{dm}{dt}\left(u_0^2 - u_2^2\right) \tag{6.7}$$

Comparing eqns (6.6) and (6.7), we require

$$(u_0 - u_2)u_1 = \frac{1}{2}\left(u_0^2 - u_2^2\right) = \frac{1}{2}(u_0 - u_2)(u_0 + u_2)$$

Hence

$$u_1 = \frac{1}{2}(u_0 + u_2), \quad \text{or} \quad u_2 = 2u_1 - u_0 \tag{6.8}$$

Also, by mass continuity (see Section 4.3) the mass flow per second, dm/dt, is given by

$$\frac{dm}{dt} = \rho u A = \rho u_1 A_1 \tag{6.9}$$

(Note that the changes in pressure are sufficiently small that the density of air ρ is essentially constant; see Example 4.3.)

Substituting for u_2 from eqn (6.8) and for dm/dt from eqn (6.9) in eqn (6.6) yields

$$P = 2\rho u_1^2 A_1 (u_0 - u_1) \tag{6.10}$$

Let $u_1 = (1 - a)u_0$, where a is called the induction factor. Then

$$P = \frac{1}{2}\rho u_0^3 A_1 [4a(1-a)^2] \tag{6.11}$$

The power coefficient C_P, which represents the fraction of the power in the wind that is extracted by the turbine, is given by

$$C_P = P \big/ \left(\tfrac{1}{2}\rho u_0^3 A_1\right) = 4a(1-a)^2 \tag{6.12}$$

Maximizing P by setting dC_P/dt to zero gives the maximum power extracted P_{max} when $a = \frac{1}{3}$, that is,

$$P_{max} = \frac{1}{2}\rho u_0^3 A_1 [16/27] \tag{6.13}$$

which is ~59% of the power in the incident wind passing freely through an area equal to that of the turbine, A_1. This limit for the power coefficient C_P of 16/27 of the incident wind power is called the Betz or Lanchester–Betz limit.

6.6 **Wind turbine blade design**

The thrust on the turbine is generated and translated into rotational energy by shaping the turbine blades as aerofoils. A wind turbine is shown in Fig. 6.6(a) and a section of a blade is shown in Fig. 6.6(b). The air speed at the turbine is u_1 and the rotational speed of the blade is v (i.e. perpendicular to the direction of air flow). The resultant velocity $\boldsymbol{u}_\alpha$ of the air relative to the blade makes an angle ϕ to the direction of the blade, given by

$$\tan\phi = u_1/v \tag{6.14}$$

The angle of attack of the wind on the blade is α. The motion of the wind over the aerofoil section gives rise to a lift force $\mathcal{L}$. The lift $\mathcal{L}$ of an aerofoil is perpendicular to the direction $\boldsymbol{u}_\alpha$ of the air flow, so the thrust T on the aerofoil (neglecting drag $\mathcal{D}$) is given by $\mathcal{L}\cos\phi$ and the power P developed by $\mathcal{L}(\sin\phi)v$ (force multiplied by velocity). Using eqn (6.14) to substitute for v gives

$$P = \mathcal{L}(\sin\phi)u_1\cot\phi = \mathcal{L}\,(\cos\phi)u_1 = Tu_1 \tag{6.15}$$

showing that the power developed equals that delivered by the thrust of the wind, when drag is neglected.

The speed v of the blade at a radius r is given by

$$v = \frac{rv_{\text{tip}}}{R} \tag{6.16}$$

where v_{tip} is the speed of the blade tip and R is the maximum radius of the turbine blade. An important parameter is the tip-speed ratio λ, defined as the ratio of the speed v_{tip} of the blade at the tip to the speed u_0 of the incident wind, i.e.

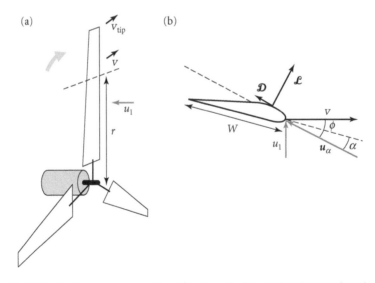

Fig. 6.6 (a) Wind incident on rotating turbine. (b) Section of turbine blade at distance r from the axis.

$$\lambda = \frac{v_{tip}}{u_0} \tag{6.17}$$

When the Betz condition ($u_1 = 2u_0/3$) is satisfied, then using eqns (6.16) and (6.17) we can express the angle ϕ in terms of r, R, and λ through

$$\tan \phi = \frac{u_1}{v} = \frac{2R}{3r\lambda} \tag{6.18}$$

From eqn (6.18) we observe that, for a given radius r, the angle ϕ is only a function of the tip-speed ratio λ, which is why λ is so significant in the design of a wind turbine. The apparent angle of the wind ϕ increases with decreasing radius r (since $\tan \phi \propto 1/r$). Turbine blades are therefore designed with a twist that increases with decreasing r in order for the angle of attack α to remain optimum. The blade width, W, also increases with decreasing r so that the component of lift generates the thrust required to maintain the Betz condition. The calculation of the width of a turbine blade is given in Derivation 6.2. The results for the width and angle of twist of a blade on a turbine with n blades are

$$\text{Angle of twist} = \phi - \alpha = \tan^{-1}[2R/(3r\lambda)] - \alpha \tag{6.19}$$

$$\text{Width} = W = 8\pi R \sin \phi/(3\lambda n) \tag{6.20}$$

EXAMPLE 6.2

A wind turbine with three blades is operating in a mean wind speed of $8\,\text{m s}^{-1}$. The turbine rotates at 15 rpm. Each blade is 40 m long. Estimate the width at the midpoint and tip of each blade.

The time τ for one revolution of the tip of a blade of length R is

$$\tau = 2\pi R/v_{tip}$$

so the number n_{rpm} of revolutions per minute (rpm) is

$$n_{rpm} = 60/\tau$$

Therefore the tip speed v_{tip} and tip-speed ratio $\lambda = v_{tip}/u_0$ are

$$v_{tip} = 2\pi R/\tau = 2\pi R n_{rpm}/60 = 2\pi(40)(15)/60 = 62.8\,\text{m s}^{-1}$$

$$\lambda = 62.8/8 = 7.85$$

From eqn (6.18) the angle ϕ the wind makes to the movement of the blade at a distance r from the turbine axis is given by

$$\tan \phi = 2R/(3r\lambda)$$

So $\phi_{tip}=\tan^{-1}[2/(3\lambda)]=4.85°$. Substituting ϕ_{tip} into eqn (6.20) gives an estimate of the width of the blade tip:

$$W_{tip}=8\pi R(\sin\phi_{tip})/(3\lambda n)=8\pi(40)(0.0846)/[3(7.85)(3)]=1.20\,\mathrm{m}$$

At the midpoint $\phi_{mid}=\tan^{-1}4/(3\lambda)=9.64°$ and $\sin\phi_{mid}=0.167$, so

$$W_{mid}=2.38\,\mathrm{m}$$

Derivation 6.2 Blade design

A wind turbine typically has three blades. The effect of a blade on the wind flow extends over a sufficient distance in the direction of the lift that the blade's reaction on the wind occurs over a large part of the whole annular area $dA=2\pi r dr$ (see Exercise 6.5). Variations in u_1 over the time for one blade to reach the position of the next blade (equal to $2\pi/(n\Omega)$, for n blades, where Ω is the angular speed of the turbine) are modelled by taking u_1 as the average wind speed at the turbine.

The total thrust dT_n on the annular area dA from n blades equals the rate of change of momentum of the wind. From eqns (6.5) and (6.9) this is given by

$$dT_n=(\rho dAu_1)(u_0-u_2)=\rho u_1^2\,2\pi r dr \tag{6.21}$$

when the Betz condition $\left(u_2=\tfrac{1}{2}u_1=\tfrac{1}{3}u_0\right)$ is satisfied.

This thrust is equal to the sum of the components of the lift $\mathcal{L}$ from each section of blade between r and $r+dr$. (We neglect the effect of drag in this discussion.) The lift $d\mathcal{L}$ is given by (see eqn (4.15))

$$d\mathcal{L}=\frac{1}{2}C_L\rho u_\alpha^2Wdr \tag{6.22}$$

where α is the angle of attack and W is the width (or chord) of the blade at r. From Fig. 6.6 the speed $u_\alpha=u_1/\sin\phi$. Equating the thrust to the sum of the components of lift and substituting for u_α gives

$$\rho u_1^2\,2\pi r dr=nd\mathcal{L}\cos\phi=\frac{1}{2}nC_L\rho u_1^2\,Wdr\cos\phi/\sin^2\phi \tag{6.23}$$

In order to satisfy the Betz condition (which maximizes the efficiency of the turbine) the width W of the blade at a distance r from the axis must therefore be given by

$$W=4\pi r\tan\phi\sin\phi/(nC_L) \tag{6.24}$$

The tip-speed ratio $\lambda=v_{tip}/u_0$, so $\tan\phi=u_1/v\equiv2R/(3r\lambda)$. Substituting for $\tan\phi$ yields

$$W=8\pi R\sin\phi/(3\lambda nC_L) \tag{6.25}$$

Table 6.1 Width and twist for an optimum turbine blade of length 24 m

Radius (m)	Width (m)	Angle (°)	
		Twist	Wind ϕ
6	2.649	12.4	18.4
12	1.377	3.5	9.5
18	0.925	0.3	6.3
24	0.696	−1.2	4.8

Since $\sin\phi$ increases as r decreases, the width of the blade increases from the tip to the root of the blade (see Fig. 6.6).

The angle of attack α is chosen to give the highest lift-to-drag ratio and is typically a few degrees. For a three-blade wind turbine, λ is often chosen to be between 6 and 10. An aerofoil generally has C_L close to 1 at the angle of attack α_0 with the largest C_L/C_D ratio. For α_0 equal to 6° and $C_L = 1$, the width W of the blade and the shape and angle of twist of the optimum blade of length $R = 24$ m are given in Table 6.1 for a tip-speed ratio $\lambda = 8$. These are calculated using eqns (6.25) and (6.18).

In practice, blade design takes account of drag as well as lift and the effects of all the sections are calculated using computer programs (blade element theory).

6.7 Dependence of the power coefficient C_P on the tip-speed ratio λ

In the discussion in Section 6.6 of the power extracted by a wind turbine, we neglected the effect of drag $\mathcal{D}$. This force acts in the direction of the wind at right angles to the lift $\mathcal{L}$ as shown in Fig. 6.6. Although the drag is small compared to the lift, its direction is nearly opposite to that of the blade's motion. Consequently the effect of the drag is enhanced by a factor of $\sim\lambda$, the tip-speed ratio, as we will now show.

As a result of the drag the rotational force is reduced and becomes

$$\mathcal{L}\sin\phi - \mathcal{D}\cos\phi = \mathcal{L}\sin\phi(1 - g\cot\phi) \tag{6.26}$$

where $g \equiv C_D/C_L$ is the drag to lift ratio of the blade's aerofoil shape. From eqn (6.18)

$$\cot\phi = 3r\lambda/(2R)$$

so the reduction varies with radius r. We can estimate the overall reduction by taking a typical radius r as $2R/3$ so $\cot\phi \sim \lambda$. The maximum power coefficient $C_{P\text{max}}$ is therefore expected to be about

$$C_{P\text{max}} \approx (1 - g\lambda)C_{P\text{Betz}} \tag{6.27}$$

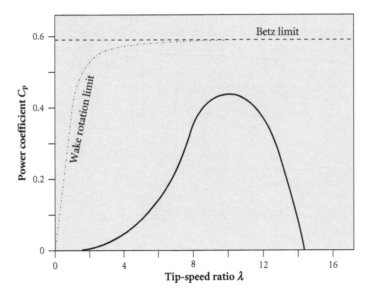

Fig. 6.7 C_P-λ curve for a high tip-speed ratio wind turbine.

For a modern wind turbine with $g \sim 1/40$ and $\lambda \sim 10$, the rotational force, and hence C_P, would be reduced by ~25%. Figure 6.7 shows a plot of the variation of C_P with λ for a turbine designed to have maximum efficiency at $\lambda \sim 10$.

EXAMPLE 6.3

A wind turbine is operating in a mean wind speed of $10\,\mathrm{m\,s^{-1}}$. It rotates at a speed of $20\,\mathrm{rpm}$. Each blade is $45\,\mathrm{m}$ long and has an aerofoil section with a drag to lift ratio of $1/50$. The blades are optimized for the mean wind speed. Estimate the maximum power coefficient and the power output at the mean wind speed. (Assume the density of air is $1.2\,\mathrm{kg\,m^{-3}}$.)

From Example 6.2

$$\lambda = 2\pi R n_{\mathrm{rpm}} / (60 u_0) = 2\pi \times 45 \times 20 \, / \, (60 \times 10) = 9.4$$

Equation (6.27) gives

$$C_{P\mathrm{max}} \approx (1 - g\lambda) C_{P\mathrm{Betz}} = (1 - 9.4/50) 16/27 = 0.48$$

Substituting in eqn (6.4)

$$P = \frac{1}{2} C_P \rho u_0^3 A_{\mathrm{turbine}} = \frac{1}{2} (0.48)(1.2)(10^3) \pi (45)^2 \approx 1.83 \text{ MW}$$

The width W and twist of a turbine's blades are designed for a particular λ. If the wind speed or the rotational angular velocity of the turbine alters then ϕ and therefore the lift $\mathcal{L}$ changes. This changes the thrust from it optimal value and C_P decreases. So we expect C_P to fall away on either side of its maximum value as seen in Fig. 6.7. At low λ the blade stalls (see Section 6.9). Above $\lambda \approx 13$ corresponds to the induction factor (see Derivation 6.1) $a > \frac{1}{2}$ and is where using the momentum change to derive the thrust on the turbine (eqn (6.5)) is invalid as it corresponds to a negative final velocity. In this region the thrust rises rather than falls and corresponds to a turbulent wake behind the turbine blades. Although the thrust T continues to rise in this region, while the speed of the wind u_1 at the turbine decreases, the power imparted to the turbine blades Tu_1 continues to fall.

Also shown in Fig. 6.7 is the theoretical maximum C_P curve, which for $\lambda \geq 4$ is close to the Lanchester–Betz limit. At lower λ the curve dips down, which is caused by the neglect of the swirling motion of the wake (see Derivation 6.3).

Derivation 6.3 The dependence of the maximum extraction efficiency on the tip-speed ratio λ

A turbine blade is designed for a particular λ and, as explained above, we expect C_P to fall away on either side of its maximum (see Fig. 6.7). This maximum value is related to the theoretical limit, which at sufficiently high λ is the Betz limit (neglecting the effect of drag). However, at low λ the theoretical maximum for C_P is decreased, as we now explain.

As wind passes through a turbine the wind acquires a swirling motion as a result of the rotation of the turbine blades. This is because angular momentum is necessarily imparted downstream since the air flow through the turbine imparts a torque to the turbine blades (see Fig. 6.8). If the speed of the blade is much greater than the speed of the wind, i.e. a high λ, then the energy associated with this rotational motion of the wind is much smaller than that associated with the momentum change of the wind in the direction of the wind (axial direction). To see this we will consider what happens to the wind when it passes through a small annular area (see Fig. 6.8).

The power extracted by the turbine from the stream-tube of wind defined by the annular area between r and $r+dr$ equals the torque dG arising from the change in angular momentum of the wind multiplied by the angular speed Ω of the turbine. At a given radius this product equals the force arising from the change in momentum of the wind in the direction of the blade motion multiplied by the speed of the blade.

The power extracted, dP, equals the loss of axial kinetic energy of the wind minus its gain in rotational kinetic energy per second. The mass flow $d\dot{m}$ equals $\rho u_1 2\pi r dr$. If the wind gains an angular velocity ω_1, which will be in the opposite sense to that of the blade, after passing through the turbine, then the rotational kinetic energy gained per second is given by $\frac{1}{2} d\dot{m} r^2 \omega_1^2$ and the change in angular momentum per second by $d\dot{m} r^2 \omega_1$. Hence the torque $dG = d\dot{m} r^2 \omega_1$ and dP is given by

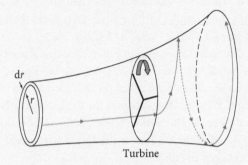

Fig. 6.8 Angular momentum of air flow after passing through turbine.

$$dP = dG\Omega = d\dot{m}\,\omega_1 r^2 \Omega = \tfrac{1}{2}\,d\dot{m}\,(u_0{}^2 - u_2{}^2) - \tfrac{1}{2}\,d\dot{m}\,r^2 \omega_1^2 \tag{6.28}$$

Defining $\omega = \tfrac{1}{2}\omega_1$ and the angular induction factor $a' = \omega/\Omega$, then

$$dP = dP_{\mathrm{B}}/(1+a') = d\dot{m}\,\omega_1 r^2 \Omega \tag{6.29}$$

where dP_{B} is the change in linear kinetic energy per second. As in Derivation 6.1, which neglected wake rotation, dP_{B} equals the change in linear momentum per second multiplied by the wind speed at the turbine (eqn (6.5)) leading to eqn (6.11), which can be expressed as

$$dP_{\mathrm{B}} = \tfrac{1}{2}\,d\dot{m}\,4a(1-a)u_0{}^2 \tag{6.30}$$

Substituting gives

$$dP = \tfrac{1}{2}\,d\dot{m}\,4a(1-a)u_0^2/(1+a') \tag{6.31}$$

Defining the radial tip-speed ratio $\lambda_r = \Omega r/u_0$ and substituting $d\dot{m}\,\omega_1 r^2 \Omega$ for dP gives

$$a'(1+a') = a(1-a)/\lambda_r^2 \tag{6.32}$$

Substituting $d\dot{m} = \rho u_1 2\pi r\,dr$ and $u_1 = (1-a)u_0$ gives

$$dP = \tfrac{1}{2}\rho(2\pi r\,dr)4a(1-a)^2 u_0^3/(1+a') \tag{6.33}$$

Maximizing dP by varying a subject to the constraint of eqn (6.30) gives $a' \ll 1$ and $a \cong \tfrac{1}{3}$, as before when neglecting wake rotation, for $\lambda_r \geq 2$. The angular induction factor $a' \approx 2/(9\lambda_r^2)$ and if we take a tip-speed ratio of $2\lambda/3$ as representative, then

$$dP \approx dP_{\mathrm{B}}/\left[1 + 1/(2\lambda^2)\right] \tag{6.34}$$

so for $\lambda \geq 4$ the maximum power that can be extracted is very close to the Lanchester–Betz limit, as shown in Fig. 6.7, in the absence of drag.

6.8 Design of a modern horizontal-axis wind turbine

A modern HAWT is illustrated in Fig. 6.3. Turbines are designed with a large tip-speed ratio λ of ~6–10 to give a higher power efficiency C_p. A larger λ also means a higher shaft speed and hence a lower torque. Lowering the torque allows the use of a smaller gearbox (if used) or smaller generator. Turbines with a large λ also have smaller width blades (see eqn (6.20)), so less blade material is required, which cuts costs. Ensuring that the blades are each wide enough to have sufficient strength means that there are only two or three blades on a modern large turbine. This can make starting difficult, but this problem can be overcome by using a starting motor.

Blades were originally made from wood; aluminium and steel were then employed. Nowadays though, fibreglass and other composite materials are increasingly used because of their high strength and stiffness coupled with low density. Carbon fibre and carbon/glass hybrid composites are also being developed. The fatigue properties of the materials used are very important because of the very large number of revolutions—typically a few times 10^8— that a turbine makes in a 30-year design lifetime.

Fatigue causes a wire that is bent repeatedly to weaken and break—the strength of the wire decreases with the number of bends. Likewise, the rotation of the blades causes the loads experienced by the turbine to change repeatedly. These changing loads weaken the structure through fatigue, which must be allowed for in the design. The fatigue of materials is discussed more in Box 6.2.

The nacelle containing the generator and control mechanisms is mounted on a tower. Most towers have a strong and, for economy, lightweight structure. As a result the natural frequency of vibration of the tower lies below the rotational frequency of the blades. When starting or stopping the turbine the shaft speed is therefore changed quickly to avoid shaking the tower.

The size of wind turbines has increased over the last 20 years. The rated power is the maximum continuous power that the turbine is designed to produce. Typical specifications in 1985 were rated power 80 kW, rotor diameter 20 m, hub height 30 m; in 1995 rated power 600 kW, diameter 46 m, height 50 m; while in 2011 a modern 5 MW HAWT turbine had a hub height at ~90 m and blades that swept out a circle of diameter ~125 m. The ratio of the annual energy yield to that which would be produced at the rated power is called the capacity factor.

The global average capacity factor in 2010 was 0.25 and generally capacity factors are in the range 0.2–0.4 and have higher values in sites with good wind such as offshore; these values correspond to rated generator powers 2.5–5 times the average power output. There is a balance between the maximum energy output and the capital cost of generators. To take full advantage of periods of high wind speed would require large but expensive generators which would have a larger output and a smaller capacity factor than a smaller generator. The choice is determined by what will give the lowest cost of electricity and is dependent on a number of factors. For example, the cost of offshore foundations has tended to favour larger turbines but their larger weight pushes costs up and it is not clear where the economic optimum lies. Recent innovations have also allowed significantly larger turbine rotor diameters to be installed on the same turbine at low wind speed sites, which increases the capacity factor and

Box 6.2 Fatigue in wind turbines

Wind turbine fatigue requirements are particularly severe because the number of load cycles is so high. For a 30-year lifetime, an 80 m diameter turbine operating for ~80% of the time at a $\lambda=8$ in a wind speed of $10\,\mathrm{m\,s^{-1}}$ will make some 2.4×10^8 rotations. This means that the maximum stresses (force per unit area) must be lower than in other structures to avoid failure through fatigue; see Fig. 6.9.

Fatigue failure is the fracture of material after it has been subjected to repeated cycles of stress changes at levels considerably below its initial static strength. The number of cycles to failure decreases as the alternating stress level increases. (The stress level can be characterized by the mean stress and its range, which is equal to $\sigma_i^{\,\mathrm{max}}-\sigma_i^{\,\mathrm{min}}$.) Fatigue involves the initiation and growth of cracks in a material under the repeated stress cycles. Discontinuities such as a sharp corner or flaws in the material are prime sites. Wind turbines installed in California in the 1980s suffered blade failures as a result of fatigue not being fully understood.

Fatigue can be quantified by using the Palmer–Milner linear damage rule (often called Milner's rule). This method breaks down the cyclic stresses that a structure undergoes into the number of cycles n_i that occur at each stress level σ_i. The total damage D_M sustained by a structure is given by

$$D_M=\sum_{i=1}^{s}(n_i/N_i)=n_1/N_1+n_2/N_2+\cdots+n_s/N_s \tag{6.35}$$

where N_i is the number of cycles to failure at the stress level σ_i. Milner's rule states that failure will occur when $D_M=1$, though factors of 2 are often found between predicted and measured lifetimes.

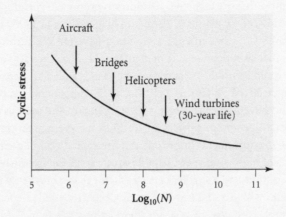

Fig. 6.9 Cyclic stress versus $\log_{10}$ (cycles to failure).
Source: Sand99.

The fatigue strength of a material is the value of the stress level σ_{max} required to cause failure after a specified number of cycles N. The results can be expressed as an S versus N plot (S–N plot), where S is the ratio σ_{max}/σ_0. The stress σ_0 is the static strength of the material. The data can be represented by the equation

$$\sigma_{max}/\sigma_0 = 1 - b\log_{10}(N) \qquad\qquad (6.36)$$

where b is a positive constant.

Equation (6.36) predicts how the stress that can be tolerated decreases with the number of cycles. The results for some fibreglass composite materials are shown in Fig. 6.10. (The data are for tensile stresses with a ratio $R = \sigma_i^{min}/\sigma_i^{max} = 0.1$; data for compressive stresses would also need to be considered.) The good quality material has a value of $b = 0.1$, i.e. the fatigue strength decreases by 10% for each decade increase in the number of cycles. The poor quality material has $b = 0.14$. As can be seen at $S = 0.2$ the good quality material has over two orders of magnitude longer lifetime. Figure 6.10 illustrates the importance of the fatigue performance of the materials used in a wind turbine.

A material with the lowest b coefficient is not necessarily the best since the static strength is also important. Increased strength could be obtained by having more fibres but with a slight increase in the value of b. Whether this change would give a better performance depends on the fatigue strength required.

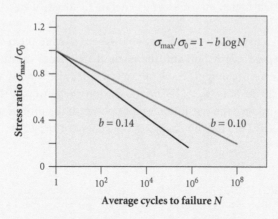

Fig. 6.10 *S–N* curves for two fibreglass composites.
Source: Sand99.

reduces the cost of electricity. For estimation a capacity factor of 0.3 can be taken but more accurate values can be obtained if details of the wind speed distribution and of the turbines are available (e.g. see Box 6.3).

The land area needed for a certain output is largely independent of the size of turbine as larger turbines need to be spaced out more (see Section 6.11), but larger turbines tend to operate in a greater mean wind speed, as their hubs are higher. Fewer turbines are required for the same output, so operation and maintenance costs are reduced; the cost per kWh therefore decreases. The price of infrastructure, such as connecting to the grid, is also reduced. Furthermore, some ridges only allow a single line of turbines, so more power is produced if they are larger. But for a given amount of capital more smaller units reduce the risk from failures. However, the reliability of turbines is now very good, with turbines available for operation ~97% of the time. (There are also small (~3%) electrical losses between the turbine and the grid connection.) In practice these considerations have tended to favour larger sizes where possible, typically 5 MW offshore and 3 MW onshore where delivery to site often limits the size. The maximum size turbine in 2011 was ~7.5 MW.

6.9 **Turbine control and operation**

In the generation of electrical energy from the wind, a wind turbine needs to be controlled to optimize its output. First the turbine needs to be oriented into the wind, which is achieved by the yaw control mechanism. The wind provides the driving torque for the electrical generator and the current flowing in the generator produces an opposing torque, the generator torque. Ignoring friction, the wind torque equals the generator torque in steady operation.

In order to optimize the aerodynamic efficiency, most modern large wind turbines tend to be variable speed variable pitch machines and generate electricity using AC–DC–AC convertors (see Box 11.1).

The output power curve for a typical variable speed variable pitch turbine is shown in Fig. 6.11. Both the speed of rotation and the pitch of the blades can be altered. This facility allows C_P (and hence the output) to be optimized when the wind is above the minimum required to operate the turbine (u_{cut-in}). The rated wind speed (u_{rated}) is such that the wind is strong enough to produce the maximum output power of the turbine generator. Generally only the turbine speed is altered, by changing the generator torque, in the region between

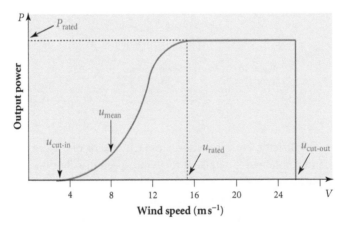

Fig. 6.11 Output power versus wind speed.

the cut-in and rated wind speeds to maintain the optimum tip-speed ratio λ. The generator torque can be varied by changing the electrical load. Above u_{rated}, the speed of the turbine is usually maintained at a constant value and the pitch of the blade is adjusted (feathering, which reduces the angle of attack and hence the lift), to reduce the wind torque and keep loads on the turbine within safe limits. The turbine can be stopped by application of the shaft brake.

For a turbine running at a constant rotational speed and where the blades have a fixed twist, the angle of attack α increases with the wind speed, because the ratio u_1/v gets larger (see Fig. 6.6). If α becomes too large then the lift drops sharply as the aerofoil stalls. This occurs at around $\alpha \sim 10°$ for the aerofoil profile shown in Fig. 4.9. This effect, called 'stall' control, provides a means of limiting the torque in high winds.

Turbines are operated typically for 65–80% of the time, depending on demand and on when the wind speed is below v_{cut-in} or above $v_{cut-out}$. The specific energy output of a wind turbine is defined as the output power per unit of swept area and is useful when comparing turbines of different size or design since the capacity factors can differ between turbines.

EXAMPLE 6.4

Calculate the average power output of a wind turbine with blades of 85 m diameter operating in wind with a mean speed of $7 \, \text{m s}^{-1}$. At this wind speed the power coefficient is 0.45. The rated output power is 2.5 MW when the wind speed is greater than $13 \, \text{m s}^{-1}$. What is the power coefficient at a wind speed of $13 \, \text{m s}^{-1}$? (Assume the density of air is $1.2 \, \text{kg m}^{-3}$.)

Substituting in eqn (6.4), we have

$$0.45 = P \Big/ \left[\tfrac{1}{2}(1.2)(\tfrac{1}{4}\pi \times 85^2)7^3 \right] \quad \text{so} \, P = 526 \, \text{kW}$$

$$C_p = 2.5 \times 10^6 \Big/ \left[\tfrac{1}{2}(1.2)(\pi \times 85^2/4)13^3 \right] \quad \text{so} \, C_p(13 \, \text{ms}^{-1}) = 0.33$$

In this example the power coefficient at the rated wind speed is about three-quarters of its value at the mean wind speed.

6.10 Wind characteristics

We know from experience that the speed of the wind in any location varies considerably with time. This variation affects the amount of power in the wind and the loads felt by the turbine. In particular, there are fluctuations in the wind speed over periods of days from changes in the weather and over periods of minutes from gusts. Averages over a ~10-minute period are used to define the steady wind speed. The shorter term fluctuations about this value are quantified by the turbulence intensity I_T, defined as the ratio of the standard deviation σ_T of the wind speed to the steady wind speed. σ_T generally increases as the steady wind speed increases, and I_T is found to depend in particular on the terrain and height. I_T increases with surface

Table 6.2 Surface roughness (z_0) values

Terrain	z_0 (m)
Urban areas	3–0.4
Farmland	0.3–0.002
Open sea	0.001–0.0001

roughness and varies approximately as $[\ln(z/z_0)]^{-1}$, where z is the height of the turbine and z_0 characterizes the terrain (see Table 6.2). Its magnitude is important in determining the fatigue loading on a wind turbine.

The steady wind is characterized by its frequency distribution $f(u)$ and its persistence. Persistence data give, for example, the number of times the wind is expected to blow for more than an hour with a speed greater than u, while $f(u)\Delta u$ gives the percentage of time that the wind speed is expected to be between u and $u+\Delta u$. The persistence of the wind is important in estimating the dependability of the generated wind power.

For sites that have an annual mean wind speed greater than $4.5\,\mathrm{m\,s^{-1}}$ the frequency distribution is often well described by the Rayleigh distribution,

$$f(u)=(2u/c^2)\exp[-(u/c)^2] \tag{6.37}$$

where $c=2\langle u\rangle/\pi^{\frac{1}{2}}$ and $\langle u\rangle$ is the average wind speed. For a Rayleigh distribution the power in the wind is given by

$$P = \tfrac{1}{2}\rho A\langle u^3\rangle = 0.955\,\rho A\langle u\rangle^3 \approx \rho A\langle u\rangle^3 \tag{6.38}$$

The Rayleigh frequency distribution for a mean wind speed of $8\,\mathrm{m\,s^{-1}}$ is shown in Fig. 6.12.

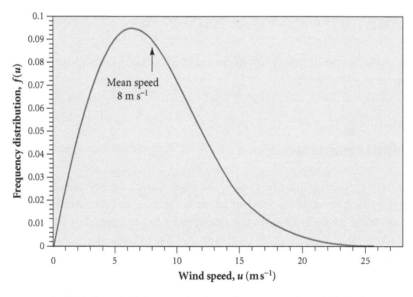

Fig. 6.12 The Rayleigh frequency distribution for a mean wind speed of $8\,\mathrm{m\,s^{-1}}$.

The wind speed u increases significantly with the height above the ground, with the speed zero at the surface. Its rate of change decreases with height as the frictional forces decrease, as shown in Fig. 6.13. A commonly used form to describe the dependence of u on height z is

$$u(z) = u_s (z/z_s)^{\alpha_s}, \tag{6.39}$$

where z_s is the height at which u is measured to be u_s, ideally as close to hub height as possible, and α_s is the wind shear coefficient, which is strongly dependent on the terrain. α_s also generally shows a large variation over a 24-hour period and can change from less than 0.15 during the day to greater than 0.5 at night. This diurnal variation arises because at night the surface temperature drops as the ground loses heat by radiation, giving a stable atmosphere with the lower part cooler than the upper part of the atmosphere. The air is not mixed and wind shear can be high. After sunrise the ground is heated by the Sun and warms the air in contact, which then rises causing mixing and reduced wind shear.

The roughness of the terrain is characterized by a surface roughness parameter z_0 and some typical values are given in Table 6.2. An approximate parametrization for the dependence of the wind shear coefficient α_s on z_0 for steady wind speeds lying between 6 and $10\,\mathrm{m\,s^{-1}}$ at a height of 10 m is

$$\alpha_s = \tfrac{1}{2}(z_0/10)^{0.2} \tag{6.40}$$

For $u_s = 8\,\mathrm{m\,s^{-1}}$ at a height of 10 m, this relation gives $\alpha_s = 0.32$, 0.13, and 0.05 for $z_0 = 1$, 0.01, and 0.0001 m, respectively. Since a typical hub height for a 1 MW turbine is 80 m, these different wind shears translate into mean hub wind speeds of 15.6, 10.5, and $8.9\,\mathrm{m\,s^{-1}}$. We can see that, for the same wind speed at 10 m, more power is produced if a turbine is mounted higher, particularly when the wind shear is large. However, large wind shears are often associated with high turbulence and hence more fatigue.

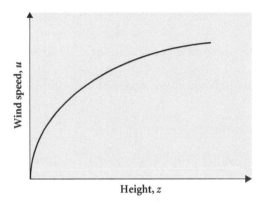

Fig. 6.13 Typical variation of wind speed with height.

6.11 **Power output of a wind turbine**

The power P_w in the wind at a given site is given by

$$P_w = \frac{1}{2}\rho A\langle u^3\rangle = \frac{1}{2}\rho A \int_0^\infty u^3 f(u)\,du \tag{6.41}$$

where u is the wind speed at the height of the turbine hub (we are neglecting the variation in wind speed over the turbine blades) and $f(u)$ is the wind speed distribution.

The average output power P_o is given by

$$P_o = \frac{1}{2}\rho A \int_0^\infty C_P(u)u^3 f(u)\,du \tag{6.42}$$

For a variable pitch variable speed turbine, C_P can be kept quite close to its maximum value over a range of wind speeds about the mean, and then reduced to keep the output at its rated value for high wind speeds. The effect of C_P varying for typical values of the mean and rated wind speeds is to make the integral roughly equal to $0.35\langle u\rangle^3$, where $\langle u\rangle$ is the mean wind speed at the height of the turbine hub. When turbines are in an array there is an array loss factor (~ 0.9), so the average output of a turbine in a wind farm is given by

$$P_o \approx 0.15\,D^2\,\langle u\rangle^3 \tag{6.43}$$

where D is the diameter of the turbine. A typical turbine spacing in a wind farm (see Section 6.12) when space is not at a premium is $8D$ (downwind) by $5D$ (crosswind), so the power density, which is the average output power per unit land or sea area P_{area} of a wind farm, is

$$P_{area} \approx 3.75\times10^{-3}\,\langle u\rangle^3\,\text{MW km}^{-2} \tag{6.44}$$

For $\langle u\rangle = 8.5\,\text{m s}^{-1}$, $P_{area}(\text{wind}) \approx 2.3\,\text{MW km}^{-2}$. Typical power densities are $\sim 2\,\text{MW km}^{-2}$ for wind farms on land and $\sim 3\,\text{MW km}^{-2}$ for farms offshore. The installed capacity (or rated power) is about $8\,\text{MW km}^{-2}$ on- or offshore.

EXAMPLE 6.5

Estimate the power output of a wind farm consisting of 25 2.5 MW turbines. The turbines' hub height is 80 m and their diameter is 90 m. The wind has an average speed of $u=7\,\text{m s}^{-1}$ and $8.1\,\text{m s}^{-1}$ at a height of 10 m and 60 m, respectively. Deduce the surface roughness parameter z_0.

Substituting in eqn (6.39) we have

$$u_{60} = u_{10}\left(60/10\right)^{\alpha_s} \text{ so } \alpha_s = \log(8.1/7)/\log6 = 0.081$$

$$u_{hub} = u_{60}\left(80/60\right)^{\alpha_s} \text{ so } u_{hub} = 8.3\text{ m s}^{-1}$$

From eqn (6.40), $z_0 = 10(2\alpha_s)^5 = 0.0011$.

Putting u_{hub} in eqn (6.43) gives an estimate for the power output P_o for each turbine in the array assuming an array loss of 0.9

$$P_o = 0.15(90)^2(8.3)^3 = 695\,\text{kW}$$

So the wind farm output is estimated to be $25 \times 695\,\text{kW} = 17.4\,\text{MW}$.

We now consider where the wind conditions favour the siting of wind turbines.

6.12 Wind farms

Good measurements over a period of time are required to determine where to site a turbine or a number of turbines, which is then called a wind farm. Experience has shown that sites onshore require an average wind speed greater than $6.5\,\text{m s}^{-1}$. Offshore wind speeds are generally higher and good sites have average wind speeds of ~$9\,\text{m s}^{-1}$ or more. Suitable locations include mountain gaps and passes, high altitude plains, exposed ridges, and coastal areas. Of particular importance in wind farms is the spacing required between the turbines. A spacing of between 3 and 10 rotor diameters, dependent on conditions, is required for the reduced wind speed downwind from a turbine to have been brought up to its original speed by gaining energy from the surrounding wind that is not slowed down. For a farm with a spacing of 7–$8D$ (downwind) by 4–$5D$ (crosswind), an array loss (the amount by which the output of the farm falls below the output of the turbines sited separately) of ~5–10% might be expected—the loss is sensitive to the layout and spacing of the turbines, the size of the array, and the distribution, in both direction and speed, of the wind. When land is at a premium a tighter spacing will generally be used. If the wind farm is on land there is a small amount of land required for the tower and service roads, but between turbines the land is available for use (e.g. for grazing cattle or growing crops).

A low wind shear reduces the differential loads on turbine blades and hence the fatigue damage, which favours flat sites, though some ridge sites have low wind shear. Offshore sea areas can provide excellent wind conditions. These areas can be out of sight of land and so have little visual impact. The turbines can be as tall as desired, and can also operate at higher rotational speeds as noise is less of a concern out at sea. Turbines can cause electromagnetic interference, and planning permission has been refused for some sites because a wind farm would affect radar coverage. Access to the electrical grid is important and some of the windiest sites are also some of the most remote. Offshore wind farms clearly need undersea cables to land, and maintenance and installation costs can be higher than for onshore farms. Floating wind turbines are under development that could be deployed in deep water, so that, for example, the wind resource off the coasts of the USA and Japan as well as in the Mediterranean could be tapped.

The power density from a wind farm with 5 MW rated turbines, each with a capacity factor of 0.3 and diameter of 125 m, would be ~2.4 MW km^{-2} or ~24 kW ha^{-1}, if spaced at 8D (downwind) by 5D (crosswind). The power density is roughly independent of turbine size as the output and area both depend on D^2, but larger turbines generally have a higher hub height and hence a higher wind speed. An example of a wind farm is the Horns Rev 2 offshore wind farm described in Box 6.3.

Box 6.3 The Horns Rev 2 offshore wind farm

The Horns Rev 2 wind farm is located in the North Sea, 30 km from the shore at Blåvandshuk in Denmark, where the depth of water is 9–17 m. There are 91 Siemens SWT-2.3-93 turbines in the farm, which covers an area of 33 km^2 and has a total capacity of 209.3 MW. The turbines' cut-in, rated (for 2.3 MW output), and cut-out wind speeds are 4, 13.5, and 25 m s^{-1}, respectively, and the rotor diameter is 93 m.

The output and coefficient of performance C_p of the turbine as a function of wind speed are shown in Table 6.3. The variable speed and pitch control of the turbine allows C_p to be close to its maximum value for the most probable wind speeds. The mean wind speed $\langle u \rangle$ at the hub height of 68 m is equal to ~9.7 m s^{-1} and the annual production in 2010 was 855.5 GWh, equivalent to 1.074 MW of continuous output. This output corresponds to a capacity factor of 46.7%, higher than more typical values of ~25–35%, and to a power density of 2.96 MW km^{-2}.

The average power output is given by

$$P_0 = \int_{u_{\text{cut-in}}}^{u_{\text{rated}}} P(u) f(u) du + P(u_{\text{rated}}) \int_{u_{\text{rated}}}^{u_{\text{cut-out}}} f(u) du \tag{6.45}$$

where $P(u)$ is the output at a wind speed of u and $f(u)$ is the wind speed distribution at the turbine hub height. The capacity factor CF of the turbine in a wind farm is given by $\eta_{\text{farm}} P_0/P(u_{\text{rated}})$, where η_{farm} is the array loss factor, typically 0.9. Approximating $P(u)$ in

Table 6.3 Output power of SWT-2.3-93 wind turbine versus wind speed

Speed (m s^{-1})	4	6	8	10	12	14	16	18
Power (kW)	98	376	914	1784	2284	2300	2300	2300
C_p (% Betz limit)	62	71	72	72	54	34	23	16

Source: Siemens data sheet.

the range $u_{\text{cut-in}} \leq u \leq u_{\text{rated}}$ by $P(u) \propto (u - u_{\text{cut-in}})$, and $f(u)$ by a Rayleigh wind speed distribution (eqn (6.37)), the capacity factor of the Horns Rev 2 wind farm is estimated to be 48.2% (Exercise 6.15*). An analytic expression for the capacity factor, calculated (see Exercise 6.15*) by assuming $P(u) \propto u^2$ in the range $0 \leq u \leq u_{\text{rated}}$, is

$$CF \approx \frac{\eta_{\text{farm}}}{x_r^2}\left[1-\exp\left(-x_r^2\right)\right] \tag{6.46}$$

where $x_r = u_{\text{rated}}/c$ and $c = 2\langle u\rangle/\pi^{\frac{1}{2}}$ is the parameter in the Rayleigh distribution formula eqn (6.37). Though only approximate, eqn (6.46) illustrates the dependence of CF on both the rated and mean wind speeds, and is a more reliable estimator for the output of a wind farm than just assuming an average capacity factor of 30%. This value for CF can be combined with the area of the farm A_{farm}, the number of turbines N_{turbine}, and their rated power P_{rated} to give an estimate of the power per unit area P_{area} of the wind farm:

$$P_{\text{area}} \approx \frac{N_{\text{turbine}}\eta_{\text{farm}}P_{\text{rated}}}{A_{\text{farm}}x_r^2}\left[1-\exp\left(-x_r^2\right)\right] \tag{6.47}$$

For the Horns Rev 2 wind farm, eqn (6.46) gives $CF = 46.2\%$ and eqn (6.47) gives $P_{\text{area}} = 2.93\,\text{MW km}^{-2}$, while eqn (6.44) gives $P_{\text{area}} = 3.42\,\text{MW km}^{-2}$. (For comparisons in other wind farms, see Exercise 6.15.)

Besides good wind conditions we also need to consider the environmental impact that a turbine, and in particular a wind farm, would have.

6.13 Environmental impact and public acceptance

Wind farms are being actively developed because they are sources of renewable energy that produce essentially no global warming nor any pollution, and also provide energy security. For contrast, the emissions of CO_2 from coal and gas power plants are ~950 and ~450 tCO_2 per GWh, respectively, while only a small amount of emitted CO_2 (~30 tCO_2/GWh) is associated with the construction and operation of the wind farms (IAEA2000), similar to the amount associated with hydro or nuclear plants (see Table 1.2). Also, there are no air pollution related deaths, and wind farms require no water. Moreover, generating electricity from wind reduces the amount of fossil fuels used in generation and so reduces CO_2 emissions.

These are all considerable environmental gains, which should not be forgotten when considering any negative aspects. Before reaching a decision on a new wind farm, a full Environmental Impact Assessment (EIA) is undertaken during which all of the positive and negative impacts of a project are considered. All aspects of the environment that could be adversely affected are identified and ways by which the significant negative impacts can be mitigated would then be developed. An example is the concern in the UK over the visual impact of a wind farm, though this is much less of an issue in Germany, China, or the Midwest of the USA. To gain high winds the turbines are often sited on ridges, so they can be very visible, and if they are in a region of natural beauty they can be felt by many, but not all, to be a blot on the landscape. Clearly if the site is remote and not visited by climbers or walkers there will be less concern. This planning consideration favours offshore installations, which can be out of sight, though they must be clear of shipping lanes and not interfere with radar installations. Where a wind farm is noticeable, experience has suggested that it is more acceptable if there are a few large rather than many small turbines. Making the local population aware of the energy and any job benefits, and also involving them in the planning process, can make planning permission easier to obtain.

Another concern that has been raised is their threat to birds—over a two-year period 183 birds were killed in the Altamont wind farm in California, while in Spain in 18 wind farms an average of 0.13 birds were killed per turbine in 2003. To put these figures in perspective, there are estimated to be over 57 million birds killed each year by cars and over 97 million killed by flying into plate glass windows in the USA, while in the UK each year 55 million birds are killed by cats. A study of power stations in Europe and the USA estimated that about 0.35 birds were killed per GWh by wind farms, compared with about 5 per GWh by fossil fuel power stations. So while care should be taken to reduce bird mortality from turbines by avoiding migratory flight paths or key habitats, the relative risks should be borne in mind.

The space that wind farms take has also been raised as an issue, but it should be noted that the land between turbines can be used for grazing or for growing crops. If we take the area occupied by a turbine, or impacted area, as D^2, where D is the rotor diameter, then the power density of a wind farm increases from ~2.5 MW km^{-2} to ~100 MW km^{-2}. To get an idea of how much land is required, we will work out the area needed to supply 10% of the UK's electricity.

EXAMPLE 6.6

Estimate the area of land required by wind farms to provide 10% of the UK's electricity demand.

The UK's electricity energy demand in 2010 was ~385 TWh per year. The number of hours in a year is 8760, so the energy demand would be met by a continuous power P of

$$P = 385 \times 10^{12} / 8760 = 43.95 \times 10^9 = 44\,\text{GW}$$

To provide 10%, i.e. 4.4 GW, then at a capacity factor of 0.3 wind farms with a rated output (or capacity) of ~14.7 GW would be required. This capacity would be provided by $(14.7 \times 10^3 / 5) = 2940$ 5-MW turbines.

A typical diameter D for a 5-MW turbine is ~125 m. So if each turbine occupied an area of $5D \times 8D$, the total area A required would be

$$A = 2940 \times 625 \times 1000 = 18.4 \times 10^8\,\text{m}^2 \approx 43\,\text{km by } 43\,\text{km}$$

The area of the UK is ~200 000 km^2, so 1840 km^2 is a small fraction (1%) of the land area, and if most of it were offshore the impact of the wind farms would be minimal.

Table 6.4 Noise levels in dB

Noise	Noise level (dB)[*]
Threshold of pain	140
Pneumatic drill at 7 m	95
Busy general office	60
Wind farm at 350 m	35–45
Rural night-time background	20–40
Threshold of hearing	0

Source: UK Department of the Environment, 1993.

[*] $I(\text{dB}) = 10 \log_{10}(I/I_0)$, where I_0 is the threshold of hearing (at 1000 Hz $I_0 = 10^{-12}\,\text{W m}^{-2}$).

Noise from wind turbines has been a concern but improvements to blade design have reduced the noise from modern turbines. Although the magnitude of noise from a wind farm is relatively low (see Table 6.4), the perception of noise is partly subjective. When the wind is blowing strongly the noise from the turbines is masked by that from the wind itself. Only when they are close to built-up areas is noise generally a concern. Offshore turbines can have a higher tip speed as noise is not a problem.

6.14 Economics of wind power

The economics of wind power depend on the capital costs to build the wind turbines, the ongoing costs to run the equipment, called operation and maintenance costs (O&M), and the revenue from the sale of the electrical energy (kWh) produced over the lifetime of the turbines (20–30 years). There are no fuel costs.

The capital costs of machines (IEA2009) range from \$1.5 million to \$2.6 million per MW in Europe, USA, and Japan, while in India and China costs are around \$1 million per MW. For land-based farms, the turbine is about three-quarters of the investment cost while for offshore projects it is about half, with the rest mainly on the foundations and cabling. Offshore costs can be more than twice those for onshore. The annual O&M costs for land-based wind farms are estimated to be ~\$12 per MWh, compared with ~\$48 per MWh offshore. The difference in capital and O&M costs for onshore and offshore wind farms is partly offset by the higher wind speeds offshore.

A modern wind turbine typically generates the amount of energy used in its manufacture in 3–10 months. Furthermore its decommissioning costs, a particular concern with nuclear power plants, are roughly covered by its scrap value. Financial payback estimates, however, are complicated and very site specific but an important factor is the effect of discounting. A simple payback estimate does not quantify what return on their investment a utility company can expect. To calculate this, we must take into account that revenue received in the future is worth less than if it were received today. For example, £100 invested today at 10% interest would be worth £110 in a year's time, so the value of £110 of revenue in a year's time would only have a present value of £100. This translation of future revenue to present value is called discounting, and the interest rate used is called the discount rate R (see Section 12.3.1, Discounted cash flow analysis). Discounting is particularly important when revenue is expected over many years, as from a wind turbine.

With the increased cumulative production of wind turbines there has been a reduction in the cost of turbines, and wind power is becoming increasingly competitive with other supplies of energy. The Energy Information Administration (EIA2016) have estimated the costs in 2016 at good sites of wind-generated electricity onshore as ~8¢ per kWh and offshore ~19¢ per kWh, compared to 6¢ per kWh for a gas-fired plant (with an effective fee of $15 per tonne CO_2 emitted) and 11¢ per kWh for a new nuclear plant. We show in Chapter 12 how to calculate the cost of producing energy and how a learning curve can be drawn relating cost to cumulative production from which future costs estimates can be made.

However, prices of turbines also depend on supply and demand: in 2003 it was a buyer's market so prices were low, by 2009 it was a seller's market and prices were high, while in 2012 it was moving back to a buyer's market. Furthermore, innovations enabling larger diameter turbine rotors to be installed on the same turbine at low wind speed sites are reducing costs. Prices are also affected by government policy on subsidies and the cost of fossil fuel generated electricity. Since there are so many factors that affect the price, the estimates of the cost of electricity given here are only indicative.

Wind power does have an additional value in saving a utility power company from using other fuels, and these direct savings are called avoided costs. Its use will also give some reduction in the requirement for conventional generating capacity. Wind energy has substantial environmental benefits, as discussed above, and these can be given a monetary value from the amount of CO_2 and other emissions saved. The external cost (environmental and health) in Europe of coal-fired generation is estimated to be 2–15c (euro cents) per kWh compared to ~0.2c per kWh for wind energy. The desire to reduce global warming can also result in requirements on utilities to use a certain percentage of renewable energy such as from wind.

As production increases, costs fall and a learning curve (see Chapter 12) can be drawn from which cost estimates can be made. The average price for wind-generated electricity was ~3.5p (UK pence) per kWh in 2003 when the global cumulative production was ~100 TWh. From the learning curve, wind would have become competitive at a price of ~2.5p per kWh with CCGT when the cumulative production reached ~400 TWh. This amount of production corresponds to approximately 150 GW of installed capacity and would have been predicted to occur in about 2010. In fact, this capacity was achieved in 2009 and the cost of wind power in areas with excellent wind conditions was estimated to be 6.8¢ per kWh compared to 6.7 and 5.6¢ per kWh for coal and gas-fired generators (Bloomberg2011).

6.15 Wind variability

Using wind power alone would require alternative generating capacity for when the wind was not blowing. But if we are considering wind providing a relatively small fraction (≲20%) of the total power on a grid (referred to as a penetration of ≤20%), its variability can be accommodated in the same way as demand variations by turning on and off conventional generating capacity (spinning reserves). For example, the variability in the output of wind farms and in the load (demand) in west Denmark is shown in Fig. 6.14 during a three-week period in February 2011. As can be seen, the load varied daily by ~0.75 GW about a mean of just over 2.5 GW. At these high levels of wind output, good interconnectivity is important so that any excess can be exported and balancing power can be imported. In west Denmark there are very strong grid connections to Germany and Norway and there is the ability to balance wind with Norwegian hydropower.

In a recent study by the IEA (IEAVTT), the estimated increase in short-term reserves had a considerable range: 1–15% of installed wind power capacity at a penetration of 10% and 4–18% of installed capacity at 20% penetration. At these levels existing reserves could provide the excess, and the increase in cost arising from wind variability was less than ~10% of the wholesale value (~€14/MWh) of the wind-generated electricity. A mitigating factor when the wind output is high is that part of the reserve requirement can be met by the other power stations which are then running low.

Connecting wind farms together over large areas helps to smooth out variations in supply and errors in the wind forecast, as well as making more reserve power supplies available. Electricity markets typically operate on the demand a day ahead and the amount of reserve depends on the forecast error in the wind power at that time; if the markets operated on a shorter timescale the amount required would decrease, though state-of-the-art wind forecasts can now predict the timing and amplitude of events to a considerable degree of accuracy 24 hours in advance (IPPR2012). The capacity credit of wind generators is the capacity of conventional generators that is no longer required to maintain a certain reliability level of generation, and rough values are 30% of the wind capacity at 10% penetration and 20% at 20% penetration.

Improvements in transmission, i.e. the grid, for better interconnectivity are very important. Adding storage, particularly for higher wind penetrations, can also be beneficial. Using plug-in hybrid electric vehicles is one possibility; producing hydrogen for longer term storage is

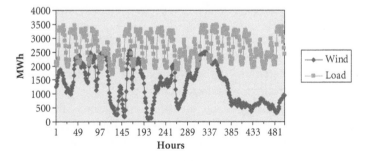

Fig. 6.14 Wind variability (output and load) in west Denmark, 1–21 February 2011. *Source:* Energinet2011.

another. Where there are appropriate resources, compressed air or pumped hydro could also be used (see Sections 11.7 and 11.8).

In a study of the global potential (Lu2009), the correlation between different regions of the contiguous USA was evaluated from a consideration of the wind resource in Montana, Minnesota, and Texas. They found that the wind was essentially uncorrelated between the three regions during the winter months, with r values of less than 0.07. In the summer months, though, there was an appreciable correlation, with r values of 0.28 (Montana–Texas) and 0.37 (Montana–Minnesota). The incorporation of other sources, in particular PV, could help smooth out the overall supply in the summer.

6.16 Global wind potential

About 0.5% (900 TW) of solar energy goes into generating wind energy and about half of this is dissipated in the Earth–atmosphere boundary layer, where there is a loss of kinetic energy through frictional dissipation. The average downward flux of kinetic energy is $\sim 1\,MW\,km^{-2}$ and provides a limit on the total global wind power that can be extracted. Over the world's unglaciated land area the flow of kinetic energy is $\sim 112\,TW$. In the presence of wind turbines, this power is divided into dissipation by boundary layer turbulence and by the turbines. The maximum amount that can be extracted by the turbines is $\sim 25\%$ (see Exercise 6.17) and of this the maximum fraction of the Betz limit that can be converted to electricity is $\sim 75\%$, which gives a limit of $\sim 20\,TWe$, a rough estimate of the technical onshore wind potential. The alteration in global surface energy dissipation from the maximum wind turbine coverage is predicted to have some effect on the climate, but for a total power generation of 2 TWe the estimated effect is insignificant—peak changes in seasonal temperatures might be $\sim 0.5\,°C$, but with almost no effect on the global mean temperature.

Higher estimates of the global potential come from considering the areas with wind speeds suitable for wind farms with a typical turbine spacing of ~ 4–$5D \times 7$–$8D$. While this spacing allows the wind speed to pick up within the wind farm, the large-scale dynamics of the atmosphere limits the maximum potential from a region. There are also practical, environmental, and social restrictions on the areas that can be exploited. The resultant potential depends on the restrictions that are applied.

The technical potential is the fraction of the gross potential when unsuitable areas, e.g. those covered with ice, have been excluded. In addition, environmental and social limitations further reduce the available area, and the constrained potential refers to the amount of technical potential taking these restrictions into account.

Two further estimates of energy supply are sometimes quoted—the practicable and the economic potential. (It should be noted that definitions of potential vary.) The practicable, or accessible, potential is that amount of the technical potential that can be utilized by a particular time. For the practicable potential, various factors need to be taken into account, such as the effects of competing land use, planning permission, grid limitations, and the rate of construction of new turbines. The economic potential is that amount of the technical potential that is economically competitive. This depends on the cost of alternative supplies and will be affected by policies such as a carbon tax.

A recent (Lu2009) estimate of the technical potential was made by Lu, McElroy, and Kiviluoma, which was similar to an earlier one by Archer and Jacobson in 2005. Lu et al. considered the output from a network of 2.5 MW turbines located in rural areas that are free from ice and unforested and with sufficient wind that the turbines operate at greater than 20% capacity. This corresponds to about 20% of the global land area. We will take 4% of this output as the constrained potential. This is the percentage of land area with suitable wind speeds that the WEC assumed could be used and was based on experience in the Netherlands and in the USA. We will apply this percentage to the offshore technical potential as well. The global onshore constrained wind potential is then estimated to be 27 600 TWh, equivalent to ~3 TWe continuous, a significant contribution to the global energy demand that is predicted to be unlikely to cause significant climatic effects and is within the limit set by atmospheric dynamics.

The resulting constrained wind potentials for the countries with the largest CO_2 emissions in 2005 are given in Table 6.5. It can be seen that for most of these countries the constrained wind potential is a significant fraction of the electrical energy consumed. For Europe, the onshore constrained potential is ~1320 TWh and the offshore is ~680 TWh, while for the contiguous USA, the corresponding potentials are 2480 TWh and 220 TWh.

Areas with great potential are in northern Europe along the North Sea, in South America near the southern tip of the continent, in Tasmania, in the Great Lakes region of North America, and along the north-east and north-west coasts of North America.

Table 6.5 The CO_2 emissions, electricity consumption, and constrained wind potentials for the countries with the largest CO_2 emissions in 2005

Country	CO_2 (billion tonnes)	Electricity (TWh)	Constrained potential (TWh)	
			Onshore	Offshore
USA	5.96	3816	2960	560
China	5.61	2399	1560	184
Russia	1.70	780	4800	920
Japan	1.23	974	23	108
India	1.17	489	116	44
Germany	0.84	546	128	38
Canada	0.63	541	3120	840
UK	0.58	349	176	248
S. Korea	0.50	352	5	40
Italy	0.47	308	10	6

Source: EIA2005.

Case study 6.1 Offshore wind potential in Europe

A study of the offshore potential for wind-generated electricity has recently been carried out (EEA 2009) and the results are summarized in Table 6.6.

Offshore, restrictions on the siting of wind farms, e.g. not too close to the shore nor in shipping lanes, reduces the potential considerably but the constrained and economic potentials just from offshore sites are still close to the total projected electricity demand in Europe in 2030. The economic potential in 2020 is estimated to be 2600 TWh offshore.

The cost of electricity (see Table 6.7) depends on the interest (discount) rate R, capital (turnkey) cost $C_{capital}$, and the operation and maintenance (O&M) costs, as discussed in Chapter 12. In this study the lifetime N is assumed to be 15 years, so using eqns (12.4) and (12.5) (see Section 12.3), we find the annual cost A_{cost} that repays the capital is

$$A_{cost} = R \times C_{capital}/[1-(1+R)^{-N}]$$

To this sum must be added the annual cost of O&M, which is estimated as a percentage p of the capital cost. The cost of electricity C_{elec} is then given by

$$C_{elec} = [A_{cost}+(p \times C_{capital})/100]/E_{elec}$$

where E_{elec} is the number of kWh produced per year.

Table 6.6 Estimated offshore wind potentials for Europe by 2030 and their share of Europe's estimated electricity demand in 2030

Potential	TWh	Share of demand
Constrained – offshore	3500	0.8
Economic – offshore	3400	0.8

Source: EEA 2009.

Table 6.7 The projected costs of wind-generated electricity on- and offshore in Europe

	2005		2020		2030	
	Off	On	Off	On	Off	On
Turnkey costs (€/kW)	1800	1000	1080	720	975	576
O&M costs (%)	4	4	4	4	4	4
Interest rate (%)	10.5	7.8	9.6	7.8	8.7	7.8
1600 load hours (€/kWh)	0.197	0.097	0.114	0.070	0.099	0.056
2500 load hours (€/kWh)	0.126	0.062	0.073	0.045	0.063	0.036

6.17 **Outlook for wind power**

The cumulative growth in global installed wind turbine capacity is shown in Fig. 6.15 and was 194 GW in 2010. There has been an average annual growth since 1996 of 28% and since 2005 of 27%. The wind power industry (GWEC2010) predicts that capacity will reach 400 GW in 2014–2015 and possibly 1000 GW by 2020, which at a capacity factor of 0.3 would be equivalent to ~300 GWe continuous. Much of this growth is driven by a desire to lower carbon emissions to reduce climate change and also improve air quality, as well as to provide energy and energy security.

In 2010 the 12 countries with largest installed capacities in GW were: China 42.3; USA 40.2; Germany 27.2; Spain 20.7; India 13.1; Italy 5.8; France 5.7; UK 5.2; Canada 4.0; Denmark 3.7; Australia 1.9; and Brazil 0.93. Growth in China and India has been accompanied by a significant increase in manufacturing capability. Deployment in North America and Europe dominates the rest of the market. There is considerable potential in the rest of Asia and also in Latin America, where it is estimated that the wind resources in Argentina could easily provide electricity for the whole region. There is a large wind potential in the northern half of Africa and, with no water requirement, wind power could help provide much needed electricity. There are also good resources in the Middle East. Offshore, there is considerable potential with the UK taking a leading role in its development.

In the IEA technology roadmap for wind power (IEA2009), 2000 GW of capacity is estimated to be installed by 2050 when it would provide 12% of global electricity, while an optimistic industry estimate is ~3500 GW of capacity by this time. We will take ~2700 GW, equivalent to a continuous output of ~800 GWe, as the accessible potential by 2050. The development of super grids (which connect energy supplies from several countries) will not only aid the growth of wind power but also the growth of other technologies and the exchange of resources, such as the output from PV farms and the trading of Danish wind

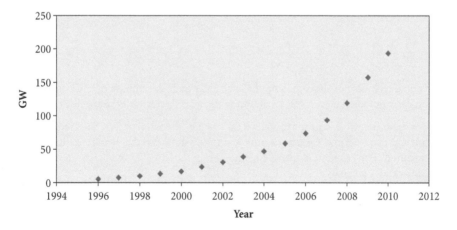

Fig. 6.15 Global cumulative wind capacity.
Source: GWEC2010.

with Norwegian hydropower. High voltage DC lines will facilitate transmission over long distances. Also required are smart grids that enable supply and demand to be better integrated. Finally better storage facilities will enable higher wind penetration.

While there is yet no globally agreed carbon price, many countries are promoting wind power through renewable energy policies as an effective way to combat climate change and provide energy. Onshore wind energy is now competitive with conventional power plants in regions where the wind is good and the cost of fossil-fuelled generation is high, as in California. With increased global capacity and the corresponding expected drop in price, wind power will become increasingly competitive.

6.18 **Conclusion**

While wind power is currently only producing a relatively small amount of electricity worldwide (2.1% in 2010), it has been expanding at an annual rate of over 25% since 1997 and it has the potential to make a significant contribution to low-carbon electricity in the long term. Improvements in the electrical grids, which would allow supplies from a wide region, would reduce the impact from the variability of wind, as would introducing smart grids that can better match supply and demand. However, the effect of wind variability has only a small impact at a penetration of less than 20%, because extra capacity has to be available to meet the variability in demand. At higher penetrations, improvements in energy storage would help, as would diversifying the sources of supply (e.g. integrating wind and solar farms).

It is crucial to expand production and deployment of wind turbines for costs to decrease, and market incentives are very important in making the technology competitive. But already (2011) wind power from the best sites is starting to be competitive with fossil fuel alternatives. The rapid growth in installed capacity is expected to continue and expand.

 SUMMARY

- Global technical onshore potential is ~20 TWe and an estimate of the constrained wind onshore potential is ~3 TWe, greater than the world's electricity usage in 2010 of 2.3 TWe.

- Power in the wind is proportional to the cube of the wind speed. Sites with wind speeds greater than $6.5\,\mathrm{m\,s^{-1}}$ onshore and $9\,\mathrm{m\,s^{-1}}$ offshore at the height of the turbine are favoured. Power extracted by a turbine is limited to 16/27 of the incident power (Betz limit) and is given by

$$P = \tfrac{1}{2}C_\mathrm{P}\rho u_0^{3}A$$

where C_P is the power coefficient, ρ is the density of air, u_0 is the wind speed, and A is the area swept out by the turbine blades. C_P is typically ~0.45 at the optimum tip-speed ratio $\lambda \equiv v_\mathrm{tip}/u_0$, where v_tip is the blade tip speed.

- Modern wind turbines are HAWTs and have maximum (rated) outputs of typically 1.5–5 MW with diameters $D=70$–125 m. On wind farms turbines are spaced at ~$(4$–$5)D\times(7$–$8)D$, with tighter spacing when land is at a premium. The annual output is typically 0.2–0.4 of the rated output, i.e. a capacity factor of 0.2–0.4. Installed capacity is currently increasing at ~25% per annum and was 194 GW in 2010, when wind provided 2.1% of the global electricity demand of ~2.3 TWe, and 239 GW in 2011.

- The accessible potential by 2050 is estimated to be ~2700 GW of installed capacity, equivalent to ~800 GWe of continuous output.

- Wind is a potentially large source of low-carbon electricity. The cost per kWh for the best wind sites is starting to be competitive with fossil fuel alternatives. Onshore wind farms have an output power density of ~20 kW ha^{-1} or 2 MW km^{-2} and much of the land between turbines is available to animals for grazing or to crops. Offshore sites have an output power density of ~3 MW km^{-2}. The installed capacity (or rated power) is about 8 MW km^{-2} on- or offshore.

FURTHER READING

Boyle, G. (ed.) (2004). *Renewable energy*. Oxford University Press, Oxford. Useful discussion of wind resources.

European Wind Energy Association (2004). *Wind energy: the facts*. EWEA, Brussels. Good overview.

Lu, Xi, McElroy, M.B., and Kiviluoma, J. (2009). Global potential for wind-generated electricity. *PNAS* **106** (27), 10933–8. www.pnas.org/cgi/doi/10.1073/pnas.0904101106 (Lu2009).

Manwell, J.F., McGowan, J.G., and Rogers, A.L. (2004). *Wind energy explained*. Wiley, Chichester. Good discussion of modern wind turbines.

Spera, D.A. (ed.) (1994). *Wind turbine technology*. ASME Press, New York. Useful information on wind turbines.

Twidell, J. and Weir, T. (2006). *Renewable energy resources*, 2nd edn. Taylor & Francis, London. Good discussion of wind characteristics.

WEB LINKS

www.awea.org American Wind Energy Association.

www.ewea.org European Wind Energy Association.

www.earth-syst-dynam.net/2/1/2011 L.M. Miller, F. Gans, and A. Kleidon, Estimating maximum global land surface wind power extractability and associated climatic consequences.

www.lorc.dk/Knowledge/Offshore-renewables-map/Offshore-wind-farms Data on offshore wind farms.

www.pnas.org/content/101/46/16115.full.pdf D.W. Keith et al., The influence of large-scale wind power on global climate.

www.windpower.org/en/ Good overview of wind power.

www.arup.com/News/2010_07_July/27_Jul_2010_Arup_and_Wind_Power_Limited_unveil_10MW_Wind_Turbine.aspx (Aero2010).

bnef.com/PressReleases/view/139 Cost of wind power (Bloomberg2011).

www.eea.europa.eu/publications/europes-onshore-and-offshore-wind-energy-potential/ (EEA 2009).

tonto.eia.doe.gov/country/index.cfm (EIA2005).

www.eia.gov/oiaf/aeo/electricity_generation.html (EIA2016).

www.energinet.dk/EN/El/Engrosmarked/Udtraek-af-markedsdata/Sider/default.aspx (Energinet2011).

www.gwec.net/ Global wind report 2010 (GWEC2010).

www.world-nuclear.org/uploadedFiles/org/climatechange/Nuclear%20Energy%20and%20GHG%20Emissions%20Avoidance%20in%20the%20EU%20(2009)%20-%20Final.pdf (IAEA2000).

www.iea.org/publications/freepublications/publication/wind_roadmap.pdf Technology roadmap, wind energy (IEA2009).

www.ippr.org/publication/55/9564/beyond-the-bluster-why-wind-power-is-an-effective-technology Good report on why wind power is an effective technology (IPPR2012).

prod.sandia.gov/techlib/access-control.cgi/1999/990089.pdf H. Sutherland, Fatigue analysis of wind turbines (Sand99).

www.vtt.fi/inf/pdf/tiedotteet/2009/T2493.pdf Design and operation of power systems with large amounts of wind power, IEA-wind, VTT research notes 2493 (IEAVTT).

 LIST OF MAIN SYMBOLS

ρ	air density	α	angle of attack
A	cross-sectional area	P	power
u	wind speed	T	thrust
$\mathcal{L}$	lift	W	width of turbine blade
$\mathcal{D}$	drag	D	diameter of turbine swept area
λ	tip-speed ratio	α_s	wind shear coefficient
C_P	power coefficient	z_0	surface roughness parameter
a	induction factor		

? EXERCISES

6.1 Calculate the power in a wind blowing with a speed of $12\,\mathrm{m\,s^{-1}}$ incident on a wind turbine whose blades sweep out an area of diameter $110\,\mathrm{m}$.

6.2 A simple drag machine (Fig. 6.16) consists of two flaps attached to a rotating belt.

The drag force F_D is given by $\frac{1}{2}C_D\rho A u_{\mathrm{rel}}^2$, where A is the cross-sectional area of the flaps, C_D is the drag coefficient for the flap, ρ is the density of air, and u_{rel} is the wind speed relative to the flap. Show that the power P_D is given by

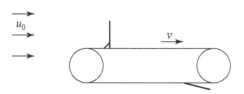

Fig. 6.16

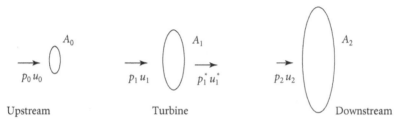

Upstream   Turbine   Downstream

Fig. 6.17

$$P_D = \tfrac{1}{2}C_D A(u_0 - v)^2 v$$

and that the maximum power $P_D(\mathrm{max})$ is given by

$$P_D(\mathrm{max}) = \tfrac{1}{2}(4/27)C_D\rho A u_0^{\,3}$$

Deduce that the power coefficient C_P for such a drag machine is equal to $4C_D/27$. The maximum value of C_D is ~ 1.5 for a cup-shaped flap so the maximum efficiency of a drag machine is 22%.

6.3 Consider stream-tubes of air (Fig. 6.17) before and after a turbine, but not across the turbine because the flow is unsteady and not streamlined.

Applying Bernoulli's principle and, noting that $p_0 = p_2 =$ atmospheric pressure and that from the conservation of mass $u_1 = u_1{}^*$, show that

$$\left(p_1 - p_1{}^*\right)\big/ \rho = \tfrac{1}{2}\left(u_0^2 - u_2^2\right)$$

$$F_{\mathrm{thrust}} = \tfrac{1}{2}\rho\left(u_0^2 - u_2^2\right)A_1$$

Hence show that the maximum thrust is given by

$$F_{\text{thrust}} = \tfrac{1}{2}\rho u_0^2 A_1 \times 8/9$$

This is similar to a circular disc of area A_1 which has a drag force

$$F_D = \tfrac{1}{2}C_D\rho u_0^2 A_1 \text{ and } C_D \sim 1$$

6.4 Using the result for F_{thrust} in Exercise 6.3, calculate the force in tonne-weight on a turbine in a wind speed of $10\,\text{m s}^{-1}$ whose blades have a radius of $50\,\text{m}$.

6.5 Deduce the form of eqn (6.2), $P \propto A\rho u^3$, by using dimensional analysis.

6.6 Show that the maximum power coefficient for two identical turbines placed one behind the other in line with the wind direction is $16/25$.

6.7 Take the extent of the disturbance of the wind in the direction of the incoming wind, i.e. parallel to the axis of the turbine, as πW, where W is the width of the turbine blade. Show that the ratio R of the time for the wind to travel a distance πW to the time for one blade to reach the position of the next blade is given by

$$R > 4\pi/(3\lambda)$$

Hence show that the reaction of the blades occurs over a large fraction of the area swept out by the blades.

6.8 Calculate, using eqn (6.25), the width and twist of the optimal turbine blade of length $R=48\,\text{m}$ for a tip-speed ratio $\lambda=10$ at radii of 12, 24, 36, and 48 m. Take $\alpha_0=5°$, $C_L=1$, and $n=3$.

6.9 The two blades of a VAWT are rotating at a speed v in a wind speed u (Fig. 6.18).

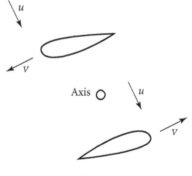

Fig. 6.18

Show from considering the direction of the wind over the blades that the turbine will rotate as illustrated no matter what direction the wind comes from.

6.10* Show that the value of a that minimizes

$$dP = \tfrac{1}{2}\rho(2\pi r dr)4a(1-a)^2 u_0/(1+a')$$

subject to the constraint

$$a'(1+a') = a(1-a)/\lambda_r^2$$

is given by the equation

$$\lambda_r^2 = (4a-1)^2(1-a)/(1-3a)$$

and that $a' = (1-3a)/(4a-1)$. Hence show that for $\lambda_r \geq 2$, $a \approx \frac{1}{3}$ and $a' \approx 2/(9\lambda_r^2)$. (The method of Lagrange multipliers states that the extremum values of a function $f(a, a')$ subject to the constraint $g(a, a')=0$ are given by the solutions to $\partial f/\partial a + \lambda \partial g/\partial a = 0$ and $\partial f/\partial a' + \lambda \partial g/\partial a' = 0$.)

6.11 Calculate the number of cycles (N) a turbine blade would make in 30 years on a wind turbine with $D=100$ m, $\lambda=10$, and a mean wind speed of 12 m s^{-1}. The data for the fatigue strength of two possible turbine materials can both be represented by $\sigma/\sigma_0 = 1 - b\log_{10}(N)$. Material x has $\sigma_0 = 100$ MPa and $b=0.10$, and material y has $\sigma_0 = 120$ MPa and $b=0.12$. Which material would you choose?

6.12 The pressure P_t acting on a wind turbine is given by $P_t = (4/9)\rho u_0^2$ (Exercise 6.3). Estimate the variation in P_t across the height of the circle swept out by the turbine blades arising at a site with a wind shear characterized by $z_0 = 0.1$, $u_s = 10$ m s^{-1} at $z_s = 10$ m, hub height$=70$ m, and $D=50$ m.

6.13 A wind farm is situated on a ridge where the wind speed distribution $f(u)$ is given by

$$f(u) = 2a^2 u \exp(-a^2 u^2)$$

where $a^2 = \pi/(4c^2)$, and c is the average wind speed at the height of the turbines. Each turbine has blades that are 45 m long and a power coefficient $C_P = 0.45$ (independent of wind speed). Calculate the average power extracted by a single turbine when $c = 8$ m s^{-1}.

The density of air is $\rho = 1.2$ kg m^{-3}; $\displaystyle\int_0^\infty u^4 e^{-a^2 u^2}\,du = \dfrac{3\sqrt{\pi}}{8a^5}$.

6.14 For a wind farm with a spacing of $8D \times 5D$, estimate the difference in energy density (MW ha^{-1}) for a farm with 20 turbines each with a rated output of 5 MW, hub height$=90$ m, and diameter$=115$ m, (a) assuming a capacity factor of 0.3; (b) using the formula $P_o \approx 0.15 D^2 \langle u(z)\rangle^3$, with $u_s(10)=9$ m s^{-1} and $z_0=0.01$.

6.15* Evaluate eqn (5.45) when the output of a wind turbine $P(u)$ is approximated in the range $u_{\text{cut-in}} \leq u \leq u_{\text{rated}}$ by $P(u) \propto (u - u_{\text{cut-in}})$, and show that the capacity factor is given by

$$CF \approx \frac{\eta_{\text{farm}}}{(x_{\text{rated}} - x_{\text{cut-in}})} \int_{x_{\text{cut-in}}}^{x_{\text{rated}}} \exp(-x^2)\,dx$$

where η_{farm} is the array loss factor, $x = u/c$, and $c = 2\langle u\rangle/\pi^{1/2}$. Show that when the approximation $P(u) \propto u^2$ in the range $0 \leq u \leq u_{\text{rated}}$ is made, the capacity factor is then given by

$$CF \approx \frac{\eta_{\text{farm}}}{x_{\text{rated}}^2}\left[1 - \exp(-x_{\text{rated}}^2)\right]$$

6.16 Data on four offshore wind farms are given in the table below:

Make estimates of the power density (MW km^{-2}) and capacity factors for these wind farms and compare them with the observed values.

Farm	P/area (MW km^{-2})	Area (km^2)	No	P$_{rated}$ (MW)	Dia (m)	u_{mean} (m s^{-1})	u_{rated} (m s^{-1})	Prod (GWh)	Cap (%)
Barrow	2.2	10	30	3	90	9	15	195.1	24.7
Egmond	1.3	27	36	3	90	8.5	15	315.2	33.3
Horns 1	3.2	20	80	2	80	9.7	16	565.8	40.4
Kentish	2.4	10	30	3	90	8.7	14	209	26.5

6.17 The momentum balance in the Earth–atmosphere boundary layer in the presence of wind turbines is given by

$$F - kv^2 - M = 0$$

where F is the rate of momentum transfer into the layer from the atmosphere (assumed constant), kv^2 is the force arising from natural turbulent dissipation in the boundary layer, with k the friction coefficient and v the mean wind speed, and M is the rate of momentum extraction by the wind turbines.

The total extracted power $P_{tot}(Fv)$ is divided between dissipation in the boundary layer (kv^3) and power extracted by the wind turbines P_{wind} ($= \frac{2}{3} Mv$ (eqn 6.6) in the Betz limit). Show that the maximum value of $P_{wind} = [4/(3^{5/2})] P_{tot}(M=0)$ and that the wind speed is then $3^{-1/2}v_0$, where $P_{tot}(M=0)$ is the power dissipation in the boundary layer and v_0 is the wind speed when no turbines are present.

6.18 The angular speed ω of a fixed pitch wind turbine is controlled by setting the generator torque $\tau_{gen} = K\omega^2$, where $K = \frac{1}{2}\rho A R^3 C_P^{max}/\lambda_{opt}^3$. The angular acceleration $d\omega/dt$ of the turbine is proportional to ($\tau_{wind} - \tau_{gen}$), where $\omega\tau_{wind} = P_{wind} = \frac{1}{2}C_p\rho A u^3$ and $\omega = \lambda u/R$. Show that the turbine will alter speed so that $\lambda = \lambda_{opt}$, provided $C_p > C_p^{(max)}\lambda^3/\lambda_{opt}^3$.

6.19 Evaluate the cost of electricity for (a) onshore and (b) offshore wind farms in 2030 if the lifetime N is 25 years and all other parameters are equal to their values in Table 6.7.

6.20 Discuss the significance of wind variability on the contribution wind power can make to providing low-carbon energy.

7 Solar energy

✔ List of Topics

- ☐ Solar spectrum
- ☐ Semiconductors
- ☐ p-n junction
- ☐ Solar cells
- ☐ Efficiency of cells
- ☐ Silicon cells
- ☐ Thin-film cells
- ☐ Light trapping

- ☐ Multilayer devices
- ☐ Developing technologies
- ☐ Solar panels and farms
- ☐ Economics of photovoltaics
- ☐ Environmental impact
- ☐ Outlook for photovoltaics
- ☐ Solar thermal power plants
- ☐ Outlook for solar thermal power

→ Introduction

The average solar power incident on the Earth is ~1000 W m^{-2} (~100 mW cm^{-2}) or about 150 000 TW. This power is far larger than the world power consumption in 2010 of ~18 TW$_{th}$. Currently ~10% of the world's power is supplied by biomass, while ~80% is derived from fossil fuels. Both are the consequence of photosynthesis, in which plants use solar energy to convert water and carbon dioxide into carbohydrates. While biomass is not necessarily a net producer of CO_2, the burning of fossil fuels definitely is. However, biomass is not a good converter of solar energy as the efficiency of biomass production is low (~0.2–2%).

A more efficient conversion (~18%) of solar energy directly to electrical power is provided by photovoltaic (PV) cells. In 2010 these provided a peak power of ~41 GWp, which could rise to ~1000 GWp by 2030 and by 2050 provide ~11% of generated electricity (IEA-PV). The price of PV cells (~$1 per Wp in 2012) is now close to being competitive in sunny regions with fossil and nuclear power for electricity supply to a national grid, and is expected to achieve grid parity in many regions by 2020. PV cells are already very competitive for applications in areas far from a grid. There has been an expansion of concentrated solar thermal power (CSTP) plants since ~2005 and in 2010 these provided a peak power of ~1 GWp; it has been estimated that CSTP might generate ~10% of global electricity by 2050 (IEA-CSP). Wp (watts peak) is defined as the output under an illumination of 1000 W m^{-2}.

We evaluate PV cells and solar thermal power plants in this chapter, and discuss biomass in the next chapter. We first look at the solar energy spectrum and at how a solar PV cell works. We see what limits its efficiency and the potential for improvement. We then look at CSTP plants, and briefly at exploiting the temperature difference that occurs in the oceans between surface and deep water as a result of solar heating.

7.1 **The solar spectrum**

The smooth spectrum shown in Fig. 7.1 is that of a black body at 5800 K. This spectral shape is close to that incident from the Sun on the Earth's atmosphere. The effect of passing through the atmosphere is to reduce the total from 1.37 kW m^{-2} for sunlight incident on the atmosphere, called AM0, to ~1.0 kW m^{-2} for that passing through a typical thickness of the Earth's atmosphere taken to be 1.5 times its height, called AM1.5. AM1.5 corresponds to sunlight incident at an angle of 48° to the vertical. The effect of absorption by water vapour, carbon dioxide, and methane is nearly all in the infrared region, corresponding to photon energies below ~1.7 eV. The energy of the photons in the visible part of the solar spectrum ranges from ~1.7 eV (0.7 μm) to ~3 eV (0.4 μm).

Solar energy reaches the Earth's surface by direct radiation (focusable by mirrors) and diffuse radiation (unfocusable). The diffuse percentage is strongly dependent on how clear the sky is, and a typical yearly average is about 30%. The total amount of radiation varies considerably with cloudiness, season, and location, from a yearly total on a horizontal surface of ~2300 kWh m^{-2} in the tropics to ~800 kWh m^{-2} by the Arctic Circle (latitude 66.5°). The average flux (watts per m^2) on a cloudy day is typically ~10% in the UK and ~50% in the tropics of the flux on a sunny day. Note that the sky would be black (except for stars and planets) in the absence of diffuse radiation and is blue because short wavelengths are scattered more than long ones by molecules and particles whose size is $\ll \lambda_{\text{visible}}$.

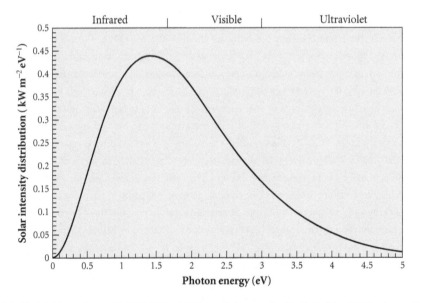

Fig. 7.1 Black-body spectrum at $T = 5800$ K. The total intensity is normalized to that of the AM1.5 solar spectrum of 1 kW m^{-2}.

EXAMPLE 7.1

Direct sunlight of average intensity $200\,W\,m^{-2}$ is incident normally on a solar cell. The area of the cell is $0.1\,m^2$. What is the total incident energy in one day in kWh and in MJ? How is this total energy altered if the sunlight falls at angle of $30°$ to the normal to the surface of the cell?

The incident power P is the intensity I multiplied by the area A of the cell. So

$$P = I \times A = 200(0.1) = 20\,W = 0.02\,kW$$

The incident energy E is the power multiplied by the time t. So

$$E = P \times t = 0.02\,(kW) \times 24\,(h) = 0.48\,kWh$$

or

$$E = P \times t = 20\,(W) \times 24 \times 60 \times 60\,(seconds) = 1.73\,MJ$$

If the sunlight falls at $30°$ to the normal to the cell, then the incident power is reduced by a factor $\cos(30°)$, as $A\cos(30°)$ is the projected area of the cell normal to the beam. The incident energy then becomes

$$E = 0.42\,kWh \text{ or } 1.50\,MJ$$

The variation of the inclination of sunlight with latitude and season is the main cause of the variation in the Sun's intensity or insolation on a horizontal surface with location.

EXAMPLE 7.2

A source of light with a wavelength of $510\,nm$ (green) of intensity $500\,W\,m^{-2}$ is incident on a solar cell. What is the incident flux of photons?

The energy of a photon $E_\gamma = h\nu = hc/\lambda$, where $hc = 1240\ eV\ nm$. So

$$E_\gamma = 1240/510 = 2.43\ eV = 2.43 \times 1.6 \times 10^{-19} = 3.89 \times 10^{-19}\,J$$

The flux F is the number of photons per square metre per second, so the intensity I is given by the flux multiplied by the energy per photon, i.e.

$$I = F \times E_\gamma$$

Hence

$$F = I/E_\gamma = 500/(3.89 \times 10^{-19}) = 1.29 \times 10^{21} \text{ photons } m^{-2}\ s^{-1}$$

7.2 Semiconductors

Photovoltaic solar cells are made from semiconductor materials. To understand how they work, we first need to realize why some materials are conductors and some insulators, before looking at semiconductors.

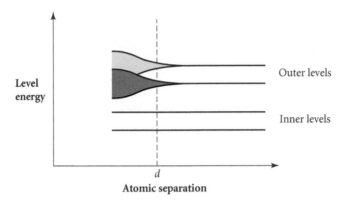

Fig. 7.2 Energy levels as a function of atomic separation: at a separation d there are two bands of levels with a band gap between them.

In a metal that conducts electricity easily, such as copper, the atoms are $\sim 10^{-10}$ m across and are separated by $\sim 2.5 \times 10^{-10}$ m. The electrons in the outer shells can move quite easily from one atom to another and so the metal conducts well. But this does not explain why some materials do not conduct and are insulators. To understand this we need to look at what states are available to the electrons in a solid. For atoms that are well separated, the electron energy levels are also well defined, as shown in Fig. 7.2. As the atoms come closer, the levels spread and overlap. At a separation of d, there are still spaces, called band gaps, between them. The effect is larger for the outer electrons which overlap most.

A single level, which corresponds to a certain spatial distribution of an electron, will spread into $2N$ states, where N is the number of atoms in the crystal. The factor 2 comes about because the electron has two spin states (spin ½ up or down). So, for example, for sodium with one electron in the outer 3s level, only half of the states in the 3s band will be occupied. When an electric field is applied, electrons start to accelerate and gain energy, and so occupy higher states. A steady drift velocity (current) is quickly attained when the number of electrons being scattered down into lower energy states equals the number being promoted up by the electric field. The electrons are scattered by defects in the crystal and by the thermal vibrations of the atoms. Since these vibrations increase with temperature we expect the conductivity to decrease if the temperature rises, and this is what is measured.

For materials where there are two electrons in an outer s shell we would expect the band to be full. On applying an electric field, no empty energy levels are available close by for electrons to be promoted into, so the material is an insulator (unless an empty band overlaps a full band, as in magnesium, when the material will be a conductor). Conductivity can only arise through electrons that have been thermally excited to the next band. This will only occur to a significant extent when the gap between the filled (valence) band and empty (conduction) band is relatively small, ~ 1 eV. Such materials are called semiconductors. We expect their conductivity to increase with temperature as more electrons will be thermally excited into the conduction band, and that is what is found. Fig. 7.3 illustrates these three cases.

A particularly important semiconductor is silicon, which as an isolated atom has four electrons in a half-filled shell. In a solid these electrons are shared (covalent bonds) between four silicon atoms, and a full valence band is formed with a 1.1 eV gap to the next band. Besides

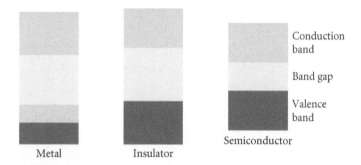

Fig. 7.3 Occupancy and band structure of a metal, an insulator, and a semiconductor.

silicon, there are a number of other semiconductor materials that are increasingly important for use in photovoltaic cells, such as GaAs, CdTe, and CuInGaSe$_2$ (CIGS).

The presence of impurity atoms can give rise to electron states within the band gap and this contamination must be reduced before the material acts as a pure semiconductor, called an intrinsic semiconductor. The development of the technique of zone-refining, in which sections of a rod of semiconductor are heated sequentially so the molten region moves from one end to the other, thereby concentrating the impurities at one end of the bar, played a key part in the development of semiconductor devices. Silicon can be produced with less than 1 part in 10^{11} impurities.

The addition, called doping, of certain impurity atoms to intrinsic material can significantly alter the conduction properties of semiconductors. If we include atoms with five outer electrons (pentavalent atoms) within a silicon crystal, then each of these atoms will have one electron which is only weakly bound and can easily be excited into the conduction band. Such atoms are called donors and the doped silicon is called n-type. Donor atoms with only three outer electrons (trivalent atoms) can gain an electron quite easily through thermal excitation of an electron from the top of the valence band, which leaves a positively charged vacancy called a hole. Such atoms are called acceptors and the doped silicon is then called p-type. A particularly important device, and the basis of the photovoltaic solar cell, can be made by forming a junction between p- and n-type material, called a p-n junction.

7.3 p-n junction

Figure 7.4 shows separate pieces of p- and n-type material and a single piece of semiconductor doped to form a p-n junction. We can see thermally excited electrons in the conduction band of the n-type and positive (i.e. absence of electrons) holes in the valence band of the p-type.

In the piece with both p- and n-type material touching, Fig. 7.4(b), electrons have diffused across the junction from the conduction band on the n-side, as a result of the concentration gradient, and these have filled the vacancies (holes) in the top of the valence band on the p-side. An electric field is set up across the junction and causes a drift of electrons which balances the electron diffusion due to the concentration gradient. (There are corresponding currents of holes.) As a result the n-side becomes positively charged while the

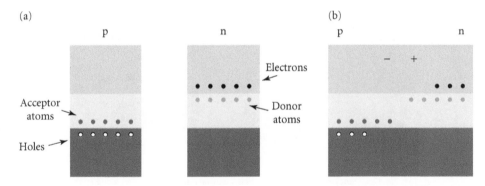

(a)

Fig. 7.4 (a) p- and n-type material. (b) p-n junction.

p-side becomes negatively charged. The energy of the electrons on the n-side is therefore lowered compared with electrons on the p-side and the junction region is normally drawn as shown in Fig. 7.5(a).

The region about the junction where there are no charges is called the depletion zone. The electrons (holes) in the conduction (valence) band on the p- (n-) side of the junction are called the minority carriers; on the n- (p-) side the majority carriers are electrons (holes).

The drift and diffusion currents are driven by opposite gradients in the concentration and electrostatic potentials. The potential relating to the concentration is called the chemical potential. When in equilibrium the sum of these potentials, called the electrochemical potential η, is a constant and is given by $\eta = q\phi + \mu$, where q and μ are the charge and chemical potential of the electron or hole and ϕ is the electrostatic potential. The value of the electrochemical potential η, in a semiconductor, also called the Fermi level ε_F, is the same on the p-side as on the n-side of the junction, as shown in Fig. 7.5(a). For a current to flow across the junction there must be a difference in the electrochemical potentials on either side of the junction.

Figure 7.5(b) shows the effect of biasing the p-region positively by V relative to the n-region. The junction is then forward-biased. (The circuit symbol for a diode is shown below the junction.) This lowers the conduction band on the p-side relative to that on the n-side of the junction by an amount $|e|V$, where e is the charge of the electron. This alters the balance between the

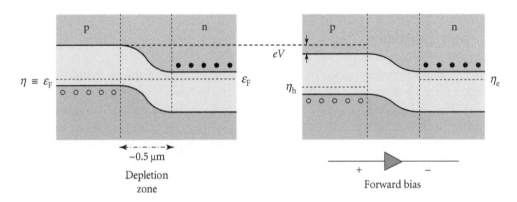

Fig. 7.5 (a) Energy levels and depletion zone in a p-n junction. (b) Effect of positive bias on a p-n junction with standard circuit symbol for a diode shown.

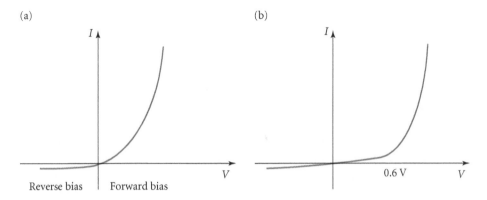

Fig. 7.6 (a) IV characteristics of ideal p-n junction. (b) Actual IV characteristics for silicon diode.

diffusion and drift currents across the depletion zone and gives a net electron (hole) current from the n- (p-) to the p- (n-) side, called the forward current. The total forward current I is given by

$$I = I_S[\exp(V/V_T) - 1] \tag{7.1}$$

where I_S is called the saturation current and $V_T \equiv kT/|e| \approx 0.026$ V at room temperature (see Derivation 7.1). I_S depends on the area of the junction and on the doping. The direction of positive charge flow is across the junction from the p- to the n-side and the ideal current–voltage characteristics given by eqn (7.1) are shown in Fig. 7.6(a).

We can see that the p-n junction acts as a diode, allowing current to flow easily only when forward-biased. Fig. 7.6(b) shows the actual characteristics for a silicon p-n junction: until there is a forward bias of ~0.6 V, conduction is less than for an ideal junction. This is caused by trapping and generation of electrons and holes within the depletion zone. But the idealized equation for the forward current is a very useful approximation, and we will now see how such a device can act as a photocell.

Derivation 7.1 Current–voltage characteristic of a p-n junction

The effect of biasing the p-region positively by V relative to the n-region lowers the conduction band on the p-side relative to that on the n-side of the junction by $|e|V$ (Fig. 7.5(b)). This alters the balance between the diffusion and drift currents across the depletion zone and gives a net electron (hole) current from the n- (p-) to the p- (n-) side, called the forward current. We can work out the size of this current by considering what happens at the edge of the depletion zone.

The energy shift of $|e|V$ has the effect of increasing the electron concentration at the edge of the depletion zone on the p-side by the Boltzmann factor $\exp[|e|V/(kT)] = \exp(V/V_T)$, where $V_T \equiv kT/|e| \approx 0.026$ V at room temperature. The excess concentration Δn_e^P at the edge is therefore equal to $n_e^P[\exp(V/V_T) - 1]$; n_e^P is the concentration of the minority carriers on the p-side when V is zero. This excess drops to zero over the distance that the electron diffuses in the p-region before recombining with holes. This concentration gradient gives rise to a diffusion current proportional to $\Delta n_e{}^p$, i.e. to $n_e{}^p[\exp(V/V_T)-1]$. Outside the depletion zone the electric field is very small and there is essentially no electron drift current. At the edge of the zone the total forward electron current is therefore equal to the diffusion current.

The current is constant throughout the device. Thus, as the diffusion current decreases, as electrons drop into the vacancies in the valence band, i.e. recombine with holes, the drift current of electrons in the valence band moving away from the junction into vacancies increases. The latter is normally described as a hole drift current towards the junction. (Minority carrier drift currents are much smaller than majority carrier drift currents.) There is a similar contribution from the minority hole diffusion and majority electron drift currents on the n-side, giving the total forward current I as

$$I = I_S[\exp(V/V_T)-1]$$

where I_S is called the saturation current (this is eqn (7.1)). I_S depends on the area of the junction and on the doping. The direction of positive charge flow is across the junction from the p- to the n-side and the ideal current–voltage characteristics given by eqn (7.1) are shown in Fig. 7.6(a).

7.4 Solar photocells

When light falls on a silicon p-n junction some of the photons can create electron–hole pairs through the photoelectric effect, in which a photon is absorbed by an electron, promoting it from the valence to the conduction band. The minimum energy that the photon must have equals the band gap E_{gap}. For silicon this is 1.1 eV and corresponds to a wavelength of ~1.1 μm, using the Einstein relation $E_\gamma = h\nu = hc/\lambda = (1.24/\lambda)$ eV, with λ in μm. Figure 7.7 shows photons incident on a junction producing electron–hole pairs.

The top metal electrode is in narrow strips to let the light fall on the junction. An anti-reflection coating increases the transmission of light into the junction. To reduce the series resistance, the top of the n-layer is more highly doped (labelled n⁺). Minimizing surface recombination at the rear contact is achieved by placing a highly doped p-layer (p⁺) just in front of the contact. This reduces the electron concentration in this region and hence the chance of recombination.

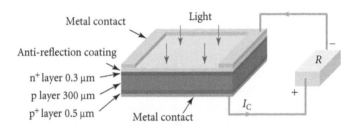

Fig. 7.7 Operation of a solar photocell.

Electrons promoted to the conduction band in the p-layer diffuse and, when within the built-in field across the depletion region, are swept to the n-side, while the holes diffuse to the p-side of the junction. This produces a reverse current I_L (as the electrons flow across the junction from the p- to the n-side), and the photocell current I_C is given by

$$I_C = I_L - I_S[\exp(V/V_T) - 1] = I_L - I_S[\exp(I_CR/V_T) - 1] \tag{7.2}$$

where the forward bias V across the junction equals I_CR. This forward bias produces a current given by the second term, eqn (7.1), flowing across the junction from the p-side to the n-side, which is in the opposite direction to the light-induced current I_L.

When R is infinite then I_C is zero and the open circuit voltage V_{OC}, by which the p-side is more positive than the n-side of the junction, is given by

$$V_{OC} = V_T \ln(1 + I_L/I_S) \approx V_T \ln(I_L/I_S) \tag{7.3}$$

since $I_L \gg I_S$. (V_{OC} is less than E_{gap}, since the maximum voltage must be less than the band gap; see Derivation 7.2.). Note $V_T \equiv kT/|e| \approx 0.026$ V at room temperature.

When R is zero then V is zero and the short circuit current I_{SC} equals I_L. For a finite resistance R, the photocell current I_C generates power P_C given by

$$P_C = I_C V = I_C{}^2 R \tag{7.4}$$

As V increases P_C increases until V is slightly less than V_{OC} as shown in Fig. 7.8.

For this small solar cell, $I_{SC} = 1$ mA, $I_S = 10^{-14}$ A. So from eqn (7.3), $V_{OC} = 0.66$ V. Note that $I_L = I_{SC}$. We can generate the curve of power P_C versus voltage by using eqns (7.2) and (7.4). First calculate I_C for a given V and then calculate P_C. Table 7.1 gives the results for a number of voltages between 0 and 0.66 V.

At the maximum power point $P_C = P_m = 0.55$ mW, $V = V_m = 0.58$ V, and $I_C = I_m = 0.95$ mA. We notice that I_m and V_m are both close to I_{SC} and V_{OC}, respectively. The fill factor (*FF*) is defined as the ratio

$$FF = P_m/(I_{SC}V_{OC}) \tag{7.5}$$

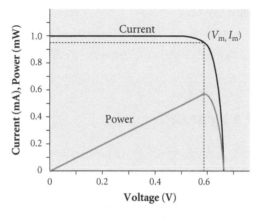

Fig. 7.8 Characteristics of a photocell.

Table 7.1 I_C in mA, P_C in mW, V in volts for cell with $I_{SC}=1$ mA, $I_S=10^{-14}$ A

V (volts)	0.10	0.30	0.50	0.55	0.58	0.60	0.62	0.64	0.66
I_C	1.0	1.0	1.0	0.98	0.95	0.89	0.77	0.51	0
P_C	0.1	0.3	0.5	0.54	0.55	0.54	0.48	0.33	0

For the example in Table 7.1 the fill factor FF equals 0.83. FF is a measure of how close the IV characteristic is to a rectangle. It is useful for quality control, with good solar cells having $FF>0.7$: typically FF lies between 0.75 and 0.85.

EXAMPLE 7.3

A photocell has a saturation current $I_S=2\times10^{-12}$ A, a short circuit current $I_{SC}=30$ mA, and an area of 1cm². Find the maximum power output, the fill factor, and the conversion efficiency of the cell. What resistance across the cell is required to give the maximum output?

From eqn (7.3) $V_{OC}=0.61$V. Calculating I for a number of voltages V below V_{OC} using eqn (7.2), and then calculating the power P from eqn (7.4) gives the following results:

V (V)	0.50	0.52	0.53	0.54	0.56
I (mA)	29.6	29.0	28.6	27.9	25.5
P (mW)	14.8	15.1	15.2	15.1	14.3

Peak power P_m is 15.2 mW when $V_m=0.53$ V and $I_m=28.6$ mA. From eqn (7.5),

$$FF=P_m/(V_{OC}I_{SC})=0.0152/[(0.61)(0.03)]=0.83$$

Approximately 100 mW cm⁻² of solar radiation falls on the Earth, so this cell has an efficiency of ~15%.

As voltage equals current multiplied by resistance, the resistance R required for maximum output is given by $V_m=I_mR$, so

$$R=V_m/I_m=0.53/0.0286=18.5\ \Omega$$

We now look at why solar cells (single junction) have efficiencies of typically only ~10% to ~30%.

7.5 Efficiency of a solar cell

The conversion efficiency is defined as the ratio of the maximum power output to the incident solar power, which for AM1.5 solar radiation (Fig. 7.1) is close to 100 mW cm⁻². This is not 100% for several reasons: one is that not all the photons have sufficient energy (>1.1 eV for a silicon cell) to produce electron–hole pairs, as can be seen in Fig. 7.9.

The photons with energies less than 1.1 eV carry 23% of the incident solar energy. Only 1.1 eV of the energy of any higher energy photons is available to produce power, with the rest lost as

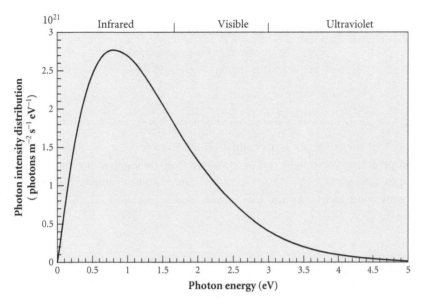

Fig. 7.9 Photon intensity for AM1.5 solar radiation (100 mW cm^{-2}), assuming a black-body distribution ($T = 5800$ K).

heat. The result is that only 47% of the incident solar energy contributes to the power with 30% going to heat. Hence 47% is the limiting efficiency from the solar spectrum for a silicon cell. A similar reduction is from the voltage factor, which is the ratio $|e|V_{m}/E_{gap} \approx 0.7/1.1 \approx 0.65$. This is the ratio of the energy given to an electron to the minimum energy required to produce an electron. The final significant loss comes about because not all of the electron–hole pairs that are produced are collected by the field across the junction; about 10% recombine.

A potentially large loss (~40%) from reflection from the front surface of the silicon can be reduced by a large amount using quarter-wavelength layers of material to act as an anti-reflection coating. The reflectance ρ between two media with refractive indexes n_{1} and n_{2} is

$$\rho = (n_{1} - n_{2})^{2}/(n_{1} + n_{2})^{2} \qquad (7.6)$$

Silicon, since it is partly conducting, has a complex refractive index which is frequency dependent and averages about 3.5. Substituting this value into eqn (7.6) gives $\rho \approx 40\%$. If we add an odd number of quarter-wavelength thick layers with a refractive index n_{1} that is intermediary between that of silicon (n_{2}) and air ($n_{0} = 1$) to the silicon, then ρ can be reduced considerably. Figure 7.10 shows a ray of light incident almost normally on such an arrangement.

The odd number of quarter-wavelength thicknesses makes the reflected components a and b out of phase. The reflectance at each surface must be equal so $(n_{1})^{2} = n_{0}n_{2}$. The effect over the range of solar wavelengths is to reduce ρ to ~6%. Multiple reflection coatings can reduce it still further, to ~1%.

There are small losses from contacts on the front surface. There is very little loss from photons not absorbed, as a result of optimization of the silicon thickness plus the addition of reflecting layers on the back of the cell (total ~3%). The overall efficiency η_{C} from multiplying all these factors together is

$$\eta_{C} \approx (0.47)(0.65)(0.9)(0.96) \approx 26\% \qquad (7.7)$$

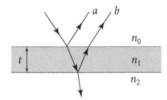

Fig. 7.10 Quarter-wavelength anti-reflection coating (the angle of incidence is exaggerated for clarity).

We can see that the efficiency is dependent on the band gap E_{gap}: decreasing E_{gap} increases the photocurrent, as more light can produce electron–hole pairs, but decreases the maximum output voltage, as $|e|V_{OC} < E_{gap}$. So there is an optimum value of E_{gap}. E_{gap}^{opt} is ~1.3 eV and the semiconductors GaAs and CdTe have band gaps close to this optimum value (see Derivation 7.2).

Derivation 7.2 Efficiency dependence on recombination and band gap

The illumination of the p-n junction creates additional electrons in the conduction band and holes in the valence band of the p-material. The electrons and holes diffuse and, as the p⁺ layer acts as a barrier to electrons, most electrons reach the depletion region where they are swept to the n-side by the internal field. When the junction is open circuit, the accumulation of electrons on the n-side reduces the internal field and the diffusion current of electrons from the n- to the p-side increases until there is a balance between this 'forward' current and the photo-generated current. The forward current of electrons recombines with holes on the p-side; hence the forward current is also called the 'recombination' current. The illumination causes a difference in the electrochemical potential $\eta_e - \eta_h$ across the junction, raising that of the electrons relative to that of the holes, and the increased number of electron–hole pairs increases the recombination rate.

The minimum recombination current and maximum open circuit voltage (V_{OC}) is theoretically obtained when the only recombination mechanism is radiative. Then the rate of radiation emitted following recombination would equal the rate of absorption of photons that generate the electron–hole pairs. This is an example of the principle of detailed balance, and the maximum theoretical efficiency that results for a single p-n junction solar cell is called the Shockley–Queisser limit.

The number of photons emitted per unit area per second with energy E to $E+dE$ into a solid angle $d\Omega$ is given by the generalized Planck distribution:

$$n(E,T,V)dEd\Omega = \frac{2}{h^3c^2} \frac{E^2}{\exp\left(\dfrac{E-eV}{kT}\right)-1} dEd\Omega \tag{7.8}$$

where T is the temperature of the material with an electrochemical potential difference $\eta_e - \eta_h = eV$ between the electrons and holes (see Exercise 7.9*).

For a solar cell in the dark ($V=0$) in radiative equilibrium with its surroundings at the same temperature T as the cell, the number of photons incident per unit area per second on the cell will be equal to the number emitted. Integrating the photon flux incident on the solar cell from an enclosing hemisphere at temperature T gives the total number N with energy E to $E+dE$ incident per unit area per second as

$$N\,dE = 2\pi \int_0^{\pi/2} n(E,T,0)\cos\theta\sin\theta\,d\theta\,dE = \pi n(E,T,0)\,dE \tag{7.9}$$

where the $\cos\theta$ factor accounts for the reduction in flux when photons are incident at an angle θ to the surface normal.

When the solar cell is illuminated and connected to a load such that the voltage across the cell is V, then the photocurrent density J through the cell is given by

$$J = e\Omega_S \int_{E_g}^{\infty} n(E,T_S,0)\,dE - e\pi \int_{E_g}^{\infty} [n(E,T_C,V) - n(E,T_C,0)]\,dE \tag{7.10}$$

where the first term is the reverse current density J_L across the junction caused by the illumination from the Sun (normal to the surface of the cell), which is taken to be that from a black body at a temperature T_S subtending a solid angle Ω_S, and the second term is the forward current density J_F through the cell at a temperature T_C and voltage V. J_F equals the difference between the recombination rate per unit area when the voltage across the cell is V and the generation rate per unit area caused by the radiation from the surroundings.

From the form of the generalized Planck distribution, the potential eV must be less than the band gap E_g for the second integral in eqn (7.10) to remain finite. For $E - eV \gg kT_C$, the forward current density is given to a good approximation by $J_F = J_S \{\exp[eV/(kT_C)] - 1\}$, where the saturation current density $J_S = e\pi \int_{E_g}^{\infty} n(E,T_C,0)\,dE$ (cf. eqn (7.2)). The factor $\exp[eV/(kT_C)]$ gives the increase in the number of electron–hole pairs, and hence in the recombination rate, over that when the cell is in the dark ($V=0$).

Near the maximum power point the forward current density J_F is well approximated by

$$J_F = J_S \exp\left(\frac{eV}{kT_C}\right) = e\pi\left(\frac{2}{h^3 c^2}\right)\left(\int_{E_g}^{\infty} E^2 \exp\left(-\frac{E}{kT_C}\right)dE\right)\exp\left(\frac{eV}{kT_C}\right)$$

$$\cong e\pi(kT_C)\left(\frac{2}{h^3 c^2}\right)F(E_g,T_C)\exp\left(-\frac{E_g}{kT_C}\right)\exp\left(\frac{eV}{kT_C}\right) \tag{7.11}$$

where $F(E,T) = E^2 + 2EkT + 2k^2T^2$. For $E_g/(kT_S) > 2$, the reverse current density J_L is closely given by

$$J_L \cong e\Omega_S(kT_S)\left(\frac{2}{h^3 c^2}\right)F(E_g,T_S)\exp\left(-\frac{E_g}{kT_S}\right) \tag{7.12}$$

The power density $\mathscr{P}$ generated by the cell is given by $\mathscr{P} = V(J_L - J_F)$ and is a function of E_g and V, and at its maximum both $\partial\mathscr{P}/\partial E_g$ and $\partial\mathscr{P}/\partial V$ are zero. Setting $\partial\mathscr{P}/\partial E_g = 0$ (see

Exercise 7.8) gives the following approximate relation between V_m, the voltage at maximum power, and E_g:

$$eV_m = E_g\left(1 - \frac{T_C}{T_S}\right) - kT_C \ln\left(\frac{\pi}{\Omega_S}\right) \tag{7.13}$$

The efficiency η of the cell is given by the ratio of $\mathcal{P}$ to the incident solar power density $\mathcal{P}_{inc}$, with

$$\mathcal{P}_{inc} = \Omega_S \int_0^\infty En(E, T_S, 0)\,dE = \frac{2\pi^4 k^4 T_S^4 \Omega_S}{15h^3 c^2} = \frac{\sigma T_S^4 \Omega_S}{\pi} \tag{7.14}$$

where σ is the Stefan–Boltzmann constant. When V_m is substituted into $\mathcal{P}$, the maximum efficiency η_m is given by

$$\eta_m = \frac{15eV_m}{\pi^4(kT_S)^3}\left[F(E_g, T_S) - \left(\frac{T_C}{T_S}\right)F(E_g, T_C)\right]\exp\left(-\frac{E_g}{kT_S}\right) \tag{7.15}$$

Taking $T_S = 5800$ K, $T_C = 300$ K, $\Omega_S = 6.8 \times 10^{-5}$, then η has its maximum value of 30% when $E_g = 1.3$ eV.

The reasons that the theoretical maximum efficiency is limited are:

- Insufficient energy: photons with an energy less than E_g contain 31% of the incident energy.
- Thermalization: photon energy in excess of E_g is lost as heat (27%).
- Thermodynamics: eV_m is less than E_g by the Carnot factor $(1 - T_C/T_S)$, which gives a 2% loss, and by $kT_C\ln(\pi/\Omega_S)$, which gives a 9% loss. The latter is of the form $T\Delta S$ and represents the energy loss caused by the increase in entropy of the photons when re-emitted.
- Emission: a small loss (1%) through re-emission associated with the forward current.

The entropy loss can be reduced by concentrating the light from the Sun. The maximum concentration possible is $\pi/\Omega_S \approx 46\,000$, which would increase the maximum theoretical efficiency to 38% when $E_g = 1.3$ eV. In practice, the increase in the generated current limits the concentration to about 1000; otherwise the loss through the series resistance of the solar cell becomes too large. The cell must also be kept cool: for silicon, the dependence on temperature is a ~1% loss in absolute efficiency for a 7 °C rise in temperature, while for Group III-V semiconductors such as GaAs the corresponding temperature rise for a ~1% loss is 20 °C (see Section 7.7.2).

In an actual solar cell, the recombination rate is much larger than that due to radiation alone, and is dominated by recombination caused by impurities giving rise to traps within the band gap. Making the approximation that this contribution to the forward current has the same exponential dependence on V as radiative recombination, so that it makes the total forward current equal to $f_i J_F$, results in V_m becoming

$$eV_m = E_g\left(1 - \frac{T_C}{T_S}\right) - kT_C \ln\left(\frac{f_i\pi}{\Omega_S}\right) \tag{7.16}$$

with the maximum efficiency still given by eqn (7.15). Silicon has $E_g = 1.1$ eV and when crystalline $f_i \sim 2000$, which decreases the maximum efficiency from 29% at an operating voltage of $V_m = 0.79$ V for $f_i = 1$ to 21% at $V_m = 0.61$ V for $f_i \sim 2000$ (see Table 7.2 for the highest conversion efficiencies for different types of solar cell).

The open circuit voltage V_{OC} is the value of V that makes $J_L = J_F = J_S \{\exp[eV/(kT_C)] - 1]\}$ and, as $J_L \gg J_S$, V_{OC} is given by $V_{OC} = V_T \ln(J_L/J_S)$, where $V_T = kT_C/|e|$, which is eqn (7.3), as

Table 7.2 Highest conversion efficiencies for different types of solar cell

Material[*]	Band gap (eV)	Efficiency (%) for cell of area	
		~1 cm²	~1 m²
GaAs	1.4	27.6	–
GaAs (multi)	1.8–0.7	42.3	–
Si (c)	1.1	25.0	21.4
Si (mc)	1.1	20.4	17.6
a-Si	~1.7	10.1	–
a-Si (multi)	~1.7–1.3	11.9	–
CdTe	1.5	16.7	10.9
CIGS	~1.2	19.6	15.7
Dye Sensitized	~1.6	10.4	–
Organic	~1.4	8.3	–

[*]c, crystalline; mc, multicrystalline; multi, multijunction
Source: Photo2011.

both I_L and I_S are proportional to the area of the cell. From eqn (7.11) the minimum saturation current density is given by

$$J_S = J_0 \exp[-E_g/(kT_C)] \tag{7.17}$$

where J_0 is given by $e\pi(kT_C)\left(\dfrac{2}{h^3c^2}\right)F(E_g,T_C)$. Actual current densities J_0^e are larger by a factor f_i, i.e. $J_0^e = f_iJ_0$. For crystalline silicon ($E_g = 1.1$ eV and $f_i \approx 2000$), $J_0^e \approx 10^{10}$ A m^{-2} $\equiv 10^9$ mA cm^{-2} and $V_{OC} = 0.67$ V. Substituting eqn (7.17) into the expression for V_{OC}, and defining $eV_g \equiv E_g$, gives

$$V_{OC} = V_g - V_T\ln\left(J_0^e/J_L\right) \approx V_g - 0.43 \tag{7.18}$$

for materials with $0.5 < E_g < 1.75$ eV and $f_i \sim 2000$.

We first look at the construction of silicon crystalline cells before looking at thin-film solar cells and at devices under development.

7.6 Commercial solar cells

There are two main types of solar cell in production today: silicon and thin-film cells. The desire to reduce the cost of cells led to the research and development of thin-film cells. Considerable progress has been made over the last 25 years, although silicon still has the largest share of the market (~85–90% in 2010). We will first describe silicon-based cells. Then we will discuss the newer thin-film technologies, first single layer devices and then multiple layer cells. At the end of this section we consider some of the technologies that are under development.

7.6.1 Crystalline silicon cells

Crystalline silicon cells are produced either as single crystal or polycrystalline (multicrystalline) cells. For single crystal cells, thin wafers, ~200–400 μm thick, are sawn from silicon crystal ingots. The technology for making these ingots has been developed for the semiconductor industry. Much of the material used for silicon solar cells used to be surplus microelectronics material, but since ~2008 there has been dedicated production of silicon for solar cells and the cost of silicon solar modules has fallen. But the wafers are still expensive and techniques are being explored to reduce these costs. One way is to use ingots of polycrystalline silicon made from casting silicon, which is cheaper but less efficient. Another way, developed to avoid the losses caused by sawing, is to make polycrystalline silicon in the form of a ribbon.

The construction of a silicon cell is illustrated in Fig. 7.7. The thickness of the silicon needs to be ~200–300 μm in order to capture most of the sunlight, as the light absorptivity of silicon is low. This is a consequence of the nature of the band gap, being indirect rather than direct: when indirect, the excitation of an electron from the valence to the conduction band has to be accompanied by a change in the vibrational state of the lattice. To use a thinner silicon layer (~50 μm) requires light trapping techniques. These scatter or reflect the light causing it to pass through the silicon layer several times. The use of highly light-absorbing semiconductor material would reduce the amount of material required considerably, which could help reduce the cost of the cell. This has been the motivation to develop thin-film solar cells.

7.6.2 Thin-film cells

There are a number of materials that have good solar light absorption: in particular, GaAs, CdTe, $CuInGaSe_2$ (CIGS), and amorphous hydrogenated silicon (a-Si:H). All of these films only need to be about ~1 μm thick, so much less material is required ($\lesssim 1\%$) than for silicon cells. Materials are only part of the cost and the challenge in thin-film technology is to develop techniques for fast deposition of films while maintaining film quality. Cells with a large area (~1 m^2) generally have more imperfections and so have lower efficiency than small-area (~1 cm^2) devices. Typical percentage conversion efficiencies that have been obtained (2011) are shown in Table 7.2.

GaAs has a band gap of 1.4 eV, close to optimal. It can withstand high temperatures as the band gap is sufficiently large to keep thermal excitation small. This enables concentrators to be used with GaAs cells (see Section 7.7.2). These focus the solar radiation on to the active cell area and can increase the flux by up to 1000 times. The active region is grown epitaxially on a very thin GaAs single crystal layer. This layer is, however, relatively expensive so only small area GaAs cells have been made. Where high performance is required, multijunction cells are now preferred, e.g. in space applications and for CPV.

CdTe, CIGS, and a-Si:H have all been made into large area ($\sim 1\,m^2$) solar cells and so are candidates for solar power plants. One promising semiconductor is CdTe, which also has a good band gap of 1.5 eV.

EXAMPLE 7.4

A 4 cm^2 GaAs solar cell has a saturation current of 4×10^{-15} mA. Under normal illumination of AM1.5 solar radiation the short circuit current is 127 mA. What is the conversion efficiency under normal and under $\times 1000$ illumination?

The open circuit voltage V_{OC} under normal illumination is given by eqn (7.3) as

$$V_{OC} = V_T \ln(I_{SC}/I_S) \approx 0.026 \ln[127/(4 \times 10^{-15})] = 0.988 \text{ V}$$

We could find the maximum output power P_m by finding I and P for different V, as in Example 7.3. However, good approximations (see Exercise 7.7) for V_m and I_m are given by

$$V_m = V_{OC}(1 + x_{OC} \ln x_{OC}) \quad \text{and} \quad I_m = I_{SC}(1 - x_{OC}) \tag{7.19}$$

where $x_{OC} = kT/\{|e|V_{OC}\} = V_T/V_{OC}$, $V_T \equiv kT/|e| \approx 0.026$ V at room temperature. So the fill factor FF is

$$FF = (1 - x_{OC})(1 + x_{OC} \ln x_{OC}) \tag{7.20}$$

For normal illumination $x_{OC} = 0.0263$. Substituting in eqn (7.19) gives

$$V_m = 0.988[1 + 0.0263 \ln(0.0263)] = 0.89 \text{ V}$$

$$I_m = 127(1 - 0.0263) = 124 \text{ mA}$$

So

$$P_m = 0.89(124) = 110 \text{ mW}$$

Under $\times 1000$ illumination the open circuit voltage increases as the short circuit current is 1000 times larger, i.e.

$$V_{OC} = 0.026 \ln[127\,000/(4 \times 10^{-15})] = 1.168 \text{ V}$$

The value of $x_{OC} = 0.0223$ and the maximum output power is when $V_m = 1.07$ V and $I_m = 124$ A corresponding to

$$P_m = 1.07(124) = 133 \text{ W}$$

Under AM1.5 illumination the solar intensity is $100\,\text{mW cm}^{-2}$, so as the area of the cell is $4\,\text{cm}^2$ a solar power of 400 mW falls on the cell. The conversion efficiency is the output power P_m over the incident solar power, i.e. $110/400 = 28\%$.

Under ×1000 illumination the incident solar power is 400 W and the output power P_m is then 133 W, giving a conversion efficiency of 33%. Increasing the illumination gives an improvement of ~18% in conversion efficiency. The fill factor remains approximately constant, changing from 0.88 to 0.895.

CdTe solar cells

A schematic diagram of the construction of a CdTe cell is shown in Fig. 7.11. The cell is fabricated on a thin sheet of glass (2–4 mm thick). This is normally coated with an anti-reflection film. The first layer on the glass is a transparent conducting oxide, usually tin or indium tin oxide, which provides a good electrical contact to the thin CdS layer. The polycrystalline CdS layer is n-doped. Its band gap is ~2.4 eV and is transparent down to wavelengths of ~500 nm. Below this wavelength light is attenuated but some is still transmitted to the CdTe, as the CdS is only ~100 nm thick.

The CdTe is polycrystalline and p-doped. Its energy gap is ~1.45 eV and is well matched to the solar spectrum. The CdTe is less doped than the CdS so most of the depletion zone lies in the CdTe layer, which is typically ~5 μm thick. A mixed metallic contact is generally used to make contact with the CdTe layer.

This is an example of a heterojunction device since the p- and n-regions are in different semiconductors. Originally, there were instabilities associated with the difficulty of making good back contacts; these are now better understood and the performance of CdTe cells is now very competitive. In 2010 the price ($ per Wp) of a CdTe module was less than that of a silicon module, and the largest planned (2011) solar farm, in China, will use CdTe panels. While the thickness of CdTe has been reduced, the availability of the rare metal Te is still an uncertainty in the economics of these panels. Two other semiconductors with promise are a-Si and CIGS. We will look at CIGS next.

CIGS solar cells

The structure of a CIGS cell is shown in Fig. 7.12. The molybdenum contact layer is sputtered onto the glass substrate, followed by the CIGS p-type layer (1–2 μm). This can be formed by first depositing an indium–gallium–selenium compound, $(\text{InGa})_2\text{Se}_3$. This layer is then

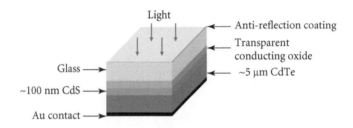

Fig. 7.11 Construction of a CdTe cell.

Light

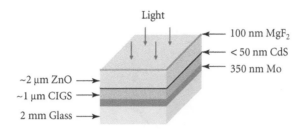

100 nm MgF_2

< 50 nm CdS

350 nm Mo

~2 μm ZnO →

~1 μm CIGS →

2 mm Glass →

Fig. 7.12 Layer structure of a CIGS cell.

reacted with Cu and Se, followed by In and Ga evaporation in the presence of Se. This technique can give a material in which the band gap varies with depth (see Multilayer thin-film cells, below). Thin CIGS layers with a uniform band gap can also be made by simultaneous deposition of all the elements. The n-type CdS layer is then put down to form the heterojunction, and the ZnO layer is deposited to provide the transparent contact. Finally a 100 nm MgF_2 layer provides an anti-reflection coating.

By altering the ratio of In to Ga, the band gap can be varied between 1.1 and 1.2 eV. For a small area device (1 cm^2) with $E_g = 1.2$ eV illuminated under AM1.5 (1 kW m^{-2}) radiation, $J_{SC} = 34.8$ mA cm^{-2}, $FF = 79.2\%$, a conversion efficiency of 19.6% has been obtained (2010). Large area modules have been made with efficiencies of ~16%. The cost of CIGS modules is close to being competitive (2010) with a-Si and CdTe modules, but the ready supply of indium is an uncertainty.

Amorphous silicon solar cells

The first amorphous silicon (a-Si) solar cell was made in 1976. a-Si is produced by electrically decomposing silane, SiH_4, together with a small amount of boron dopant. The hydrogen provides additional electrons that combine with dangling bonds in the a-Si producing an intrinsic semiconductor, a-Si:H. The material is a silicon–hydrogen alloy with 5–20 (atomic)% of H and the resultant band gap is ~1.7 eV. The disordered arrangement of atoms, together with the hydrogen, also gives a high optical absorption, which allows the device to be only ~1 μm thick. Figure 7.13(a) shows the layout of an a-Si solar cell and Fig. 7.13(b) the current–voltage characteristic of a small area device.

The cell is built upon a glass substrate with the transparent indium tin oxide providing the electrical contact to the p-doped region of a hydrogenated amorphous SiC layer. The SiC has a larger band gap than a-Si and so allows a larger fraction of the solar radiation through to the intrinsic hydrogenated a-Si layer (a-Si:H). Below the a-Si:H there is an n-doped region with a contact layer.

In a crystalline silicon cell, electrons and holes produced by light within a few diffusion lengths of the junction contribute to the photocurrent. But in the p- and n-regions of the a-Si the diffusion lengths are quite small and most of the contribution is from electron–hole pairs produced in the a-Si:H region. The field between the doped regions causes the holes to flow to the p-region and the electrons to the n-region. The layers form a single p-i-n junction, where i stands for the intrinsic a-Si:H layer.

A problem with such cells is that their efficiency decreases under illumination. This effect, first seen by Staebler and Wronski, results from the creation of metastable defects. The effect

(a)

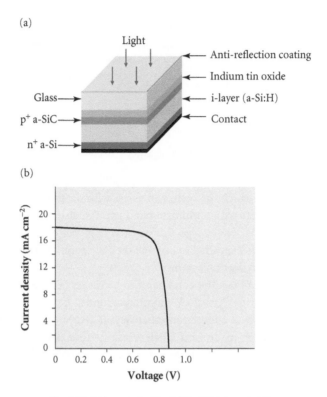

(b)

Fig. 7.13 (a) Layout of a-Si cell. (b) a-Si IV characteristic.
Source: NREL98.

has been reduced to a ~20% decrease by diluting the silane with hydrogen and by optimizing the growth conditions. Large area single junction modules have now been produced with an efficiency of ~10%. In 2010 the cost of a-Si modules was competitive with that of Si modules, and their market share (~5%) is similar to that of CdTe modules (~5%).

Thin-film multicrystalline silicon

By changing the deposition conditions it has proved possible to make good quality multicrystalline (mc) Si films as well as a-Si:H layers. But for these to be useful the probability of absorbing the light in the thin mc-Si layer must be improved. This is possible by using light trapping. This method is described in Box 7.1 and can lead to an increase in path length of ~$4n^2$, where n is the refractive index of the material. For silicon n is ~3.5, so this increase of ~50 can enable relatively thin layers (~10 μm) of mc-Si to be used.

The efficiency of a thin-film cell can be increased by making a multilayer device consisting of several p-i-n junctions on top of one another. We will now consider multilayer thin-film cells.

Multilayer thin-film cells

A multilayer device can utilize different regions of the solar spectrum. For example, we will consider a two-layer cell with a wide band gap material as the upper and a narrow band gap as

Box 7.1 Light trapping in thin films

Silicon has a low absorption, so light trapping is required if thin layers are to be used. This allows layers to be made by deposition, which uses less material. There is also less recombination in these thin layers, which increases the open circuit voltage and hence the efficiency. One way of enhancing the light absorption is to texture the top surface of the silicon and make the back surface reflective, as illustrated in Fig. 7.14.

The texturing of the top surface can cause the light to make two reflections. This reduces the overall amount reflected, but more important is the increase in path length for light within the silicon. Silicon has a high refractive index of $n \sim 3.5$, so if silicon is in a medium with a refractive index of unity then only light incident within a cone of half-angle $\theta_c = \sin^{-1}(1/n)$ will escape. Otherwise the light is totally internally reflected, as shown in Fig. 7.14(a).

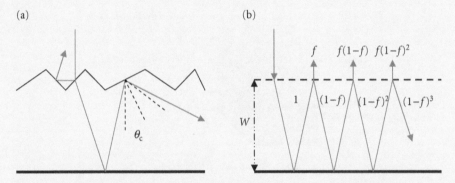

Fig. 7.14 (a) Textured top and reflective back. (b) Schematic of multiple reflections.

Light passing through the top surface of the silicon undergoes refraction and as the surface is irregular it acts as a diffuser. For an ideal diffusing surface the intensity per unit solid angle $B(\theta)$ is related to the incident intensity I_0 by

$$B(\theta) = (I_0/\pi) \cos \theta \quad \text{(Lambert's law)} \tag{7.21}$$

After each reflection the fraction f of light that escapes is given by

$$f = \int_0^{\theta_c} (1/\pi) \cos \theta \, 2\pi \sin \theta \, d\theta = \sin^2(\theta_c) = 1/n^2 \tag{7.22}$$

which is the fraction of light that falls within the critical angle θ_c. Figure 7.14(b) shows the amounts lost and intensities remaining after each reflection (NB The actual light is scattered in all directions). The mean distance D that light travels between each reflection off the back surface is given by

$$D = \int_0^{\pi/2} (1/\pi) \cos \theta (2W/\cos \theta) 2\pi \sin \theta \, d\theta = 4W \tag{7.23}$$

where W is the depth of the silicon layer and $(2W/\cos\theta)$ is the distance travelled by light reflected at an angle θ.

The mean path P travelled is given by summing the fraction of light undergoing $1, 2, 3 \ldots$ reflections, each multiplied by $4W$, which from Fig. 7.14(b) is

$$P = 4W + (1-f)4W + (1-f)^2 4W + \ldots = 4W/f = 4n^2 W \qquad (7.24)$$

where the identity $(1-x)^{-1} \equiv 1 + x + x^2 + \ldots \ (x < 1)$ has been used to give the sum of the geometric series. Since n for silicon is ~3.5, this path length is ~50 times longer than that with no texturing of the front surface and no back reflective surface. In practice, this technique allows mc-Si films of thickness of the order of 10 μm to have good absorptivity. One way the front surface can be textured is by using an acid etch.

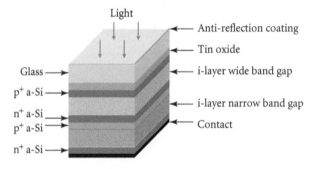

Fig. 7.15 Two-layer a-Si solar cell.

the lower. High energy photons will be absorbed in the upper layer while lower energy light will be transmitted by the upper and be absorbed in the bottom layer. The construction of such a two-layer a-Si solar cell is shown in Fig. 7.15.

The layers are in series so the current through both layers is the same and is limited by the layer producing the smallest current. The layers must be matched and one possible combination would be a-Si:H (E_g ~1.7 eV) with a-SiGe:H (E_g ~1.1 eV). The latter is made by co-depositing Ge and Si using a mixture of SiH_4 and GeH_4. An alternative for the wide band gap material is a-SiC made with CH_4 and SiH_4. Another possibility for the lower layer is microcrystalline Si with light trapping.

A four-layer device with band gaps of 1.8, 1.4, 1.0, and 0.7 eV has a theoretical efficiency greater than 50%.

EXAMPLE 7.5

A two-layer a-Si solar cell has band gaps of 1.7 eV and 1.1 eV. The photocurrent density from photons with energies between 1.1 and 1.7 eV is 19 mA cm^{-2}, while that from photons with energies greater than 1.7 eV is 21.3 mA cm^{-2}. Compare the output power for this two-layer solar cell with that of a simple silicon solar cell ($E_g = 1.1$ eV). Assume the open circuit voltage of each layer, and of the silicon cell, is given by $V_g - 0.4$ V. Take the fill factor to be 0.8.

The two layers are in series so the voltage across the cell will be the sum of the voltages across each layer. The short circuit current density through the layers is that from the lower layer, 19 mA cm^{-2}, as it is the smaller. The open circuit voltage is the sum of V_{OC} from each layer, so

$$V_{OC} = (1.1-0.4)+(1.7-0.4)=2.0\,\text{V}$$

The maximum power P_m is given by eqn (7.5):

$$P_m = FF \times V_{OC}I_{SC}=0.8(2.0)19=30.4\,\text{mW cm}^{-2}$$

The simple silicon solar cell has a short circuit current given by

$$I_{SC}=19+21.3=40.3\,\text{mA cm}^{-2}$$

The open circuit voltage is $(1.1-0.4)=0.7\,\text{V}$. The output power is therefore

$$P_m=0.8(0.7)40.3=22.6\,\text{mW cm}^{-2}$$

We can see that the efficiency is significantly higher in the two-layer cell.

7.7 Developing technologies

7.7.1 Electrochemical cells

There has been considerable research on trying to produce solar cells using a lower cost technology than the vacuum evaporation or crystal growing techniques required in the production of silicon or thin-film cells. In 1991 dye-sensitized solar cells (DSSC) were invented by M. Gratzel and B. O'Regan and these have attractive features.

In a DSSC a stack of titanium dioxide (TiO$_2$) nanoparticles ~20 nm in diameter are coated with dye molecules and the whole is immersed in an electrolyte (Fig. 7.16(a)). Light is absorbed by the dye and electron–hole pairs are produced. The electrons go into the conduction band of the n-type semiconductor TiO$_2$ and diffuse to a transparent conductive substrate. The holes in the dye molecule are filled by electrons from negatively charged ions in the electrolyte that come from the opposite electrode. The electron to ion transfer at this electrode is catalysed by a layer of platinum or carbon.

The energy levels involved are shown in Fig. 7.16(b) and the maximum potential difference (open circuit voltage) is given by the difference in the potential of the conduction band in TiO$_2$ and the potential of electrons bound in the ions in the electrolyte (the redox potential). The recombination of electrons and holes is much slower than the injection of the electrons into the conduction band of TiO$_2$. So diffusion of the electrons to the conducting substrate and the refilling of the holes by electrons from the ions in the electrolyte occur. By this means the electrons and holes produced by the solar radiation are separated. Unlike in a p-n junction, the separation is not brought about by an internal electric field.

The percentage of light absorbed by the dye molecules attached to the TiO$_2$ nanospheres is very small, but the large surface area of all the nanospheres increases the absorption considerably. TiO$_2$ is cheap and the technology shows promise: currently conversion efficiencies of

(a)

(b)

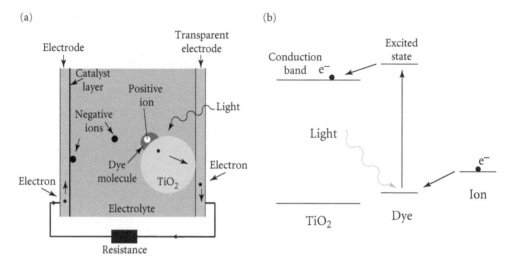

Fig. 7.16 (a) DSSC. (b) Level structure in DSSC.

laboratory cells exceed 10%, though long-term stability needs to be established. Research on using solid state hole conductors to avoid the possibility of leakage of the electrolyte, thereby making the devices simpler to manufacture and more robust, is underway but as of 2011 the efficiency of these devices was only ~5%. However, cells can be manufactured that are coloured and translucent, which makes them very attractive for integrating into buildings.

Research on nanocrystals of lead sulphide and lead selenide, called **quantum dot** absorbers, has indicated that such crystals might enhance the performance of the TiO_2 cells if used in place of the dye molecules. When illuminated with light these nanocrystals produced three electrons on average per photon absorbed, rather than just the one in a p-n junction cell, and would give a larger solar photocurrent and greater efficiency. Such crystals are called quantum dots when their size is comparable with that of the orbital radius of electron–hole excitations (excitons) within the crystal.

7.7.2 Concentrated photovoltaics (CPV)

We can avoid the need to make a large PV array by using reflectors or lenses to concentrate the light onto a single smaller area device. Silicon-based cells with 27% conversion efficiency at an intensity up to 100 times that of normal sunlight (100 Suns) and GaAs multijunction cells with 42% efficiency at 500 Suns have been developed. The reflective material for the concentrators is much cheaper than the solar cell, but the concentrators need to track the movement of the Sun, which increases the cost. Fresnel lenses and parabolic reflectors have been used but as of 2011 the market impact of CPV was small.

We expect an improvement in conversion efficiency with illumination as the open circuit voltage is expected to increase from eqn (7.3) above. This relation predicts that V_{OC} would be 0.18 V larger at 1000 Suns, which for silicon would correspond to a 25% increase in both V_{OC} and efficiency (see Example 7.4).

An interesting method for concentrating and converting sunlight that is attracting research is to load a sheet of transparent material (glass or plastic) with a suitable dye and a wavelength shifting material that causes the sunlight to be absorbed and re-emitted at a longer wavelength. A significant fraction of the emitted light is then channelled by total internal reflection to the edges of the glass where it is absorbed by photocells. No light is lost by re-absorption because of the wavelength shift. The challenge is to improve the efficiency of the device, which is called a luminescent solar concentrator.

7.7.3 Organic and organo-metallic semiconductor solar cells

In the search for cheap solar cells there has been considerable R&D on organic thin-film solar cells, which might be produced at low cost by being fabricated on a flexible plastic substrate. Stability needs to be improved but conversion efficiencies have now exceeded 8%. If printing technology could be used then vast areas of solar cells might be produced quickly and cheaply (cf. the printing of newspapers).

Recently (2013), organometal trihalide perovskite semiconductors have been found by H. Snaith et al. at Oxford University that conduct the electrons and holes produced when they absorb light over sufficiently long distances that a p-i-n type of solar cell can be made. The intrinsic perovskite ($(CH_3NH_3)PbX_3$, where X is Cl, Br, or I) layer is only 330 nm thick . The cells' efficiency is over 15% and their open-circuit voltage is 1.07 volts. The cells should be cheap to manufacture and, provided they prove to be stable, they could be a significant advance in solar cell technology.

7.7.4 Thermo-photovoltaic cells (TPC)

Solar radiation can be focused onto an intermediate absorber rather than directly onto a photocell. The absorber re-emits the solar energy as thermal radiation, which is detected by photocells surrounding the absorber. The thermal radiation is from a surface at a lower temperature than the Sun, but the radiation can in principle be utilized more efficiently. Thermal radiation can also be generated by burning gases. The application of TPCs is discussed in Box 7.2.

Box 7.2 Thermo-photovoltaic cells (TPC)

Solar radiation is first focused onto an intermediate absorber, which re-emits the solar energy as thermal radiation. The temperature of the emitter is typically 1500 K, rather than ~6000 K for the Sun, and the mean photon energy is ~0.45 eV compared with ~1.8 eV for solar radiation. The photocurrent density J_L ($A\,m^{-2}$) produced by a photocell with band gap $E_g \equiv eV_g$ is proportional to the incident radiation intensity $\mathcal{P}_{inc}$ ($W\,m^{-2}$). Assuming the thermal radiation is that of a black body, then J_L also depends on the ratio $x \equiv E_g/(kT)$. Combining eqns (7.12) and (7.14) (see Derivation 7.2), for $x > 2$ J_L is given to a good approximation by

$$J_L = \mathcal{P}_{inc}\left(\frac{15}{V_g\pi^4}\right)xe^{-x}(x^2+2x+2)$$

(7.25)

The open circuit voltage V_{OC} depends slightly on J_L (see Derivation 7.2) and can be estimated by the relation

$$V_{OC} = V_g - V_T \ln(J_0/J_L) \tag{7.18}$$

where the empirical constant $J_0 \approx 10^{10}$ A m^{-2}.

The amount of energy emitted per unit area per second $\mathscr{P}_E$ by a black-body radiator is given by the Stefan–Boltzmann law:

$$\mathscr{P}_E = \sigma T^4 \tag{7.26}$$

The value of the Stefan–Boltzmann constant σ is 5.67×10^{-8} W m^{-2} K^{-4}. The incident radiation intensity $\mathscr{P}_{inc}$ is proportional to $\mathscr{P}_E$, with the proportionality constant dependent on the geometry of the device.

While the fraction of photons with $E_\gamma > E_g$ is lower than for a solar spectrum, the use of an intermediate absorber allows the energy of the photons with $E_\gamma < E_g$ that are transmitted through the cell not to be lost. These lower energy photons are returned to the emitter by a reflective layer on the back of the photocell. At the emitter their energy is absorbed and then re-emitted. The efficiency of conversion can be further improved by using selective emitters that radiate in a fairly narrow band, rather than with a broad thermal spectrum. An example is the rare earth oxide Er$_2$O$_3$, which emits in a band around 1500 nm. This is matched to photocells with an $E_g < 0.8$ eV.

In order to reduce the amount of thermal radiation with $E_\gamma < E_g$ falling on the photocell and heating it, a filter can be added. The filter can absorb long wavelength radiation or, by using dielectric quarter-wavelength coatings, reflect it back on the absorber. A schematic of a TPC system is shown in Fig. 7.17; only one photocell of the several that would surround the emitter and filter is shown. As the mean photon energy is only 0.45 eV, a

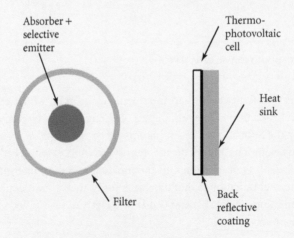

Fig. 7.17 Schematic of a TPC system; only one photocell is shown.

low band gap photocell is used, such as GaSb with $E_g=0.72$ eV, or Ge with $E_g=0.66$ eV. The predicted efficiency of such a system using GaSb cells is ~25%.

A further advantage of using a TPC is that the emitter can be heated by burning gas after dark. Or the TPV system can be run entirely with gas heating. A ceramic radiator burning methane can operate at ~1500 K and the total system can provide both heat and electricity. One possibility for producing methane is an anaerobic digester (see Chapter 8). A TPC system would then provide heat and power in regions well away from a grid, such as in some rural parts of developing countries. The lack of moving parts makes TPC attractive as maintenance should be low.

EXAMPLE 7.6

A methane heated radiator operates at a temperature of 1500 K. The cylindrical heater is 0.25 m long and 0.02 m in diameter. The TPCs surrounding the radiator form a cylinder 0.08 m in diameter. The photocells are GaSb ($E_g=0.72$ eV) with a fill factor of 0.7. Estimate the electrical output of the system. Assume that the emitted radiation is that from a black body.

From eqn (7.26) the emitted intensity from the radiator

$$\mathscr{P}_E=2.87\times10^5 \ \mathrm{Wm^{-2}}$$

The geometry is cylindrical so

$$\mathscr{P}_{inc}=\mathscr{P}_E\times\left(r_{radiator}/r_{photocell}\right)=\mathscr{P}_E/4=7.18\times10^4\,\mathrm{Wm^{-2}}$$

The ratio

$$x = E_g/(kT)=0.72\times1.6\times10^{-19}/(1.38\times10^{-23}\times1500)=5.57$$

so using eqn (7.25)

$$J_L=1.44\times10^4\,\mathrm{Am^{-2}}$$

From eqn (7.18)

$$V_{OC}=0.72-0.35=0.37 \ \mathrm{V}$$

The current produced by the photocells is the current density multiplied by the area, i.e.

$$I_L=J_L2\pi r_{photocell}L_{photocell}=1.44\times10^4(2\pi\times0.04)0.25=905 \ \mathrm{A}$$

Finally eqn (7.5) gives the power output

$$P_m=FFV_{OC}I_L = 0.7(0.37)(905)=0.23 \ \mathrm{kW}$$

The radiant power P_E is

$$P_E=\mathscr{P}_E2\pi r_{radiator}L=2.87\times10^5(2\pi\times0.01)0.25=4.5 \ \mathrm{kW}$$

The conversion efficiency will be higher than $P_{\mathrm{m}}/P_{\mathrm{E}} = 5\%$, as no allowance has been made for photons returned to the radiator and for the use of selective emitters. This TPV system could provide both heating and power to a home.

7.8 Solar panels

Solar panels are built up from single units. For example, a silicon cell made from a thin textured multicrystalline wafer ($\sim$125 μm) might be 125 cm^2 in area with $V_{\mathrm{OC}} \sim 0.6$ V, and a panel might contain about 40 such cells connected in series with an output voltage of 24 V. When illuminated with solar radiation of 1000 W m^{-2}, the 0.5 m^2 area panel would provide $\sim$75 W, corresponding to 15% conversion efficiency. This output under 1000 W m^{-2} illumination is termed the watts peak (Wp) output, and the cost of panels is often given in terms of $ per Wp or € per Wp. Figure 7.18 shows how the module (panel) cost per Wp has decreased and how the volume has increased since 1976. (The bump around 2008 was due to a shortage of silicon.)

The amount of power that a solar panel will produce over a year depends on how much sunlight the location receives. A 1-kWp array of panels gives an output of 1 kW when the solar intensity is 1 kW m^{-2}. The yearly output from a 1-kWp array is therefore numerically the same as the annual amount of solar energy per square metre. A 1-kWp array will provide $\sim$1800 kWh y^{-1} in southern California, 850 kWh y^{-1} in northern Germany, and 1600–2000 kWh y^{-1} in India and Australia. As 1 kW continuous for a year is equivalent to 8760 kWh, we see that we obtain 10–20% of the peak output of a panel, dependent on location. The global average of the capacity factor is 0.12 (2010) and this is expected to rise to 0.17 by 2050 with improvements to the technology.

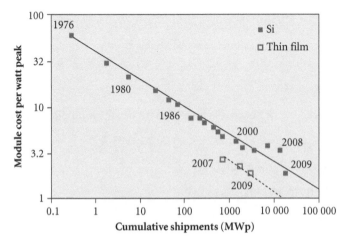

Fig. 7.18 Module cost dependence on cumulative shipments of Si and thin-film PV panels. *Source:* Solarcell.

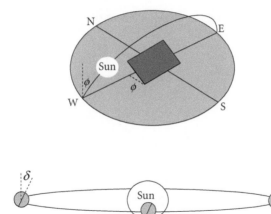

Fig. 7.19 Optimum orientation of a fixed solar panel.

7.8.1 **Solar farms**

Many large photovoltaic solar farms are now operational globally. One example is the Conergy 24 MWp farm near SinAn, a city south-west of Seoul in South Korea, which covers an area of 670 000 m² and produces 35 000 MWh annually. The panels are mounted on a single-axis tracking system, which increases the output relative to a fixed orientation by ~20%. The capacity factor of the farm is given by the ratio of the annual output to the annual peak output and equals 16.7%. The power density is 6.0 MW km^{-2}.

Figure 7.19 illustrates the optimum orientation for a fixed solar panel when located at a latitude of ϕ. The panel points south and its normal is tilted at an angle of ϕ to the vertical. When the Sun is at an equinox (21 March or 21 September) the Earth's axis is perpendicular to the direction of light from the Sun, and the Sun rises and sets as shown in Fig. 7.19(a).

The tilt (declination) δ of the Earth's axis to the plane of the Earth's motion around the Sun, which is illustrated in Fig. 7.19(b), varies from 0° to ±23.5°, lowering the path across the sky in winter and raising it in summer. For a panel oriented towards the equator, and with its slope angle equal to its latitude, the intensity I on the panel is

$$I = S\cos\delta\cos H \tag{7.27}$$

where S is the solar intensity on a surface perpendicular to the direction of the Sun's rays (the direct normal irradiance, DNI), H is the angle that the Sun has moved in its plane of motion since noon, and δ is the declination. The angle H is given by $H = 15°(t-12) = \pi(t-12)/12$ radians, where t is the time in hours.

EXAMPLE 7.7

Fixed silicon solar panels with an efficiency of 15% are located at a latitude of +35°. The active area of panels is 10 000 m². Estimate the maximum daily output in kWh of the solar farm in

September, when the number of hours of sunshine each day is 7. Take the intensity of the Sun as $1\,\mathrm{kW\,m^{-2}}$.

The optimum orientation is for the solar panels to be pointing south and tilted by the angle of latitude, in which case eqn (7.27) holds. In September $\delta\sim0$, so the total solar radiation R in one day is given by

$$R=A\times\int_{a}^{b}S\cos[\pi(t-12)/12]\mathrm{d}t$$

where $A=10^4\,\mathrm{m^2}$ is the panel area, and a and b are the times the Sun rises and sets. In this case, $a=8.5$, $b=15.5$, and $S=1\,\mathrm{kW\,m^{-2}}$. Integrating gives

$$R = A\,(12S/\pi)\,[2\sin(3.5\pi/12)]=10^4\,(12/\pi)\,1.587=6.06\times10^4\,\mathrm{kWh}$$

If the panels tracked the Sun, then $R=7\times10^4\,\mathrm{kWh}$.

7.8.2 Distributed generation with photovoltaics

Solar panels are also increasingly being used on homes, with any surplus electricity being fed into the grid. An average British household uses ~3000kWh of electricity per year and the available roof area per house is about $10\,\mathrm{m^2}$. On this area we could install ~1 kWp, which would provide in the UK around 750 kWh per year, which is about a quarter of the total consumption. The rooftop area available in the EC could provide some 1000 GWp, but presently subsidies (e.g. feed-in tariffs) would be required for this to be economic.

However, in developing countries solar power could be much more cost-effective. While a typical household in a developed country consumes about 3500 kWh a year, an estimate of the basic electricity needs for three people in a dwelling in a developing country is ~200 kWh, which could be provided by a 100 Wp system in regions of high irradiance.

EXAMPLE 7.8

A solar panel is made up of 40 silicon cells in series, each of area $0.01\,\mathrm{m^2}$, open circuit voltage 0.6 V, and fill factor 0.7. The short circuit current density of a panel under AM1.5 illumination is $400\,\mathrm{A\,m^{-2}}$. In the UK there is about $750\,\mathrm{kWh\,m^{-2}\,y^{-1}}$ of solar radiation. If an area of $8\,\mathrm{m^2}$ is available on a house, estimate the amount of energy per year that could be provided by solar panels.

AM1.5 illumination gives $1\,\mathrm{kW\,m^{-2}}$. An annual amount E_S of $750\,\mathrm{kWhm^{-2}}$ is equivalent to a continuous illumination $\mathcal{P}_{inc}$ given by

$$\mathcal{P}_{inc}T=E_S$$

where T is one year (i.e. 8760 h). Substituting gives $\mathcal{P}_{inc}=85.6\,\mathrm{Wm^{-2}}$. The short circuit current density J_{SC} is proportional to the intensity of solar radiation so

$$J_{SC}=(85.6/1000)400=34.2\,\mathrm{A\,m^{-2}}$$

A solar panel has an area of $0.4\,m^2$ and $V_{OC}=40\times0.6=24\,V$. Its short circuit current $I_{SC}=J_{SC}\times0.01=0.342\,A$.

With $8\,m^2$ of solar panels the total short circuit current $I_{SC}=0.342(8/0.4)=6.84\,A$. The average output power P_{out} is given by eqn (7.5) as

$$P_{out}=FF\times I_{SC}V_{OC}=0.7(6.84)24=115\,W=0.115\,kW$$

The amount of energy E produced in a year is given by the product of the power by the time, so

$$E = 0.115(8760)=1007\,kWh\approx 1\,MWh$$

Solar panels are often used with battery storage. This allows operation of equipment at night, but the battery can also provide a load that is quite close to optimal. The voltage across a battery remains reasonably constant while being charged. The power provided by the panel is the product of the battery voltage and the current produced by the solar panel. This power can be quite close to the maximum power point over a wide range of current and hence solar intensity. For example, a 12 V battery requires a charging voltage between 12 and 15 V. This voltage can be provided by a 30-cell silicon module with an insolation varying between 0.2 and $1\,kW\,m^{-2}$. (See Exercise 7.9.)

Solar panels can provide power in remote locations, for example for telecommunications equipment and lighting, and also for small electronic devices. Where AC power is required, an inverter is used that converts the DC output to AC. Resistive loads give a voltage proportional to the current and so are not well matched to a solar panel supply. For such loads a DC–DC converter is used, where the input DC voltage is close to the optimal voltage for the panel.

7.9 Economics of photovoltaics (PV)

We can see from Fig. 7.18 that the price per Wp of silicon panels has dropped significantly over the last 30 years and also that thin-film panels (e.g. CdTe) are now (2010) cost-competitive. There has been a >20% reduction in cost for every doubling in global production of silicon panels (learning curve effect), with the total price (module plus system cost) falling from ~$8 per Wp in 2004 to ~$4 per Wp in 2010. The module (panel cost) is about 40–50% of the total cost, the remaining amount being taken up with the cost of inverters, controllers, connectors, mounting racks, and installation.

Grid parity, when the cost of PV electricity (see Example 7.9) equals that from the national grid, has already been achieved in some sunny regions, e.g. in California and Italy, where the cost of fossil-generated power is high, and is expected to be achieved in many regions by 2020 (IEA-PV). The IEA (IEA2010) have estimated that in sunny parts of the world the cost of electricity is ~21–33¢ per dollar (corresponding to a 5–10% discount rate; see Section 12.3.1), while in regions of low insolation costs are around 60¢ per dollar. In California (Ca2010), costs are ~13–22¢ per dollar, with capital costs of $3600–5000 per kWp and capacity factors of 0.2–0.3. These costs assume a financial recovery period of 20 years. However, if the efficiency of solar panels only decreases at 0.5% per annum, the output will be still >75% of their rated capacity after 50 years, which would decrease the cost of PV electricity significantly.

Electricity charges (tariffs) to consumers vary considerably globally with the highest (2011) now exceeding 20¢ per kWh. Government subsidies and policies also affect the market. Feed-in tariffs, whereby utilities pay a guaranteed price ($ per kWh) to solar power producers, have been effective in promoting the PV market, e.g. in Germany, making it competitive and allowing the cost of a PV installation to be paid off within a few years.

The installation of PV systems in many sunny rural areas away from any grid can also now provide the most economic and dependable power supply. Moreover, the cost of extending a grid is very high and the grids may not be very reliable. Diesel generators are an alternative but while they have a relatively low initial capital cost compared with PV panels they have significantly higher operating costs, making PV particularly suitable for low power applications such as telecommunications and lighting.

In 2007 the market shares for remote industrial (5%) and remote local (4%) PV installations were much less than that for grid-connected applications (90%). (The remaining 1% is for items such as calculators.) In grid-connected systems any excess supply in the day can be exported to the grid, while any shortfall during the night is imported. Battery storage systems can be included to give security in case of power cuts.

EXAMPLE 7.9

The capital cost of manufacturing and installing an array of solar panels that will produce 1 kWp is $4000. The annual solar energy density in the location where the panels will be installed is 2000 kWh m^{-2}. Calculate the cost of electricity per kWh. Take the lifetime of the solar panels as 30 years and assume the discount rate is 6%.

As the area of solar panels produces 1 kWp, then the annual amount E_{elec} of electricity produced will equal (numerically) the annual solar energy density, i.e. $E_{elec} = 2000$ kWh.

The capital cost $C_{capital}$ is $4000, the discount rate R is 6%, and the lifetime N is 30 years, so using eqns (12.4) and (12.5) (see Section 12.3) we find the annual cost A_{cost} that repays the capital from

$$C_{capital} = A_{cost}[1-(1+R)^{-N}] / R$$

So

$$A_{cost} = R \times C_{capital}/[1-(1+R)^{-N}] = 0.06(4000)/[1-(1.06)^{-30}] = \$291$$

The cost of electricity C_{elec} is given by

$$C_{elec} = A_{cost}/E_{elec} = 291/2000 = 0.15 = 15¢.$$

7.10 Environmental impact of photovoltaics

Solar PV power in operation produces no pollutants, no greenhouse gases, and is a safe way of generating electricity. It is visually unobtrusive; there are no moving parts, which reduces maintenance and also results in no noise pollution. This means that planning permission is generally straightforward. PV cells can also be integrated into buildings (building-integrated PV or BIPV), which can save money as smaller quantities of conventional building materials are required: the modules can replace tiles and windows. A large area is required to produce MWs of power, but this can be on rooftops. One square kilometre will produce an average annual generation of ~10–40 MW, but PV is ideal for distributed power generation not requiring a grid.

In production, some hazardous materials such as Cd and As are used, but the quantities are small. With effective safeguards and regulations the risks can be kept very small and acceptable. Energy is required to manufacture the modules and is mainly from fossil fuels, so there will be an associated CO_2 emission. For the energy to manufacture not to be significant then the time required to produce this amount of energy should be small compared with the lifetime of the solar cell, which is about 30 years. This time is called the energy payback time (EPBT) and was only 1–2 years in 2011. The emissions associated with manufacturing PV systems are 57 gCO_2eq/kWh (see Table 1.4), compared with the global average of 600 gCO_2eq/kWh. In 2008, two-thirds of electricity generation was from fossil fuels.

7.11 Outlook for photovoltaics

The global capacity of photovoltaic systems has grown at an average yearly rate of over 40% since 2000 with a total installed capacity by the end of 2010 of 41 GWp. The production of PV panels has increased from 0.93 GWp in 2004 to 20.6 GWp in 2010, with over two-thirds now made in China, where the cost of labour and materials is lower. In 2010, ~17 GWp were installed, with over half in Germany and Italy and most of the rest in the Czech Republic, Japan, USA, France, and China. Grid parity has already been achieved in some sunny regions and is expected to be achieved in many regions by 2020 (IEA-PV). The current dominant technology is silicon-based cells, but the cost of some thin-film technologies is now close (see Fig. 7.18), in particular CdTe and a-Si, and with increased production and the ready supply of certain elements (such as Te) their share of the market could increase.

With an average capacity factor of 0.12, the contribution of 41 GWp PV to global electricity production was 0.2% in 2010. While installations connected to the grid are expected to be the main market, off-grid installations in Asia and Africa will be important. By 2020 North America, China, and India are expected to dominate, and by 2030 Africa, the Middle East, and Latin America will also be significant areas. With the increase in deployment in sunny regions and improvements in performance, the capacity factor is expected to improve to ~0.17 by 2050. By this time the IEA technology roadmap (IEA-PV) estimates the global share of PV electricity production could be ~11%, a substantial contribution to providing energy and reducing carbon emissions. Solar power also helps provide energy security, being immune to fuel supply and costs.

While solar power is only available during the day, this coincides with the peak load in industrialized societies. Solar power can also complement wind power where both are available, since the wind tends to be stronger at night and in winter. It is ideal for distributed generation and there are huge rural areas where many millions are without electricity, e.g. in India.

However, the development of storage is very important, as electricity is needed in the evening. Batteries couple well to the output of PV panels but costs and capacities need to be improved.

The variability of solar power means that to cope with a large fraction of solar generated electricity in a grid, extra capacity, storage (see Section 11.8), and demand management will be required. Smoothing with super grids that connect the supply from several countries and smart grids that are designed to cope with a variable supply will help. In particular, generating electricity with solar power in countries with high irradiance and using HVDC (high voltage direct current) transmission to distribute the power, as in the DESERTEC scheme, which also uses a combination of renewable sources distributed across a large region, looks like a promising plan. Figure 7.20 shows this ambitious scheme for a super grid across Europe and the Middle East and north Africa region which would enable hydro, solar, wind, and geothermal power to

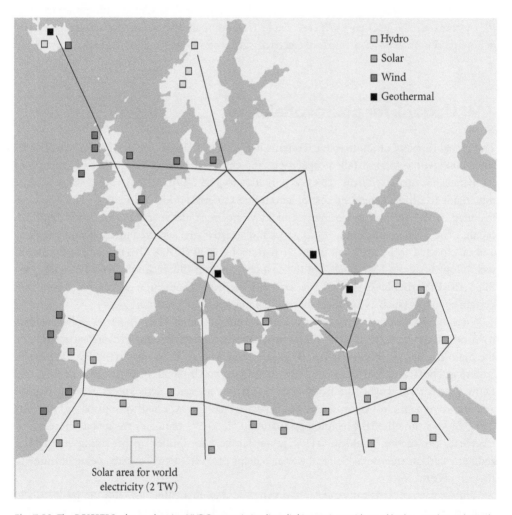

Fig. 7.20 The DESERTEC scheme showing HVDC transmission lines linking regions with good hydro, geothermal, wind, and solar sources. Also shown is the area of a solar farm (with present technology) required to provide the current global electricity demand of ~2 TW.
Source: Desertec.

be transmitted across Europe, helping to reduce the variability of renewable power and provide access to the best sources.

One way to combine solar electricity generation with storage, and so provide power at night, is to use thermal plants heated by solar radiation. We will now describe this emerging technology.

7.12 Solar thermal power plants

Innumerable methods have been devised for using solar radiation to provide heating in houses and offices, a few of which have been described in Chapter 2, but only since 2006 has there been renewed interest in and deployment of large-scale solar power plants. The first commercial plants began operation in California in the 1980s and an early example was at Barstow in California where in the 1980s and 1990s a large array of mirrors was used to direct the Sun's rays onto a tank on top of a central tower. Oil and then molten salt (at over 500 °C) were used to transfer heat to a boiler for a conventional thermal power plant that produced 10 MW. The molten salt was also used as a store of heat. Another large plant in the Mojave Desert in California used parabolic trough collectors to provide heat for an 80-MW plant from $\sim$500 000 m^2 of collector area. A tank containing oil was heated to $\sim$390 °C and an average conversion efficiency of 18% was achieved.

However, the decrease in the cost of fossil fuels in the 1990s and the withdrawal of incentives stifled any significant growth until 2006 when government initiatives in Spain and in the USA revitalized the market. By 2010 there was globally 1 GW of capacity with a further 15 GW in planning, with parabolic trough systems accounting for the greatest share of the CSTP market.

7.12.1 Concentrated solar thermal power (CSTP) plants

On clear days, which are often found in hot, semi-arid regions typically lying between latitudes 15° and 40° N and S of the equator, about 80–90% of the solar radiation reaches the Earth directly, without significant scattering, and can be concentrated by mirrors or lenses onto small areas. To be currently economic, the DNI needs to above 1900–2100 kWh m^{-2} y^{-1}. The power density of a CSTP plant without storage, when the capacity factor is $\sim$20%, is $\sim$25 MWe km^{-2}.

Coupling CSTP systems with thermal storage improves their availability and as a result their competitiveness: one estimate gave 85% (80%) of the cost with no storage for 5 (10) hours' storage. The development of super grids using HVDC transmission would also increase the potential of CSTP. For example, the large amount of sunlight in the Middle East and north Africa could be used to supply much of the energy needs of these regions as well as providing a significant amount of electricity for Europe (e.g. the DESERTEC scheme; see Fig. 7.20). The cost of production of electricity by CSTP plants in north Africa has been estimated to be $0.21 per kWh and that of transmission to southern Europe at $0.02–0.06 per kWh (IEA).

Besides parabolic trough systems, there are solar towers like that at Barstow, typically with capacities of 100–250 MW, and parabolic dishes that focus the Sun's rays onto heat engines that couple directly to generators. These dish systems, generally providing ~10 kW, avoid the need for cooling water, and provide the highest conversion efficiencies. Hybrid systems have also been developed in which either solar radiation or hot gases from the burning of fossil fuel or biomass are used to heat the engine. These improve the availability of the plant.

We now describe a solar–Stirling CSTP dish system, which uses a thermal engine based on the Stirling cycle.

7.12.2 Solar driven Stirling engines

The Stirling engine was conceived in 1816 by the Revd Dr Robert Stirling, who may have thought it a safer alternative to a steam engine, whose boiler (if poorly constructed) could explode. In his engine, a gas is sealed in a cylinder and alternately heated and cooled. In the process it drives a piston, which in turn drives a generator. The heat supply is external and the cold side can be an air-cooled heat exchanger. The internal pressures are lower than those in a steam engine. But it is slow to warm up, less compact than a steam engine, and requires precise machining, and as a result was never competitive.

However, it has several attractive features: it is quiet; it can be made to run very reliably; it has a high thermal efficiency; it has a completely external heat supply; and it has no emissions. As a result, it has already found important applications in submarines and in space. It is now being developed to provide low-carbon electricity with either solar radiation or biomass providing the energy input. (More details on the Stirling cycle are given in Box 7.3.) Typical generator sizes are 10–25 kW and these systems could be used to provide electricity in arid rural regions far from the grid as they do not require a water supply, and could be combined with thermal storage to enable electricity generation at night.

Box 7.3 Stirling engine

An ideal Stirling cycle is represented on a T–V diagram in Fig. 7.21(a). In Fig. 7.21(b) a schematic is shown of how the sealed gas is moved from the heated to the cooled cylinder through a porous matrix, called a regenerator, that acts as temporary heat store or supply. The basic principle of the cycle is that more work is produced when a gas expands from V_1 to V_2 at T_h than it takes to compress the gas from V_2 to V_1 at T_c, where T_h and T_c are the temperatures of the hot and cold sides of the sealed engine, respectively. Stirling's brilliant innovation was to store and reuse the heat transferred during the constant volume parts of the cycle.

The cycle starts with the left-hand piston at the right of the hot cylinder and the right-hand piston about two-thirds of the way along the cold cylinder. At the end of step 1 heat Q_1 will have been absorbed from the regenerator in the isochoric (constant volume) process. In step 2 heat Q_2 is absorbed, the gas expands, and the left-hand piston moves fully left, doing work W_2 in the isothermal process at the temperature T_h of the hot cylinder. Step 3 is isochoric and heat Q_3 is rejected to the regenerator. Finally in step 4 work W_4 is done compressing the gas and rejecting heat Q_4 to the cold cylinder in the isothermal process at T_c.

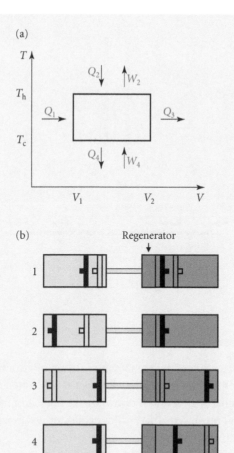

Fig. 7.21 (a) Stirling cycle. (b) Schematic of Stirling engine operation.

The values for the heat flows Q_1 and Q_2 are given by

$$Q_1 = \alpha Nk(T_h - T_c) \quad Q_2 = W_2 = NkT_h \ln(V_2/V_1) \tag{7.28}$$

where $\alpha(\alpha = C_v/R)$ for a monatomic gas is 1.5 and for a diatomic gas is ~2.5. $\Delta U = Q_2 - W_2 = 0$, as we are assuming an ideal gas, so $U = U(T)$. Likewise

$$Q_3 = \alpha Nk(T_h - T_c) \quad Q_4 = W_4 = NkT_c \ln(V_2/V_1) \tag{7.29}$$

Assuming the regenerator is 100% efficient (i.e. assuming all of $Q_3 = Q_1$ is absorbed by the regenerator and then subsequently absorbed by the gas), then the efficiency ε is given by

$$\varepsilon = (W_2 - W_4)/Q_2 = (T_h - T_c)/T_h = \varepsilon_C \tag{7.30}$$

where ε_C is the efficiency of a Carnot cycle. Actual Stirling engines are more complicated thermodynamically, but very good thermal efficiencies (~40%) can be achieved, with regenerator efficiencies of over 95% possible.

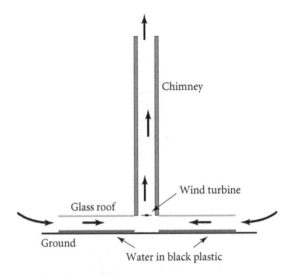

Fig. 7.22 Solar chimney.

7.12.3 **Solar chimneys**

Another method of utilizing solar energy in areas of strong solar radiation, and where land is readily available, is to construct a large solar chimney as illustrated in Fig. 7.22. Warm air is produced under a large area of glass and is drawn up the high chimney. A prototype was built in Spain in the 1980s and produced 50 kW with a collector 240 m in diameter. The chimney was 195 m tall and 10 m in diameter.

A large, ambitious project being planned in Australia is to build a giant solar tower some 1000 m tall with a diameter of 150 m; the collector would be about 5000 m in diameter. Black plastic pipes full of water placed on the ground under the glass will heat up during the day and will give out their heat to the air during the night, providing output throughout 24 hours. The output power would be 200 MW, sufficient power for ~200 000 homes. The conversion efficiency is ~2% but, where land is cheap and there is good sunshine, the cost per kWh is estimated to be quite competitive at ~$0.15 per kWh. The principle of the solar chimney is explained in Box 7.4.

Box 7.4 Solar chimney

The updraught depends on the height of the chimney. A simple model of the chimney shows how a pressure difference arises through the difference in the density ρ of the heated air within the chimney at temperature T_i and that outside at temperature T_o. Figure 7.23 is a schematic of the chimney and the air pressure variation within and outside.

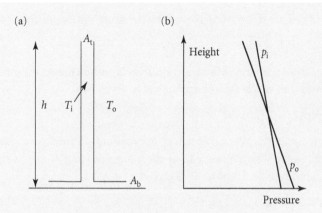

Fig. 7.23 (a) Solar chimney schematic. (b) Pressure variation in solar chimney.

The change in air pressure with height is given by the equation for hydrostatic equilibrium (see eqn (4.3)),

$$\mathrm{d}p/\mathrm{d}z=-\rho g \qquad (7.31)$$

which equates the change in pressure $\mathrm{d}p$ across a vertical element of thickness $\mathrm{d}z$ to the pressure exerted by the weight of air contained in the element.

Making the assumption that the air inside the chimney is all at T_i and that outside at T_o, then

$$\Delta p_o=\rho_o gh \quad \text{and} \quad \Delta p_i=\rho_i gh \qquad (7.32)$$

where the small change in ρ with Δp is neglected in comparison with the much larger relative change due to the difference in temperature between inside (T_i) and outside (T_o) the chimney.

Since ρ_i is less than ρ_o, there is less of a pressure drop within the chimney than outside, which gives rise to a pressure that is lower at the bottom and higher at the top inside than outside the chimney (Fig. 7.23(b)).

These differences in pressure drive air in at the bottom and out at the top, and this process, called the stack or chimney effect, is used to provide natural ventilation in buildings. The overall pressure difference Δp driving the air flow is given by

$$\Delta p=(\rho_i-\rho_o)gh=\rho(T_i-T_o)/T_o)gh \qquad (7.33)$$

We can estimate the flow of gas through an opening (orifice) using Bernoulli's principle. Consider an air stream of cross-sectional area A_1 flowing slowly with velocity u_1, which then passes through a smaller opening A_2 (corresponding to A_b and A_t in Fig. 7.23(a)). By conservation of mass, its velocity through the opening will be u_2, where

$$u_2=u_1 A_1/A_2$$

The pressure difference $p_1 - p_2 = \Delta p$ is given by Bernoulli's principle (eqn (4.2)):

$$\Delta p / \rho = \tfrac{1}{2} (u_2{}^2 - u_1{}^2) \approx \tfrac{1}{2} u_2{}^2 \tag{7.34}$$

if $u_1 \ll u_2$, i.e. $A_1 \gg A_2$. Energy will be lost because of turbulence as the air flows through the opening, and empirically the velocity u_2 will be given by

$$u_2 = C_T (2\Delta p / \rho)^{1/2} = C_T [2(\Delta T / T_o) g h]^{1/2} \tag{7.35}$$

where C_T is a constant < 1. We can obtain an approximate estimate of the power P generated by the prototype solar chimney in Spain using this value for u_2 as the wind speed incident on a wind turbine. Its power P is given by

$$P = \tfrac{1}{2}\varepsilon A \rho u^3 \approx 0.1 D^2 [2(\Delta T / T_o) g h]^{3/2} \tag{7.36}$$

where $\rho = 1.25 \ \mathrm{kg\,m^{-3}}$, the efficiency $\varepsilon \sim 0.4$ for a wind turbine (see Chapter 6), $A = \pi D^2 / 4$ is taken as the cross-sectional area of the chimney whose diameter is D, and we have taken $C_T \sim 0.8$. For the prototype, $D = 10$ m and $h = 195$ m. For $\Delta T = 20$ K and $T_o = 300$ K this gives an output power of ~ 40 kW. For the solar chimney planned for Australia, $h \sim 1000$ m and $D \sim 150$ m and if we take $\Delta T = 30$ K and $T_o = 300$ K then $P \sim 200$ MW.

These estimates are rather rough, as the model that the speed of air flow up the chimney is given by that through an opening is very approximate. However, they do show how a considerable amount of power can be generated by a very tall solar chimney.

7.12.4 Ocean thermal energy conversion (OTEC)

The oceans collect a huge amount of solar radiation. As a result a vast amount of heat energy is stored within the top 100 m of the oceans, where the temperature is $\sim$20–25 °C higher than deep below the surface. Ideas on the exploitation of this resource, called ocean thermal energy conversion (OTEC), have been developed (see Box 7.5). The small temperature differences, however, result in low thermal efficiencies. To date there are significant technical challenges to be met and the process is not economically competitive.

Box 7.5 Ocean thermal conversion (OTEC)

The oceans cover a huge area of the Earth and absorb an enormous amount of energy per day. Approximately 1000 MW falls on a square kilometre of sea. The result is a small temperature difference of $\sim$20–25 °C between the surface and water below $\sim$1000 m. Using this temperature difference to drive a heat engine is the basis of OTEC. The temperature of the top $\sim$100 m of the sea is roughly constant. It then decreases until it is $\sim$5 °C at $\sim$1000 m, and below that depth it remains approximately the same.

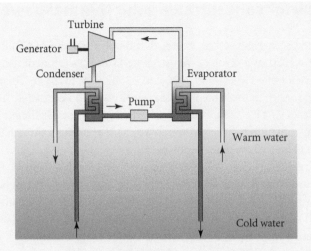

Fig. 7.24 Schematic of a possible OTEC system.

A schematic of a possible OTEC system is shown in Fig. 7.24. From the second law of thermodynamics there is a limit on the output power P_{out} given by

$$P_{out} < \eta_{Carnot} P_{flow} \tag{7.37}$$

where η_{Carnot} is the Carnot efficiency,

$$\eta_{Carnot} = (T_h - T_c)/T_h = \Delta T/T_h \tag{7.38}$$

and P_{flow} is the heat flow. If the pump draws a volume Q of seawater per second, and the specific heat capacity and density of the seawater are C and ρ, respectively, then

$$P_{flow} = \rho C Q \Delta T \tag{7.39}$$

So, the output power is limited to

$$P_{out} < (\rho C Q/T_h)(\Delta T)^2 \tag{7.40}$$

Note the quadratic dependence on ΔT, which favours regions with large ΔT. Seawater has $\rho \approx 1000\,\text{kg m}^{-3}$, $C \approx 4200\,\text{J kg}^{-1}\,^{\circ}\text{C}^{-1}$ so $\Delta T = 20\,^{\circ}\text{C}$ and $P_{out} = 1\,\text{MW}$ requires $Q > 0.18\,\text{m}^3$ $\text{s}^{-1} \equiv 180$ litres per second.

Such a system would need large and expensive equipment: big pumps and very efficient heat exchangers. Encrustation by marine organisms (bio-fouling) is a serious problem. The manufacture of long, large-diameter cold water pipes is also difficult. Dredging up the cold water from the deep will release some CO_2 dissolved in the seawater. But the amount is less than 1% of that generated burning coal to produce the same amount of energy. An OTEC system, however, may affect the climate and the fauna and marine life of the region. OTEC is not yet economically competitive, but with global warming and technical developments, it is closer to becoming a viable option in some tropical regions.

7.13 Outlook for concentrated solar thermal power (CSTP)

In their accelerated scenario, the IEA expect that the deployment of CSTP could accelerate in the period 2010–2020, as a result of industrial R&D and government incentives for CSTP, with a global capacity of ~150 GW by 2020 and ~1000 GW by 2050. The capacity factor is expected to increase with improvements in storage. By 2050 an estimated 10% of global electricity could be from CSTP plants. While the technology is not dependent on the supply of any rare elements, obtaining sufficient water will be a challenge, though hybrid water–air cooling systems will help alleviate the problem.

Besides generating electricity, CSTP plants can be used for desalination, for which there is expected to be a growing demand with the increasing scarcity of water in many hot regions. CSTP could also be used for solar-assisted reforming, in which methane (natural gas or biogas) is reacted with water to produce hydrogen and carbon dioxide (see Section 11.17). (NB The CO_2 could also be captured and stored to reduce carbon emissions.)

Currently, large trough CSTP plants cost $4200–8400 per kW capacity with the cost of electricity $0.2–0.3 per kWh, dependent mainly on the amount of sunlight (IEA-CSP). Storage increases the cost of the plant but increases the capacity factor, with the result that the cost of electricity is similar to that from plants without storage. However, the output is more reliable and, as prices come down with increased production, CSTP plants are expected to become more competitive with fossil-fuel plants. The IEA have projected that the cost of electricity from CSTP could be as low as ~$0.05 per kWh by 2050, though this will require considerable investment and favourable policies.

SUMMARY

- Solar radiation is a huge resource, offering more than 5000 times current world power consumption. The solar intensity on a clear sunny day is ~1000 W m^{-2}.

- Solar cells are currently (2010) mainly (85–90%) silicon semiconductor p-n junction devices. Their output current density J_C is given by

$$J_C = J_{SC} - J_S[\exp(V/V_T) - 1]$$

where J_{SC}, the short circuit current density, is proportional to the light intensity, J_S is the saturation current density (~4×10^{-9} A m^{-2}), and $V_T = kT/|e| \approx 0.026$ V.

- The maximum output power density $\mathcal{P}_m$ of a photocell is given by

$$\mathcal{P}_m = FF \times J_{SC} V_{OC}$$

where V_{OC} is the open circuit voltage and FF is the fill factor, typically ~0.8.

- Silicon solar cells have, under AM1.5 illumination, V_{OC} ~0.6 V, J_{SC} ~400 A m^{-2}, and FF ~ 0.7. Output power density $\mathcal{P}_m$ ~180 W m^{-2}, equivalent to a conversion efficiency of ~18%.

- Solar PV power density is ~100–400 kWe ha^{-1} or 10–40 MWe km^{-2}.

- Large area thin-film cells, in particular amorphous silicon, CuInGaSe$_2$ (CIGS), and CdTe, are starting to become cost-competitive, though their market share is small (~10% in 2010). There is R&D on new technologies, such as electrochemical, organic, and thermo-photovoltaic systems. Large CSTP thermal plants are also being developed.

- The cost of generating electricity by photocells has dropped significantly over the last 25 years and is now near grid parity in some sunny regions and very competitive in rural sunny regions away from the grid.

- Solar power as a source of low-carbon electricity has great potential and, with strong investment and favourable policies, PV and CSTP could provide ~20% of global electricity by 2050. The corresponding accessible potentials are PV ~500 GWe and CSTP ~500 GWe.

FURTHER READING

Boyle, G. (ed.) (2004). *Renewable energy*. Oxford University Press, Oxford. Good qualitative discussion of solar photovoltaic technologies.

Herrera, D.K. and Aguirre, C.A. 2012. *Solar power for communities*. Nova Science, Hauppage, NY. Useful background for communities contemplating solar power schemes.

Progress in Photovoltaics: Research and Applications (2011). **19**: 84–92. Solar cell conversion efficiencies (Photo2011).

Progress in Photovoltaics: Research and Applications (2011). **19**: 286–93. Fundamental losses in solar cells.

Rosenberg, H.M. (1988). *The solid state*. Oxford University Press, Oxford. Good introductory textbook for condensed matter physics.

Sze, S.M. (1985). *Semiconductor devices*. Wiley, New York. Good textbook on the physics of photocells.

Twidell, J. and Weir, T. (2006). *Renewable energy resources*. Taylor & Francis, London. Provides useful information about photovoltaics.

Wurfel, P. (2005). *Physics of solar cells*. Wiley, New York. More advanced textbook on the physical principles of solar cells.

WEB LINKS

www.eere.energy.gov/topics/solar.html Good overview of solar cell technologies.

www.greenpeace.org/international/en/publications/reports/Solar-Generation-6/ Good summary of PV potential.

www.solarthermalworld.org/sites/gstec/files/REN21_GSR2011.pdf Renewables 2011, Global status report.

en.wikipedia.org/wiki/Solar_cell Good summary of solar cells.

www.desertec.org/SolarEnergycc.ch/ (Desertec).

www.iea.org/publications/freepublications/publication/csp_roadmap.pdf IEA CSP technology roadmap (IEA-CSP).

www.iea.org/publications/freepublications/publication/pv_roadmap.pdf IEA PV technology roadmap (IEA-PV).

www.nrel.gov/docs/legosti/fy98/23733.pdf T. Brog, Thin film CdTe PV modules (NREL98).

www.iea.org/Textbase/npsum/ElecCost2010SUM.pdf Projected costs of generating electricity (IEA2010).

www.energy.ca.gov/reti/ California Energy Commission (Ca2010).

solarcellcentral.com/cost_page.html (Solarcell).

LIST OF MAIN SYMBOLS

V	voltage	FF	fill factor
I	current	V_{OC}	open circuit voltage
P	power	I_{SC}	short circuit current
$\mathscr{P}$	power per unit area	J	current density
R	resistance	I_S	saturation current
V_T	thermal motion voltage ($\equiv kT/|e|$)	I_L	light-induced current
I_S	saturation current	E_g	band gap energy
I_C	photocell current	V_g	band gap voltage
p^+	highly doped p-material	Wp	watts peak
n^+	highly doped n-material		

EXERCISES

7.1 Blue light of wavelength 475 nm falls on a silicon photocell whose band gap is 1.1 eV. What is the maximum fraction of the light's energy that can be converted into electrical power?

7.2 Sunlight of intensity 600 W m^{-2} is incident on a building at 60° to the vertical. What is the solar intensity, or insolation, on (a) a horizontal surface; (b) a vertical surface?

7.3 When it is heated to temperatures of about 1700 K, approximately 20% of the radiation emitted by the rare earth oxide ytterbium oxide, Yb_2O_3, is in a narrow band around 1000 nm. Would this source of radiation be suitable for a Si or a GaAs photocell? (Band gaps: Si 1.1 eV, GaAs 1.4 eV.)

7.4 The intensity I_{AM0} of solar radiation incident on the Earth's atmosphere (AM0) is given to a good approximation by

$$I_{AM0} = \sigma T^4 \Omega_S / \pi \ \mathrm{W\,m^{-2}}$$

where Ω_S is the solid angle subtended by the Sun and σ is the Stefan–Boltzmann constant. Find T given that the intensity I_{AM0} is 1367 $\mathrm{W\,m^{-2}}$ and $\Omega_S = 6.8\ 10^{-5}$ sr.

7.5 Explain why some materials are insulators, some conductors, and some semiconductors.

7.6 Plot the current through an ideal p-n junction which has a saturation current of 10^{-11} A, for bias voltages of −0.4 to +1.0 V in 0.2 V steps.

7.7* The current–voltage relation for an ideal diode is given by

$$I = I_L - I_S[\exp(V/V_T) - 1]$$

where $V_T \equiv kT/|e| \approx 0.026$ V at room temperature. By differentiating the power P, given by $P = VI$, with respect to V and equating the derivative to zero, show that the maximum power P_m occurs when

$$(1 + V/V_T)\exp(V/V_T) = I_L/I_S + 1$$

By approximating $(1 + V/V_T)$ by V_{OC}/V_T, where $V_{OC} = V_T \ln(I_L/I_S)$, and $(I_L/I_S + 1)$ by I_L/I_S, show that V_m and I_m are given by

$$V_m = V_{OC}(1 + x_{OC} \ln x_{OC}) \quad \text{and} \quad I_m = I_{SC}(1 - x_{OC})$$

where $x_{OC} = V_T/V_{OC}$. Note $I_L = I_{SC}$.

7.8* The density of electrons dn_e with energy ε to $\varepsilon + d\varepsilon$ near the bottom of the conduction band is given by $dn_e = D(\varepsilon) f_e(\varepsilon)\, d\varepsilon$, where $D(\varepsilon)$ is the density of electron states and $f_e(\varepsilon) = \{\exp[(\varepsilon - \eta_e)/(kT)] + 1\}^{-1}$ is the Fermi–Dirac (or Fermi) distribution, which is the probability that an electron is in the energy range ε to $\varepsilon + d\varepsilon$. The electrochemical potential $\eta_e \equiv \varepsilon_F$, the electron Fermi energy, and is the value of ε for which $f_e(\varepsilon) = \frac{1}{2}$.

Show for $D(\varepsilon) = a(\varepsilon - \varepsilon_C)^{1/2}$ and $(\varepsilon_C - \eta_e) \gg kT$ that $n_e = N_C \exp[-(\varepsilon_C - \eta_e)/(kT)]$, where N_C is the effective density of states at the bottom of the conduction band, given by $N_C = a(kt)^{3/2}(\pi/4)^{1/2}$.

A corresponding expression holds for the number of holes n_h near the top of the valence band: $n_h = N_V \exp[-(\eta_h - \varepsilon_V)/(kT)]$. Hence the number of electron–hole pairs $n_e n_h$ is proportional to $\exp[-eV/(kT)]$ when there is a difference in the electrochemical potentials $\eta_e - \eta_h = eV$. (η_e and η_h are also called quasi-Fermi levels, as the electrons and holes are separately in thermal equilibrium with the semiconductor material.)

7.9* The rate at which photons are absorbed by electrons in the conduction band of a semiconductor with energy E_1 and are excited to unoccupied states with energy E_2 in the valence band is given by

$$dr_a = M^2 D_{12} f_1 (1 - f_2) n(E)\,dE$$

where M^2 is the transition probability, D_{12} is the density of states, and $n(E)dE$ is the current density (number per unit area per second per unit solid angle) of photons with energy $E \ (= E_2 - E_1)$ to $E + dE$; f_1 is the probability that the initial state is occupied and $(1 - f_2)$ is the probability that the final state is empty.

These photons also induce transitions of electrons in the valence band with energy E_2 to unoccupied states in the conduction band with energy E_1 at a rate

$$dr_i = M^2 D_{12} f_2 (1 - f_1) n(E) dE$$

There are also spontaneous transitions of electrons between the same states at a rate

$$dr_s = c M^2 D_\gamma D_{12} f_2 (1 - f_1) dE$$

where $D_\gamma = 2E^2/(h^3 c^3)$ is the density of emitted photon states per unit solid angle.

In thermal equilibrium, $dr_a = dr_i + dr_s$. Show, by substituting the separate Fermi distributions for electrons in the conduction and valence bands (see Exercise 7.8*), that $n(E)$ is given by the generalized Planck distribution, eqn (7.8).

7.10* The power density generated by a solar cell is given by $\mathcal{P} = V(J_L - J_F)$, where J_L and J_F are the reverse and forward current given by eqns (7.12) and (7.11), respectively. By evaluating the equation $\partial \mathcal{P}/\partial E_g = 0$, show that the voltage V_m at maximum power is then given by eqn (7.13).

7.11 (a) A silicon photocell has an area of $4 \, \text{cm}^2$ and is illuminated normally with AM1.5 solar radiation. The short circuit current is $160 \, \text{mA}$ and the saturation current is $4 \times 10^{-9} \, \text{mA}$. Calculate the maximum power output and the corresponding load resistor. (b)* What is the output power when the load resistor is 10% higher than the optimum value?

7.12 A household uses 4000 kWh of electricity in a year. Estimate what area of solar panels would be required to produce 1000 kWh of electricity. The insolation in the region is $800 \, \text{kWh m}^{-2} \text{y}^{-1}$.

7.13 A 30-cell silicon solar panel has a saturation current density $J_S = 10^{-7} \, \text{A m}^{-2}$. Show that this panel could be used to charge a 12 V battery by calculating the peak power voltages V_m for insolation values of 0.2, 0.4, 0.6, 0.8, and $1.0 \, \text{kW m}^{-2}$. An insolation of $1 \, \text{kW m}^{-2}$ gives a short circuit current density of $400 \, \text{A m}^{-2}$.

7.14 In a region where the solar insolation is $1800 \, \text{kWh m}^{-2} \text{y}^{-1}$, estimate the area of solar panels that would be required to produce 100 MW of electricity.

7.15 A reasonable approximation to the dependence of the short circuit current density J_{SC} on the band gap E_g for $0.5 < E_g < 1.8$ eV under AM1.5 illumination is

$$J_{SC} = (80 - 34 E_g) \, \text{mA cm}^{-2}$$

A three-layer multijunction solar cell has an upper layer with $E_g = 1.8$ eV. (a) Determine the optimal band gaps for the lower two layers. (b) Calculate the output power under AM1.5 solar illumination, assuming a fill factor of 0.8 and an open circuit voltage for each layer given by $V_{OC} = V_g - 0.4$ V. (c) What is the conversion efficiency?

7.16* A thin-film silicon solar cell has a thickness W. The upper surface is polished flat and has an anti-reflection coating. On the back surface there is a perfectly diffusing reflective coating. Show that light will have an effective path length within the silicon of $(4n^2 + 1)W$, where n is the refractive index for silicon.

7.17 A CIGS photocell of area $10\,cm^2$ and band gap $1.5\,eV$ is illuminated with laser light of wavelength $800\,nm$. The photocell has a saturation current of $10^{-7}\,mA$. The light power is $150\,W$. Use eqns (7.3), (7.19), and (7.20) for V_{OC}, I_m, V_m, and FF to estimate the conversion efficiency of the photocell.

7.18 What are the advantages and disadvantages of single crystal compared with thin-film solar cells?

7.19* A solar cell with an open circuit voltage of $0.4\,V$ utilizes quantum dot photon absorbers. These absorbers emit two electrons when the energy of the photon $E_\gamma > 1.6$ eV and only one when $0.8 < E_\gamma < 1.6$ eV. Compare the conversion efficiency under AM1.5 illumination with that of a p-n junction solar cell with a band gap of 0.8 eV and an open circuit voltage of 0.4 V. Assume that both cells have the same fill factor.

7.20* In a TPV system, the central cylindrical emitter is surrounded by two concentric quartz cylinders. The quartz cylinders transmit radiation with wavelength $\lambda < \lambda_{max}$. Wavelengths with $\lambda > \lambda_{max}$ are absorbed and re-emitted in all directions. Surrounding the quartz cylinders are the photocells.

Show that only a third of the radiant energy with $\lambda > \lambda_{max}$ from the central emitter is transmitted by the quartz cylinders to the photocells.

7.21 A solar farm consisting of fixed silicon photovoltaic panels is located at a latitude of $25°$, where the average number of hours of direct sunlight is 10 per day. The panels have an efficiency of 18% and a total active area of $3 \times 10^4\,m^2$. Estimate the maximum average power output of the solar farm.

7.22 In a Stirling engine the regenerator has an efficiency of ε_R, i.e. the heat input Q_1 required in eqn (6.32) equals $(1 - \varepsilon_R)Q_3$. Show that the efficiency ε_S of the Stirling engine operating between temperatures T_h and T_c is given by

$$\varepsilon_S = \varepsilon_C / [1 + (1 - \varepsilon_R)\alpha\varepsilon_C / \ln(V_2/V_1)]$$

where ε_C is the Carnot efficiency $(T_h - T_c)/T_h$ and $\alpha = 1.5$.

Calculate the efficiency for $T_h = 325\,°C$, $T_c = 75\,°C$, $\varepsilon_R = 0.5$, and $V_2/V_1 = 5$.

7.23 What would the capital cost and installation charges need to be for 1 kWp of solar panels for the cost of electricity to be 20¢ per kWh? The insolation is 1500 kWh m^{-2} y^{-1}. Take the lifetime of the panels to be 25 years and the discount rate to be 7%. Neglect any maintenance charges.

7.24 The capital cost of manufacturing and installing a 1 kWp array of solar panels is €7000. The annual solar energy density in the location where the panels will be installed is 1800 kWh m^{-2}. (a) Calculate the cost of electricity per kWh. Take the lifetime of the solar panels as 35 years and assume the discount rate is 5%. (b) What will be the cost of electricity if there is an annual maintenance charge of €100?

7.25 An OTEC system is proposed for a region where the temperature difference between the surface and the deep water is $25\,°C$. The pumps have a capacity of 100 litres per second. Estimate the power output if the overall efficiency is 50% of the theoretical maximum.

7.26 A solar chimney is proposed to provide 100 MW of power. The temperature difference of the air inside and outside the chimney is predicted to be $35\,°C$. Estimate the height of the solar chimney required if the diameter of the chimney is $100\,m$.

7.27 An organic photocell with an efficiency of 10% is printed on flexible sheet. Estimate the area required to produce an an average annual electrical output of 100 GW in a sunny region and compare with the area printed in a year of a country, state, or city newspaper available in that region.

7.28 Discuss the advantages and disadvantages of the DESERTEC scheme for distributing solar power to Europe.

7.29 Describe the relative merits of photovoltaic solar farms and CSTP plants.

8 Biomass

Introduction

Plants derive their energy to grow from the Sun's radiation. The primary process is photosynthesis, in which carbon dioxide and water are converted to carbohydrate and oxygen. Animals eat plants and other animals, and the whole of animal and plant material is called biomass. The burning of biomass or of materials derived from biomass is an important source of energy in the world, providing about 10% of the world's requirements.

In an energy context biomass refers to plant- and animal-derived material such as straw, logs, dung, and crop residues that are used either directly or indirectly to produce heat, electricity, or fuels. The fuels that are produced are often called biofuels. The attraction of biomass as a source of energy is that it is potentially low-carbon, since the amount of CO_2 released in its combustion has been previously removed from the atmosphere when CO_2 was converted by photosynthesis into making the plant material. However, CO_2 is generated in the land clearance, planting, fertilizing, harvesting, and transport of the biomass; but, provided this is low and we renew the biomass consumed, biomass can be a sustainable low-carbon source of energy.

We first look at photosynthesis and estimated crop yields and see that the overall efficiency of conversion of solar radiation is rather low at around 0.5%. Large areas of land are therefore required to produce significant amounts of energy. We then look at how biomass is currently used as a source of energy. Finally we discuss its future potential, both in providing energy and also in producing alternative liquid biofuels to the fossil fuels petrol (gasoline) and diesel.

8.1 **Photosynthesis and crop yields**

In photosynthesis carbon dioxide and water are converted to oxygen and carbohydrate (sugar).

$$CO_2 + H_2O + h\nu \rightarrow O_2 + [CH_2O]$$

where $h\nu$ represents light quanta (photons) and $[CH_2O]$ stands for carbohydrate. The products are ~5 eV per carbon atom higher in energy, which corresponds to ~16 MJ kg^{-1} for pure carbohydrate. The amount of energy per kilogram depends on the degree of oxidation of the carbon, with zero available when fully oxidized as CO_2, ~16 MJ kg^{-1} as carbohydrate, and the maximum available when fully reduced as CH_4 at 55 MJ kg^{-1}.

In photosynthesis a minimum number of eight photons, each with ~1.8 eV, so 14.4 eV in all, are needed to produce one O_2 molecule and one C atom fixed in carbohydrate that stores ~4.8 eV of energy. In sunlight, photons at the red and blue ends of the visible spectrum are absorbed by the chlorophyll pigment in the leaves of the plant. More green light is therefore reflected from the leaves, which is why leaves appear green. In a leaf the pigment molecules are close together and when an absorbed photon excites an electron this electron can transfer to an adjacent molecule rather than de-exciting and emitting a photon. In this way, the energy of the electron can be used to form molecules in a series of complex chemical reactions, together called the electron transport pathway. The maximum efficiency of this process is ~33% (4.8/14.4). But only ~50% of the energy of the solar photons can be absorbed in photosynthesis because of the spread in wavelengths of solar radiation. Together with losses (~25%) from the leaves, mainly reflection and transmission, the maximum theoretical efficiency is ~12%, and ~10% has been achieved under optimal conditions in a laboratory.

But, in the field the conversion efficiency is much lower since only about one-third of the solar radiation falls during the growing period. Only approximately one-fifth of the radiation lands on the leaves. About 60% of that is converted to biomass with the remaining ~40% used up in sustaining the plant through respiration. Multiplying these factors together gives us a rough estimate for the overall average annual efficiency of ~0.5%:

$$\text{Efficiency} \approx (0.12)(0.33)(0.2)(0.6) \approx 0.5\%$$

In Europe the annual amount of solar energy is ~1000 kWh m^{-2}. This is equivalent to 10 GWh ha^{-1} or 36 TJ ha^{-1} (1 ha $\equiv 10^4$ m^2). The available energy for biomass production is therefore ~180 GJ ha^{-1}, so the yield of carbohydrate biomass, which requires ~16 GJ t^{-1}, is approximately

$$\text{Biomass yield} \approx 10 \text{ t ha}^{-1}\text{y}^{-1} \cong 180 \text{ GJ ha}^{-1}\text{y}^{-1} \cong 5 \text{ kW}_{th} \text{ ha}^{-1} \equiv 0.5 \text{ MW}_{th} \text{ km}^{-2} \qquad (8.1)$$

The yield and energy content depend on the plant and growing conditions and should only be taken as a rather approximate estimate for the dry biomass yield. But we will find these values useful in estimating how much land area is required for biomass supply. The equivalent continuous thermal power of ~0.5 MW$_{th}$ could generate typically 0.2 MWe of electrical power.

Most plants have three-carbon (C_3) based compounds produced in photosynthesis as part of the Calvin cycle. Some tropical plants, e.g. sugar cane and maize, initially produce a four-carbon (C_4) based compound. The C_4 compound raises the CO_2 concentration in the leaf, which is then fixed in a Calvin cycle. In high light levels and high temperatures (tropics) these C_4 plants have greater conversion efficiency, while at lower temperatures and lower light levels (temperate regions) C_3 plants have a greater efficiency. For example, the yield of sugar cane, a C_4 plant, can be ~100 $t\,ha^{-1}y^{-1}$, though ~75% of that mass is water and the yield of sugar is typically ~10 t $ha^{-1}\,y^{-1}$. Algae, which are C_3 plants, can also have a high conversion efficiency, potentially producing an order of magnitude more biofuel per hectare than other sources; however, the challenge is to find robust strains of algae that can be harvested cost-effectively.

Respiration is the reverse reaction to that occurring in photosynthesis and proceeds by enzyme-catalysed reactions that convert carbohydrate plus oxygen to carbon dioxide and water with the release of about 5 eV per carbon atom. Respiration occurs all the time in humans and is the source of energy for human activity, such as walking. In plants respiration occurs mainly at night but can occur during the day if there is little available water. Under these conditions forests can be a source rather than a sink of carbon dioxide.

Further conversion of carbohydrates to oils occurs when certain plants ripen, and these oils provide a more compact form of energy storage, typically ~38 $MJ\,kg^{-1}$, the details of which are described in Box 8.1.

Box 8.1 Energy storage in plants

Plants synthesize carbohydrates from CO_2 and H_2O using the energy from sunlight (photosynthesis; see Section 8.1). The carbohydrates are part of a plant's structure and provide a store of energy. The simplest carbohydrates are sugars or monosaccharides, which have the composition $(CH_2O)_n$. Glucose, $C_6H_{12}O_6$, is the commonest plant sugar and is called a 6-sugar as it contains six carbon atoms. The glucose molecule can exist in several forms, in which the atoms have different bonding and orientations, called 'structural isomers'; in particular, as α-glucose and β-glucose. These two hexagonal ring forms are illustrated in Fig. 8.1. The upper illustrations indicate that the hexagonal

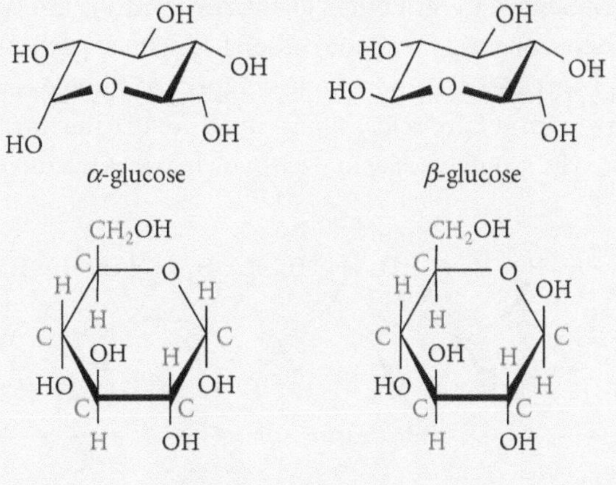

Fig.8.1 α- and β-glucose.

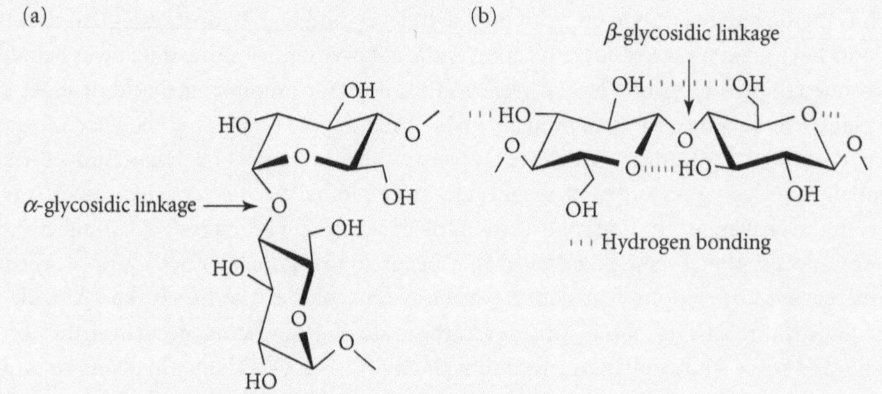

(a)   (b)

α-glycosidic linkage ⟶

β-glycosidic linkage

··· Hydrogen bonding

Fig.8.2 (a) Amylose. (b) Cellulose.

rings are not actually flat in nature. Also, the carbon and hydrogen symbols are left off the upper diagrams for clarity.

Glucose forms a disaccharide by a condensation reaction,

$$C_6H_{12}O_6 + C_6H_{12}O_6 \rightarrow C_{12}H_{22}O_{11} + H_2O \qquad (8.2)$$

that is the reverse of a hydrolysis reaction. For example, maltose is formed by linking one α-glucose and one β-glucose and cellubiose by linking two β-glucose molecules. Further condensation reactions convert glucose to polysaccharides, in particular to starch and cellulose. These biopolymers store energy and provide bulk and structure in a plant.

The differences in the bonding in starch and cellulose significantly affect their structure, with starch much more amorphous than cellulose, which forms fibrous bundles. Amylose, which is a component of starch, is a polymer of α-glucose molecules linked by α-glycosidic bonds, while cellulose is a polymer of β-glucose molecules linked by β-glycosidic bonds. The linkages in amylose and cellulose are illustrated in Fig. 8.2. The hydrogen bonding gives increased stability and leads to long, straight chains. These chains can hydrogen bond with each other, giving rise to strong micofibrils.

The sugars are a store of energy with the carbon in a state of partial reduction; on combustion ~16 MJ kg^{-1} is released. A more compact form of storage is afforded by further reducing the sugars to form fatty acids, whose structure has the form shown in Fig. 8.3. (The H of the carboxylic group is easily ionized, which makes the molecule an acid.)

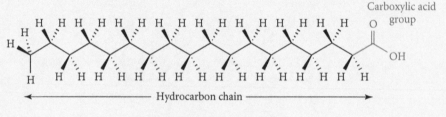

Carboxylic acid group

Hydrocarbon chain

Fig.8.3 Structure of a fatty acid.

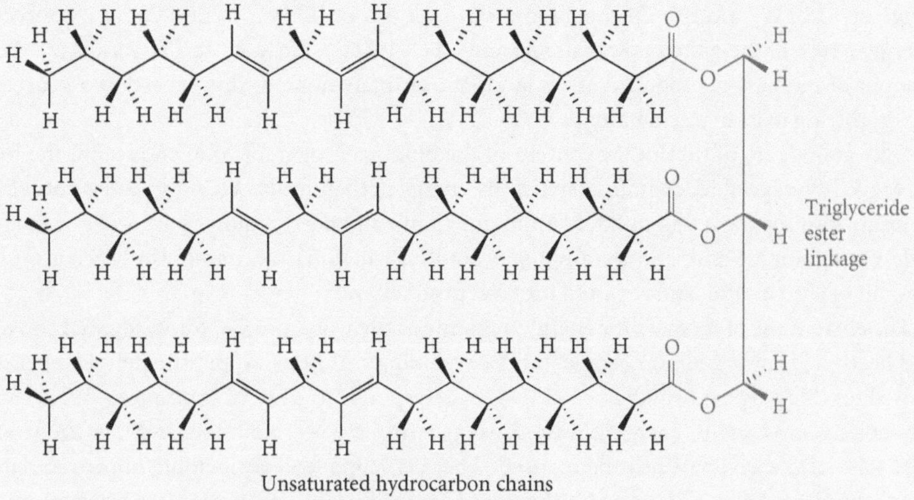

Fig.8.4 An unsaturated triglyceride, trilinolein.

The commonest length of the hydrocarbon chain lies within 12 to 24 carbon atoms. These molecules have a very low solubility in water, and they and their derivatives are called lipids. The carbon is almost fully reduced and has the highest ratio of H:C when there are no double bonds between the carbon atoms; such fatty acids are called saturated. The heat of combustion is therefore much higher than for carbohydrates, with typically 38 MJ kg^{-1} released.

A common storage molecule is a triglyceride, which is a fatty acid ester: three fatty acids joined to a glycerol molecule with the removal of three water molecules. This transformation of carbohydrate to triglyceride occurs when certain plants, e.g. olives, ripen. The reverse process occurs when a seed starts to grow (germination) with the hydrolysis (uptake of water) of 1 gram of oil producing ~2.7 grams of carbohydrate. Shown in Fig. 8.4 is an unsaturated triglyceride, trilinolein.

In this fat, the double bonds cause the chain to kink at the indicated positions, with the result that the intermolecular bonding is reduced as the molecules cannot pack together so closely as in a saturated triglyceride. As a result, such naturally occurring unsaturated fats tend to be liquids at room temperature, while the saturated fats tend to be solids. For example, olive oil is composed of triglycerides made up mainly of the isomers oleic acid (55%–85%), a monounsaturated acid (only one double bond), and linoleic acid (~9%), a polyunsaturated acid (more than one double bond).

8.2 Biomass production and use

The mass of plants and animals on land produced each year is about 1.2×10^{11} t, and contains ~5×10^{10} t of carbon. An average of ~15 MJ kg^{-1} is stored in biomass, making the annual amount of bioenergy produced equal to ~2×10^{21} J. This is the amount of energy that would be generated in a year by a power output of ~65 TW$_{th}$, which is several times the global power usage of ~18 TW$_{th}$ (2010). Of this biomass we use ~2.4×10^9 t y^{-1} ($\equiv 1.5$ TW) as a source of energy, ~10% of the global power usage, and ~11×10^9 t y^{-1} for food. In the oceans a similar amount of biomass is produced as on land, but virtually none of this is used as a source of energy, owing to its inaccessibility.

Only about 12% of the energy content of the biomass grown for food ends up in the food we eat. While vegetables contain ~60% of the energy in their feedstock, only ~5% is contained in meat, with the meat and milk of ruminants (lamb and beef) requiring an order of magnitude more biomass and land than pig and poultry, which has an important bearing on the amount of agricultural land required for food production.

The current use of biomass for energy is mainly as fuelwood (~55%) for residential cooking and heating for over 2 billion people in the developing countries. Approximately 1 kg per day is used for this purpose, which at ~10 MJ kg^{-1} corresponds to ~120 W continuous. When used for cooking, most of this energy is wasted, as open fires are very polluting and inefficient with only ~5% of the available heat being used. There is scope for considerable improvement in health and in energy efficiency with the use of more efficient ovens. Biomass provides about 20% of the energy consumption in developing countries, while in industrialized countries it provides on average 3%.

The residues from agricultural produce have an estimated energy content of about 11 EJ (1 EJ = 10^{18} J) per annum ($\cong 0.6$ TW continuous). But only a small fraction of this is currently utilized for energy production. For example, biogasse, the residue from sugar cane, is used in sugar factories as a fuel for producing electricity and hot water.

EXAMPLE 8.1

Estimate the amount of electricity in kWh that could be produced annually from a biomass-fired power station that burns biomass grown over an area of 10 km by 10 km.

Equation (8.1) gives an estimate of 5 kW ha^{-1}. This is equivalent to 0.5 MW per square kilometre, so the thermal power from $10 \times 10 = 100$ square kilometres is 50 MW. Taking the efficiency of the thermal power station as 40% would give an output of 20 MWe. In one year the amount of electricity would be $(2 \times 10^4) 8760 \approx 1.75 \times 10^8$ kWh.

We now concentrate on the commercial use of biomass as a source of energy, either directly for heat and power or indirectly as feedstock for the production of liquid and gaseous biofuels. We will then discuss the environmental impacts and end with a discussion of the economics and potential of biomass.

8.3 Biomass for heat and power

Heat and power are produced from two main sources of biomass: agricultural and municipal wastes, and energy crops. These have rather low energy content per kilogram compared with fossil fuels and relatively low density, making them bulky and expensive to transport. Economic use for energy production therefore generally requires the biomass source to be readily available, e.g. a waste dump or factory residue. For this reason bioenergy production is currently often combined with crop production or as a useful way of disposing of organic waste, both municipal and agricultural. Biomass supply in the future, though, lies in dedicated energy crops such as willow or switch grass. It is only by planting large areas with such crops that sufficient biomass will be produced.

The utilization of biomass is varied. We concentrate on a few of the major ways in which biomass can contribute to energy production, all of which can have significant savings in greenhouse gas emissions compared with fossil fuels. One of the most widespread uses is to take advantage of natural decomposition of organic waste by anaerobic bacteria. Anaerobic digestion is used in developing countries for heating and cooking and in industrialized countries to provide gas for small power units.

On a larger scale there is the combustion and, to a small extent, the gasification of biomass to produce electricity and heat. There is also the conversion of biomass to liquid biofuels to replace oil-based fuels. In North and South America there is large-scale production of bioethanol through fermentation and more recently, particularly in Europe, a significant increase in the manufacture of biodiesel from plant oils. We first look at anaerobic digestion and then at combustion and gasification, before discussing the production of liquid biofuels.

8.3.1 Anaerobic digestion

Anaerobic digestion (AD) is the decomposition of organic matter by bacteria in the absence of air. Bacteria break down the organic matter and produce a gas consisting of methane (~65%) and carbon dioxide (~35%), with traces of other gases. The gas has a calorific value of ~17–25 MJ m^{-3} at STP (standard temperature and pressure, which is 0 °C and 100 kPa $\equiv$ 1 bar) and the conversion efficiency is typically 40–60%. Anaerobic digestion occurs naturally, e.g. in compost heaps, and is the source of marsh gas. It takes place in landfill sites over a period of years, with peak methane production typically occurring after 10 years, and in purpose-built digesters, where the process occurs at higher temperatures (30–60 °C) after only a few weeks. In the latter the residue can be used as a fertilizer. In comparison with aerobic digestion, in which organic matter plus oxygen is converted to a residue plus carbon dioxide and water (as in combustion but at much lower temperatures), anaerobic digestion produces considerably less residue (~5–10 times less) as well as a gas that can be used to produce power. Anaerobic digestion also occurs in cows and is a source of a significant amount of methane in the atmosphere: CH_4 production from cows and other ruminants is estimated to account for about 5–10% of the global warming associated with greenhouse gas emissions! The biochemical processes involved are explained in Box 8.2.

Anaerobic digestion is widely used in Asian villages, where the biogas is used for heating and cooking. In China, an estimated 50 million homes use biogas. A Chinese fixed-dome digester is illustrated in Fig. 8.5. The technology for larger scale anaerobic digestion of sewage and industrial sludges and of waste water is well developed. The biogas potential within the EU from animal manure, energy crops, and waste is ~40 megatonne of oil equivalent (Mtoe) compared to a production of 5.9 Mtoe in 2007. In municipal waste digesters the organic fraction of the waste needs to be separated out, which can be costly.

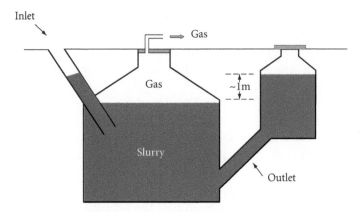

Fig.8.5 A Chinese fixed-dome anaerobic digester.
Source: AD04.

Box 8.2 Biochemical processes in anaerobic digestion

Anaerobic digestion consists of several processes in which organic matter is transformed by anaerobic bacteria to a gas consisting mainly of CH_4 and CO_2 and a residue called sludge or biosolids. Anaerobic digestion results in a significant fraction of the original biomass energy being stored in the methane, as the reactions that occur only give out a small amount of heat. These reactions are catalysed by enzymes, which are typically complex proteins whose action depends on their molecular shape. These enzymes increase the reaction rate by lowering the activation energy (the energy required to start a chemical reaction).

In anaerobic digestion there are three stages: hydrolysis, acidification, and methane production. The large molecules in the organic matter are initially broken down by hydrolysis, which is a reaction that causes molecules to break apart when water is added, i.e.

$$AB + HOH \rightarrow A\text{-}OH + H\text{-}B \tag{8.3}$$

The hydrolysis of cellulose and protein produces fatty acids, amino acids, and glucose. These are then converted to organic acids such as ethanoic (acetic) and butanoic acid, by acidogenic and acetogenic (acid- and acetate-forming) bacteria, and H_2 and CO_2 are produced, e.g.

$$C_6H_{12}O_6 \text{ (glucose)} + 2H_2O \rightarrow 4H_2 + 2CH_3COOH \text{ (ethanoic acid)} + 2CO_2 \tag{8.4}$$

In the final stage (methanogenesis) bacteria digest the products of the acidification stage and produce methane via reactions such as

$$CH_3COOH \rightarrow CH_4 + CO_2 \quad \text{and} \quad 4H_2 + CO_2 \rightarrow CH_4 + 2H_2O \tag{8.5}$$

The acidification and methane production stages are symbiotic as the H_2 and acetate inhibit the bacteria producing them, so their consumption by the methane producing bacteria is beneficial. If the production were complete then the overall process would amount to

$$C_6H_{12}O_6 \rightarrow 3CH_4 + 3CO_2 \tag{8.6}$$

The efficiency of energy conversion is very high, with ~90% of the stored energy in glucose being stored in the produced methane.

An anaerobic digester consists of a feedstock holder, digestion tank with mixing system, biogas and residue recovery, and, where necessary (e.g. in northern Europe), heat exchangers to maintain the optimum temperature for the bacteria to produce the biogas (~30–60 °C). Small digesters are mainly used for heat production while the larger units are used for electricity generation, with power outputs of ~2 MW.

In the EU the dumping of biodegradable waste in landfill sites is being reduced. Although the methane can be utilized to provide energy, recovery is variable, with typically only 50% used. Since methane is a very potent greenhouse gas, some 23 times more than carbon dioxide, reducing its emission is important. Moreover, there can be pollution from leachates from biodegradable waste in the landfill sites.

EXAMPLE 8.2

Calculate the energy efficiency of the conversion of carbohydrate to methane in anaerobic digestion.

The molecular weights of glucose $C_6H_{12}O_6$ and methane CH_4 are 180 and 16, respectively, so 180 kg of glucose is converted in anaerobic digestion to 48 kg of methane. The heats of combustion of glucose and methane are ~16 MJ kg^{-1} and ~55 MJ kg^{-1}, respectively. The stored energy in 180 kg of glucose is therefore ~2880 MJ, while in 48 kg of methane it is ~2640 MJ, so the conversion efficiency ε_{AD} is given by

$$\varepsilon_{AD} \approx 2640/2880 \approx 92\%$$

8.3.2 Combustion and gasification

Energy from biomass can be produced by direct combustion or by combustion after gasification. Pelletization of the biomass prior to combustion is required before either co-firing with fossil fuels or direct combustion, which are the most developed technologies. Gasification plants are currently not competitive because of their high investment, maintenance, and operating costs; however, gasification has the potential to be cheaper and more efficient. Combined heat and power (CHP) production gives the highest efficiency and, for power plants with an output of $\gtrsim 2$ MW, steam (Rankine) cycle generators are used. For outputs in the range of 0.2–2 MW, organic Rankine cycle (ORC) plants are preferred, while for small generators of $\lesssim 100$ kW Stirling engines show the best potential. The potential in Europe and in the UK is discussed below.

Case Study 8.1 Biomass for heat and power in the UK and Europe

Biomass in Europe could provide ~1000 TWh per year of heat and power by 2020. This would be about 10% of the expected demand. In a UK study by the UK Department of Trade and Industry (DTI), two main biomass energy chains were identified: the combustion and gasification of short rotation coppice willow, miscanthus, and straw, and the gasification, anaerobic digestion, and combustion of municipal solid waste and sewage sludge (see Fig. 8.6). The conversion technologies involved in combustion and gasification require significant investment and have commercial risks for the more advanced technologies. So, a secure biomass supply and a market for the energy are required, which will require economic incentives.

The study concluded that short rotation coppice willow has the best potential and could provide ~9% of the UK electricity by 2020 using large-scale power plants, assuming one million hectares were planted. This would agree roughly with our estimate of ~5 kW ha^{-1}, which would give ~5 GW$_{th}$ total output, ~5% of the UK electricity demand. Although one million hectares is a substantial land area of 100 km by 100 km, it is a small fraction of the 17.5 million ha of agricultural land in the UK.

A similar contribution of 15% to the electricity and heat production within the OECD countries has been suggested based on woody biomass, which is biomass from forestry and farming. The plan assumes that the power demand in the industrialized countries in the OECD will double by 2020. It requires the use of a quarter of the agricultural residues and the putting aside of 5% of crop, farm, and woodland area for the growing of woody biomass. The total area of such land in the OECD is over 1500 million ha, so 5% is over 75 million ha. This area has been estimated to provide sufficient fuel to generate 200 GW of power, enough for over 100 million homes. This is in agreement with our rough estimate of 5 kW ha^{-1}, which would give a power of 150 GW. The study concludes that there need be no conflict between land use for biomass and for food production at

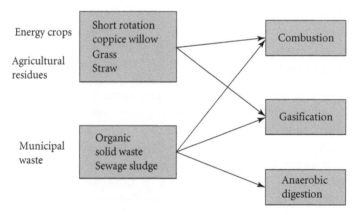

Fig.8.6 Biomass sources in the UK for 2020.
Source: UK2020.

this level of power generation. However, in the UK and in Europe there has been no significant growth since ~2005 in the area planted with dedicated energy crops, primarily because the cost of the energy produced is not competitive and there is also uncertainty over biomass policy.

Gasification can be brought about by anaerobic digestion, as described above, or by burning the biomass in a reduced supply of air. One of the earliest ways that biomass was converted to produce a more useful fuel was the burning of wood with a reduced air supply to yield charcoal. Charcoal burns at a much higher temperature than wood and the higher temperatures that could be achieved using charcoal significantly advanced the extraction of metals from ores. Modern methods of biomass conversion by thermochemical processes concentrate on the gaseous and liquid products and in particular on gasification.

The first industrial use of gasification was in the production of coal gas or producer gas from heating coal in the presence of steam. With the advent of natural gas this production ceased but a similar process is involved in the gasification of biomass. The different thermochemical processes that occur are described in Box 8.3. The gas that results from burning biomass in a reduced air supply consists of CO, H_2, CO_2, CH_4, and nitrogen (from the air) and is called producer gas.

Producer gas has a low calorific value of 1000–1200 kcal m^{-3} at STP, which is equivalent to 4.2–5.0 MJ m^{-3} (STP). But producer gas burns cleanly with low emissions. The total conversion efficiency can be ~60–70%. Each kilogram of air-dried biomass (10% water content) yields about 2.5 m^3 of producer gas, equivalent to ~12 MJ kg^{-1}, agreeing with our rough estimate of 15 MJ kg^{-1} of biomass. (Biomass has a higher output if it is low in water content, since water requires 2.3 MJ kg^{-1} to evaporate.) Producer gas can be used to provide heat, or in internal combustion engines and in gas turbines for electricity generation.

Liquid fuels in the form of oils can also be derived from heating biomass in the absence of air, a process called pyrolysis, but this has so far been less economic, producing fuels with too low a calorific value to substitute for petrol (gasoline) or diesel. Producer gas can also be purified to produce synthesis gas or syngas, which is carbon monoxide and hydrogen. Syngas can be used to synthesize chemicals and fuels such as methanol using catalysts.

Box 8.3 Thermochemical processes in gasification

Various thermochemical processes occur at the same time in a gasifier. We will consider wood as an illustration. Drying of the wood occurs at ~150 °C, which drives off the water as steam. The dried wood is decomposed by heat in the absence of air, a process called pyrolysis. This occurs between 150 °C and 700 °C and produces gases, in particular CO and H_2, liquids (oils), and a solid residue (charcoal). These products are oxidized by reacting with the oxygen in air which is supplied to produce CO_2 and H_2O by the reactions

$$C + O_2 \leftrightarrow CO_2 + 394\,MJ\,kmol^{-1} \quad \text{(A kmol of carbon is 12 kg.)} \tag{8.7}$$

$$H_2 + \tfrac{1}{2}O_2 \leftrightarrow H_2O + 242\,MJ\,kmol^{-1} \tag{8.8}$$

which occur between 700 °C and 2000 °C. [NB Throughout Chapter 8 Q-values $(=-\Delta H)$ are given for chemical reactions.] These gases are reduced by the following reactions:

$$CO_2 + C \leftrightarrow 2CO - 172\,MJ\,kmol^{-1} \tag{8.9}$$

$$C + H_2O \leftrightarrow CO + H_2 - 131\,MJ\,kmol^{-1} \tag{8.10}$$

$$C + 2H_2 \leftrightarrow CH_4 + 75\,MJ\,kmol^{-1} \tag{8.11}$$

$$CO_2 + H_2 \leftrightarrow CO + H_2O + 41\,MJ\,kmol^{-1} \tag{8.12}$$

which are predominantly endothermic, i.e. requiring heat. These reducing reactions therefore occur in a lower temperature range, between 800 °C and 1100 °C.

Of particular importance in determining the yield of these reactions is the reaction temperature, with an increase in temperature at constant pressure favouring the products of an endothermic reaction. This is an example of an important rule (Le Chatelier's principle), which follows from the second law of thermodynamics, and states

> When a reaction in equilibrium is perturbed, the equilibrium is altered in the direction that reduces the perturbation.

This means that it is important that the temperature should not fall too far in the reduction reactions, otherwise the yield of CO and H_2 will be reduced. By altering the conditions, hydrogen production can be enhanced and this can be used in fuel cells to provide electricity generation with high efficiency (see Chapter 11).

8.3.3 Municipal solid waste

A considerable amount of waste is generated per household in industrialized countries: about one tonne annually. The combustion of municipal waste with the use of the heat for electricity production or for space heating is a useful way of disposing of this waste, particularly with the decreasing availability of landfill sites. Globally in 2010 there was 221 TWh of electricity and heat generation using municipal and solid waste. The waste often needs processing before it can be used as fuel. Although the fuel is mainly organic, the combustion is not carbon-neutral, since some of the material is derived from fossil fuels, typically 20–40%. Such an analysis, called life-cycle analysis (LCA), works out the amount of CO_2 (and other gaseous emissions) per kWh of energy produced (see Section 1.4). While the burning of agricultural wastes gives less than 30 g per kWh, municipal solid waste (MSW; also called energy from waste, EfW) gives ~360 g per kWh compared with ~970 g per kWh from coal and ~450 g per kWh from a natural gas CCGT power plant. The typical energy content of MSW is ~10 MJ kg^{-1}.

Besides providing heat and power, biomass can provide fuels that reduce our dependence on oil-based fuels such as petrol (gasoline) and diesel. We now look at how biomass has been used for transport fuel production and its future potential for such fuels.

8.4 Liquid biofuels

While sustainable electric power can be provided by other renewables, biomass is a direct renewable source of carbon-based fuels and chemicals. In 2010 ~86×10^9 litres of bioethanol and ~19×10^9 litres of biodiesel were produced annually in the world, up from 16×10^9 litres of biofuels in 2000. Besides the environmental benefit from lower CO_2 emissions, biofuels produced within a country can provide energy security and economic development. Less petroleum needs to be imported and there is more immunity against international political disturbances affecting oil supplies. Energy and agricultural policies can be coupled with biocrops, aiding rural development as well as energy resources. Biofuels typically burn cleanly but life-cycle analysis shows that several feedstocks for biofuels only give quite small net savings on CO_2 emissions compared to petroleum-based fuels, when the energy used for farming and production of the biomass feedstock and for conversion is included. So there is a need for a considerable improvement both in economic competitiveness and in CO_2 savings. For those biofuels for which the CO_2 savings, given by the fossil energy replacement ratio (FER),

FER = (energy supplied to customer)/(fossil energy used)

are small, the biofuel is essentially only replacing fossil fuels. The FER is also referred to as the energy budget or energy balance.

Transport accounts for ~19% of global energy consumption and ~23% of energy-related CO_2 emissions. Nearly all of this energy comes from oil and, as the growth in car usage in developing countries is significant, non-fossil based transportation fuels could help reduce a huge source of CO_2.

There are two important liquid biofuels: bioethanol and biodiesel, the first derived from sugar-containing plants, the second from oil-containing plants. In 2010, these made up 2.7% of global transport fuels. We look first at how the biological fermentation of sugar- and starch-containing plants is used to produce bioethanol.

8.4.1 Bioethanol

The glucose in sugar cane or sugar beet, or in starch, which contains α-glucose polymers, can easily be extracted. These feedstocks for bioethanol, however, require good quality soil and plentiful amounts of water and can therefore be in competition with food crops. They are the only ones currently commercially developed and are called first-generation biocrops. In 2010, the largest producers of bioethanol were the USA, primarily from corn, accounting for

~55% of global production, and Brazil (where there is a long-established programme providing ~40% of the fuel for their cars, from sugar cane), accounting for ~30%. About 3% of the world's grain supply was used as part of the feedstock.

Vehicles can run using petrol blended with up to ~20% bioethanol, and an increasing number of cars are available that can use higher percentages. In 2010 there were more than 50 countries that had minimum blending percentages or targets.

Bioethanol from sugar feedstocks

Sugar from sugar-containing plants can be directly fermented by yeast or bacteria, which reduce the carbohydrate to ethanol and produce CO_2:

$$C_6H_{12}O_6 \rightarrow 2C_2H_5OH + 2\,CO_2 + 0.4\,MJ\,kg^{-1} \tag{8.13}$$

As the heat released is so small, nearly all the energy stored in the sugar is stored in the alcohol. The ethanol has a much higher heat of combustion ($29.7\,MJ\,kg^{-1}$) than glucose ($15.6\,MJ\,kg^{-1}$). It is sufficiently high that it can be used as a substitute for petrol (gasoline), which has a heat of combustion of ~$45\,MJ\,kg^{-1}$. Almost half the weight of glucose (molecular weight, MW = 180) is converted to carbon dioxide (MW = 44) and the maximum conversion efficiency by weight to ethanol is 51%. The estimation of the heat of combustion of a biofuel is shown in Box 8.4.

Box 8.4 Heat of combustion of biofuels

We can estimate the heat of combustion of a biofuel by assuming all the carbon and hydrogen are fully oxidized when the fuel is burned. The two reactions are

$$C + O_2 \rightarrow CO_2 + 32.8\,MJ\,kg^{-1}\,C \tag{8.14}$$

$$2H + \tfrac{1}{2}O_2 \rightarrow H_2O + 142.9\,MJ\,kg^{-1}\,H \tag{8.15}$$

The heat of combustion H_c of a fuel composed of C, H, and O is therefore given by

$$H_c = 32.8(C) + 142.9(H - O/8)\,MJ\,kg^{-1} \tag{8.16}$$

where C, H, and O are the fractions by weight of carbon, hydrogen, and oxygen. The oxygen atoms in the fuel are assumed to combust with hydrogen atoms and form water, hence the factor $(H - O/8)$ in eqn (8.16). Using this formula we would estimate H_c in $MJ\,kg^{-1}$ for carbohydrates ($[CH_2O]$) as 13.1 (~15.5), for ethanol (C_2H_5OH) as 29.5 (29.7), and for octane (C_8H_{18}) as 50.2 (47.9), where the measured values are in brackets. So, except for carbohydrates, this simple estimate (Dulong's formula) is accurate to ~5%.

Olive oil is composed of triglycerides: mainly 55–85% oleic acid, a monounsaturated acid (only one double bond), and ~9% linoleic acid, a polyunsaturated acid (more than one double bond). These two acids are structural isomers, and the composition of the triglycerides is $C_{57}H_{96}O_6$. Using the formula above, the heat of combustion is estimated to be $39.3\,\mathrm{MJ\,kg^{-1}}$, compared with $\sim 40\,\mathrm{MJ\,kg^{-1}}$ measured. This is only $\sim 20\%$ less than that released ($47.9\,\mathrm{MJ\,kg^{-1}}$) when octane, the principal component of petrol (gasoline) is burned:

$$C_8H_{18}+12.5O_2 \rightarrow 8CO_2+9H_2O \tag{8.17}$$

In this process, $114\,\mathrm{kg}$ of octane produces $352\,\mathrm{kg}$ of carbon dioxide, just over three times as much by weight.

The heat of combustion is also called the higher heating value (HHV) and assumes that any water vapour is condensed. If the water vapour produced in combustion of a fuel is not condensed, as in a car engine, then the available energy, called the lower heating value (LHV), is less than the HHV by about 5–10%. Values of LHV are given for a number of fuels with the conversion factors on p. ii.

EXAMPLE 8.3

Estimate the heat of combustion H_c of ethanol.

Ethanol has the chemical composition C_2H_5OH. Its molecular weight (MW) is therefore

$$MW = 2 \times 12 + 6 \times 1 + 16 = 46$$

The fraction by weight of carbon C is therefore 24/46, that of hydrogen $H = 6/46$, and that of oxygen $O = 16/46$. Substituting these values into eqn (8.16) gives

$$H_c = 32.8(24/46) + 142.9(6/46 - 2/46) = 29.5\,\mathrm{MJ\,kg^{-1}}$$

The FER for bioethanol from sugarcane is good, at ~ 8. The waste, bagasse, can be used to provide heat, which improves the ratio.

The largest producer of bioethanol from sugarcane is Brazil, where there has been a successful programme designed to provide a significant source of fuel for cars and reduce their dependence on oil. Bioethanol has been described as a sustainable advanced biofuel owing to its associated carbon dioxide emissions being low. A report in 2010 by the US Environmental Protection Agency estimated a 61% reduction in greenhouse gas (GHG) emissions from a life-cycle analysis that included indirect emissions from changes in land use (see Box 8.5).

Case Study 8.2 Bioethanol production from Brazilian sugarcane

---

Brazil is the world's largest producer of ethanol from sugarcane (21 billion litres in 2011), with a programme started in the 1970s following the oil price shocks and which by 2004 was using 6.2 million ha, 1.7% of the country's total arable land. Brazil adopted a policy that all cars must be able to run on a mixture of petrol (gasoline) and bioethanol, with a minimum percentage ($\sim 20\%$) of bioethanol in the fuel. The price of bioethanol in

2011 was competitive with that of petrol. The majority of bioethanol is produced in the subtropical south-eastern region of Brazil far from the Amazon rainforest. Roughly half the sugarcane is used for bioethanol and the rest for sugar; the relative amounts can be adapted to market conditions.

Sugarcane is a perennial grass that produces cane stalks 3–4 m in height and about 5 cm in diameter. It is a C_4 plant, very efficient at photosynthesizing (up to ~2%), that grows in tropical or temperate regions where there is plentiful rainfall (>60 cm y^{-1}). Genetic improvements to the sugarcane varieties can make them more resistant to disease and improve productivity. The yield of sugarcane is around 65 t ha^{-1}; every six years the sugarcane is replanted from cuttings. About 15% of the plant's mass is sugar and 15% fibre, with the remainder water, so there is ~10 t of sugar per hectare. After the cane juice containing the sugar has been extracted, the residual fibre, called bagasse, is used as a fuel in the production of the alcohol. The energy ratio FER is good at 8–10, with a typical breakdown per tonne of sugarcane being: agricultural phase 190 MJ, industrial processing 40 MJ, ethanol 2000 MJ, and bagasse 300 MJ, giving an FER of 10.

Although most of the growth of sugarcane plantations is on degraded and pasture lands, it is important that this does not indirectly cause deforestation and an associated large release of carbon dioxide as cattle grazing land is expanded. Another concern has been pollution from the burning of the crops prior to harvesting, which is used to kill venomous animals and remove the sharp leaves, making it easier to cut the stalks manually. Mechanized harvesting avoids the need for burning; it is taking over from manual harvesting, which, it is planned, will stop by 2014. However, mechanization will have a big impact on employment, since there are about 1 million people employed in the sugarcane business, many of these migrant workers, with about half as sugarcane cutters.

The huge abundance of pasture land and favourable temperature and rainfall conditions, together with the required infrastructure, make Brazil a special case for the production of bioethanol from sugarcane. It is clearly possible for Brazil to provide a significant fraction of its fuel for transport (currently 20% by energy content and ~40% of the fuel for cars) and do so with relatively low carbon dioxide emissions, and without a damaging impact on food production. However, care is needed to avoid short-term economic hardship due to mechanization, and to avoid causing any significant release of carbon dioxide through indirect change in land use.

Bioethanol from starch feedstocks

In the production of bioethanol from starch, which is contained in corn, the glucose is in the form of a biopolymer. The glucose molecules are linked together by α-glycosidic bonds. These bonds are easily broken apart using human and animal enzymes. The enzymes catalyse the decomposition by hydrolysis of starch to glucose, which can then be fermented to produce bioethanol. However, a considerable amount of energy is required in the production of the corn.

The corn produced in the USA is mainly for animal feed, with some for bioethanol production. Energy is needed in producing the bioethanol, and on the farm for planting, harvesting, and making fertilizer. The resulting FER for corn-ethanol is positive but small at ~1.2–1.4. As the FER is small, the corn-ethanol does not reduce CO_2 emissions significantly, but it does reduce the amount of petrol and hence oil required, though about 50% more bioethanol is required for the same stored energy.

However, US government subsidies for corn-ethanol have meant that farmers have used more of their corn crops to produce bioethanol, so that there is less for animal feed, and corn and food prices have risen as a result. So while corn-based ethanol can displace oil, we ideally need a glucose-containing feedstock that is cheap both to produce and to process, is not in competition with food production, and has a good FER.

Moreover, the amount of bioethanol produced in the USA ($\sim 5 \times 10^{10}$ litres y^{-1}) is only equivalent to ~6% of the petrol that is consumed. To produce another 6%, which would be ~1 EJ y^{-1} of bioethanol, would require ~1.5×10^7 ha. This area is ~8% of USA cropland. So we would like a plant that can grow on marginal land unsuitable for food crops. Starch-rich plants that will grow on degraded soil, such as the cassava plant, are possibilities. The large group of cellulose-based plants, such as switchgrass, could also provide a suitable feedstock. However, extracting the glucose from cellulose is less straightforward than from starch, and as a result cellulosic feedstocks are called second-generation biocrops.

Bioethanol from cellulosic feedstocks

Cellulosic feedstocks, such as wood and grasses, contain mainly cellulose, hemicellulose, and lignin, and are also called lignocellulosic-based feedstocks. Lignin is a biopolymer rich in phenolic components that confer stiffness and make up about a tenth to a quarter of the biomass. It is the part of a plant that fossilizes and becomes coal. These plants can grow on marginal areas unsuitable for food crops. Switchgrass, for example, is a deep-rooted perennial that prevents soil erosion and can restore degraded land. Switchgrass needs only a small amount of fertilizer or pesticide and uses water efficiently. As a result the costs of production can be low.

Cellulose, which is the largest component of biomass (40–60%), is a biopolymer (polysaccharide) of glucose, as is starch, and consists of bundles of long chains of glucose molecules bonded together by β-glycosidic linkages. The fibre bundles are strong because of a high level of hydrogen bonding between the glucose chains, and are resistant to cleavage. The other component of biomass, hemicellulose (20–40%), interlinks the cellulose and is mainly made up of the 5-sugar xylose. The hemicellulose and lignin enclose the cellulose bundles and protect them from microbial attack (Fig. 8.7).

The hydrolysis of cellulose to glucose is less straightforward than in starch-containing plants because of the hemicellulose and lignin that encase the cellulose. Pretreatment with dilute acid combined with heat and pressure is used to separate the hemicellulose and lignin and expose the cellulose for hydrolysis. The hydrolysis can be acid-catalysed but this process is expensive as it can require pressure vessels and significant amounts of energy.

Enzyme hydrolysis is currently being actively pursued. This process was first noticed in the Second World War when a fungus was found that attacked cotton clothes and tents. The

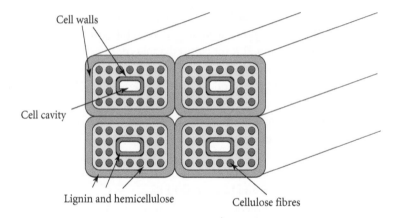

Cell walls

Cell cavity

Lignin and hemicellulose

Cellulose fibres

Fig.8.7 Cellulose, lignin, and hemicellulose in a plant (schematic).

fungus was providing cellulose enzymes. The process requires low temperatures and can be carried out at atmospheric pressure. Energy requirements are low and reaction vessels are cheap to construct. The conversion efficiency is quite high. But currently enzymes are too expensive and act too slowly for good economy. A cocktail of three cellulose enzymes—endoglucanase, exoglucanase, and betaglucosidase—was discovered that showed good performance that might reduce costs. Enzyme hydrolysis (saccharification) and fermentation can also be combined to reduce the inhibition of enzyme action by sugar build-up.

Yeasts that ferment 6-sugars have been known since antiquity but only recently have yeasts been developed that can ferment 5-sugars such as xylose, the main constituent of hemicellulose. A genetically engineered *E. coli* microorganism has been developed by the University of Florida which can ferment both 5- and 6-sugars. Bacteria are also under investigation as their speed of fermentation can be minutes rather than the hours required by yeast.

The successful development of bioethanol production from lignocellulosic feedstocks, such as wood or grass that are less expensive than corn, will significantly reduce the cost of bioethanol, provide a good FER, and could make it competitive with fossil-fuel derived petrol. It should be noted that, as the energy consumption associated with a crop becomes less fossil-fuel based, then the associated FER will improve.

8.4.2 Biodiesel from plant oils

The production of biodiesel is growing rapidly, particularly in Europe. Essentially zero in 1995, the amount produced globally by 2003 was 1.5×10^9 litres per year and by 2010 was 19×10^9 litres per year. Use of 100% biodiesel occurs in some European countries such as Germany, but it is also used in a blend (5–25%) throughout North America and Europe. Biodiesel is made by the chemical transesterification of vegetable oils from oilseed crops such as rapeseed and sunflower, or from other sources such as waste cooking oil.

Diesel demonstrated his engine in Paris in 1900 using peanut oil, but the availability and cheapness of diesel fuel derived from fossil fuel oil meant vegetable oils were not used. While diesel engines can run on pure vegetable oil, their viscosity is rather high and transesterification produces a lower viscosity fuel that starts more easily. Fatty acid methyl esters (FAME) are the product and biodiesel is also called FAME. The process involves a relatively simple reaction of the oil with either methanol or ethanol using sodium or potassium hydroxide as a catalyst; the chemical processes involved are described in Box 8.5. The efficiency of the process is high (>97%) and requires ~10% by weight of alcohol; the resultant FER can be quite good (see Table 8.1).

Under development is an alternative treatment (hydroprocessing) of vegetable oils that produces hydrocarbons more similar to those found in diesel derived from fossil fuels. The triglyceride molecules (see Box 8.5) in the vegetable oils are converted to hydrocarbons by reacting the oil with hydrogen at a raised temperature and pressure in the presence of a catalyst. The product is called renewable diesel.

The diesel engine, like the petrol engine, normally has a four–stroke cycle: intake, compression, expansion, and exhaust. But in a diesel engine it is just air that is taken in and compressed, while in a petrol engine it is a fuel–air mixture. This means that the compression ratio of intake to compressed volume, corresponding to the piston being at the bottom and top of the cylinder, respectively, can be much higher in a diesel than in a petrol engine, typically 20:1 as compared to 9:1. This is because the compression must be limited in a petrol engine to avoid pre-ignition (knocking), as the temperature of the fuel–air mixture rises as it is compressed. In a diesel engine fuel is injected into the hot (~900 °C) compressed air whereupon it ignites and rapidly expands. (It is a very fast burn, not an explosion.) The maximum temperature (and pressure) is higher in a diesel than in a petrol engine and as a result the efficiency is higher, as would be expected from the maximum theoretical thermodynamic efficiency given by the Carnot formula, $(1 - T_{min}/T_{max})$.

Box 8.5 Transesterification of plant oils

Triglycerides can be used neat in diesel engines but better starting is obtained in cold weather by lowering the viscosity. This can be done by mixing the triglyceride with a solution of methanol and sodium hydroxide, the sodium hydroxide acting as a catalyst:

$$
\begin{array}{cccccc}
CH_2OOR_1 & & catalyst & & CH_2OH & \\
| & & \downarrow & & | & \\
CHOOR_2 & + \; 3CH_3OH & \Leftrightarrow & 3CH_3OOR_x & + \;\; CHOH & \quad (8.18) \\
| & & & & | & \\
CH_2OOR_3 & & & & CH_2OH & \\
\textit{Vegetable oil} & \textit{Methanol} & & \textit{Biodiesel} & \textit{Glycerin} &
\end{array}
$$

The methanol and sodium hydroxide form sodium methoxide plus water, and the sodium methoxide then successively converts the triglyceride to methyl esters plus glycerin. The first step can be represented by

$$NaOH + CH_3OH \rightarrow NaOCH_3 + HOH \qquad (8.19)$$

$$HOH + NaOCH_3 + (-CH_2OOR_1) \rightarrow (-CH_2OH) + CH_3OOR_1 + NaOH \qquad (8.20)$$

This process continues until all three methyl esters have been formed. The catalyst, sodium hydroxide, is not used up in the reaction, and the triglyceride molecules (glycerin esters) have been converted to methyl esters, hence the description of the process as transesterification.

In a modern diesel engine, the air taken into the cylinder is pre-compressed so that the final maximum pressure is higher. This increase in air (oxygen) allows more fuel to be injected and more power to be obtained. This process, called turbocharging, makes the performance of diesel engines comparable to that of petrol engines. Their efficiency is greater at ~40%, as compared to ~30% for car engines, while for large diesel engines their efficiency can be ~50%. Better control of the fuel burning has also reduced emissions from diesel engines.

A summary of the main processes involved in producing liquid biofuels from biomass is given in Fig. 8.8. Of particular importance is the yield of biofuel per hectare, and we will see that this limits how much can be produced easily without impacting on normal agricultural use of the land.

8.4.3 Liquid biofuel yields and energy budgets (FERs)

We expect roughly $10\,t\,ha^{-1}\,y^{-1}$ of dry carbohydrate and as the maximum conversion efficiency by mass to alcohol is ~50%, then assuming ~40% in practice, the annual yield for bioethanol is about ~$4\,t\,ha^{-1}$ or $5000\,litres\,ha^{-1}$. Typical values for the FER and biofuel yields from corn, sugarcane, rapeseed, cassava, jatropha, palm oil, and switchgrass are shown in Table 8.1. For switchgrass, the lignin is a by-product that cannot be fermented but, like bagasse from sugarcane, can be used as a fuel, which helps improve the FER values.

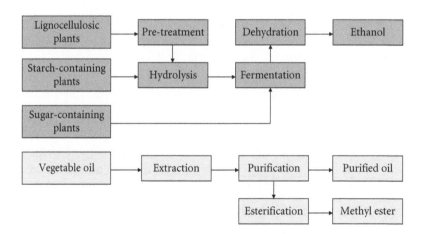

Fig.8.8 Main processes involved in producing biofuels from biomass.

Table 8.1 Typical values of FER, biofuel yield, suitability for cultivation on degraded land, water requirement, and time to replace the CO_2 associated with converting land to biofuel production, for a number of bioethanol and biodiesel feedstocks

Feedstock	FER (location)	Biofuel (litre ha^{-1})	Cultivation on degraded land?	Water requirement	Replacement of CO_2 (y)
Corn	1.34 (USA)	3400	no	high	~50
Sugarcane	8 (Brazil)	6000	no	high	~20
Rapeseed	2.3 (EU)	1000	no	high	~50
Cassava	9 (Thailand)	~3000	yes	low	~0
Jatropha	6 (Thailand)	530	yes	low	~0
Palm oil	9 (Malaysia)	3750	no	high	~100
Switchgrass	5 (USA)	2800	yes	low	~0

Source: WorldBank2009.

EXAMPLE 8.4

A palm tree plantation produces 5 tonnes of oil per hectare per year. (a) What area of plantation would be required to displace 5×10^{11} litres of petrol per year, which is the USA annual consumption, with biodiesel? (b) What would be the amount of carbon emissions displaced? (Density of petrol is $0.73 \, \text{kg litre}^{-1}$.)

(a) Petrol has a LHV of ~43.5 MJ kg^{-1}, so there are ~32 MJ per litre. The energy content of 5×10^{11} litres of petrol is therefore 1.6×10^{13} MJ. An area A ha of palm trees would produce $5000A$ kg of oil per year. The conversion efficiency to biodiesel is close to 100% and biodiesel has a LHV of ~38 MJ kg^{-1}. The efficiency of diesel cars is ~4/3 times that of petrol cars. So the area required is given by

$$5000 \times 38 \times A \times 4/3 = 1.6 \times 10^{13} \quad \text{so} \quad A = 63 \times 10^6 \, \text{ha} = 0.63 \times 10^6 \, \text{km}^2$$

(b) Assume that petrol combusts like octane. Equation (8.17) shows that 114 kg of octane produces 352 kg of CO_2. This is equivalent to $(12/44) \, 352 = 96$ kg of C. So 1 kg of octane produces 0.84 kg of carbon on combustion. Petrol has an FER of 0.83, so 1 kg of petrol produces roughly 1 kg of carbon. A volume of 5×10^{11} litres of petrol has a mass of 3.7×10^{11} kg. This amount would produce ~0.37 Gt of carbon.

The energy content of the biodiesel is 1.2×10^{13} MJ. The FER is 9 (Table 8.1), so 1.33×10^{12} MJ of fossil energy $\equiv 3.1 \times 10^{10}$ kg of petrol (gasoline) $\equiv 0.03$ Gt of carbon is used. So

$$C_{\text{displaced}} \approx 0.37 - 0.03 \approx 0.34 \, \text{Gt of carbon per year}$$

8.5 Environmental impact of biomass

The production of a significant amount of energy from biomass requires a large area of land, and land clearance and deforestation are major environmental concerns. The amount of CO_2

released in the process can be many times the annual CO_2 savings from using the biomass or biofuel produced on the land instead of fossil fuels. Plants and soils are huge stores of carbon, holding about 2.7 times the amount in the atmosphere, and clearing the land by fire results in a large release of CO_2, with more given off over many subsequent years from the decomposition of plant matter. The biomass planted on the cleared land can after several years replace this lost carbon if fossil-fuel replacement ratios are greater than one.

However, the replacement time can be many decades, as shown in Table 8.1, with the estimated times for the conversion of grassland, savannah, tropical forest, and peatland ranging from a couple of decades to 400 years. The conversion of marginal cropland, though, results in no significant release, arguing strongly that these are the areas that should be developed for bioenergy.

Jatropha, cassava, and sweet sorghum can all grow on marginal land and do not need much water. The cassava and sorghum plants are rich in starch and can be easily converted to ethanol, and the extraction of oil from jatropha is straightforward. These are all suitable biofuel feedstocks, and China is encouraging their use on non-arable land so that there will not be competition with food production.

The use of residues and wastes can also have little environmental risk and there is considerable potential from forestry and timber waste. The combustion of wood gives much less SO_2 than coal and hence less acid rain. Large areas of energy crops may reduce biodiversity; however, forestry energy crops can have a greater variety of wildlife and flora than arable or pasture land. It should be noted that biocrops, like all crops, can be vulnerable in bad weather, so reserves and a range of biomass supplies is prudent.

Perennial lignocellulosic crops on marginal land could also provide a significant source of biomass with low associated emissions. They can reduce erosion, improve water retention, and help rural economies, but they are not yet economic. In the long term, aquatic biomass (microalgae) could be a huge source, with an estimated very high yield of ~25 000 litres ha^{-1} of biodiesel. [Microalgae can consist of ~50% oil.] This is equivalent to ~850 GJ ha^{-1} y^{-1}, so 60 million ha of arid land covered with algae ponds could produce enough biodiesel to displace 1.2 billion tonnes of conventional diesel, the global consumption in 2009; this area is only 8% of the land area of the contiguous USA. Algae require a supply of concentrated carbon dioxide to increase their production rate. Open ponds are susceptible to contamination by airborne organisms but genetically modified strains may be more robust, though so far costs are uncompetitive.

While algae might provide ~850 GJ ha^{-1}, equivalent to 2.7 MW$_{th}$ km^{-2} continuous power, typical biomass yields are 150 GJ ha^{-1} for energy through combustion and 90 GJ ha^{-1} for liquid biofuels, so the area required to produce a significant fraction of a country's energy demand is very large. In the USA, primary energy demand in 2009 was 90 EJ, and to produce 5% from biomass combustion would require ~30 Mha, which is 4% of the contiguous USA land area or about a sixth of the cropland area. To displace 25% of the petrol consumption of ~16 EJ y^{-1} the land area required would be similar; hence the contribution from first-generation biomass is limited. However, the potential from agricultural residues and wastes is considerable and would have no impact on cropland.

Besides emissions associated with land clearance there are nitrous oxide emissions from crop residues and emissions associated with the use of fertilizers. When all these are taken into account, some biomass feedstocks result in an increase in emissions compared with the use of fossil fuels. Hence the choice of biomass and its location is vital if biomass is to be an effective low-carbon energy source that does not adversely affect food supply, biodiversity, and the world's forests. Improving the agricultural land-use efficiency in developing countries would help accommodate the demands for global biofuel and global food production.

Regulation and certification can help, but it is essential that complete LCAs are made and required for all biomass projects. Moreover, the availability of water could be the determining factor, as by 2030 demand is expected to be over $6500\,\mathrm{km^3\,y^{-1}}$ while the sustainable supply is only $\sim 4200\,\mathrm{km^3\,y^{-1}}$.

8.6 Global potential and economics of biomass

Currently the main feedstocks for biomass heat and power generation are forestry, agricultural, and municipal residues and wastes, and there are large amounts of these resources untapped. First-generation crops for energy could be expanded but as these use food crops, these could affect food supply and prices. In the medium term, lignocellulosic crops on marginal and surplus land could be the main source of biomass. The technical potential resource taking sustainability issues into account has been estimated to be 200–500 EJ ($\equiv 6.3$–$15.9\,\mathrm{TWy}$) (WEC2010). Residues and wastes account for 50–150 EJ, with the rest from energy crops and improved agricultural productivity. In the longer term aquatic biomass (algae) could provide a significant amount.

The estimated biomass demand could be as great as 250 EJ by 2050 but the actual contribution from biomass will be very dependent on cost-competitiveness and on greenhouse gas emission policies. Estimates vary considerably—for example, values from residues range between 4.4 and 24 EJ per year by 2030. Cost is vital and in 2010 a production cost of biomass of $4 per GJ was the upper limit for a significant contribution from biomass. This upper cost corresponds to 1.1¢ per $\mathrm{kWh_{th}}$ or approximately 3.3¢ per kWh, assuming a 33% thermally efficient power plant.

The main use of modern biomass is for the production of heat, and is often cost-competitive with fossil fuels. For example, in California in 2010 the cost from fluidized-bed combustion of biomass (see Section 3.13) was very similar to that from fossil-fuel generation, at 10¢ per kWh. The most efficient utilization is for combined heat and power (CHP), where efficiencies can be as high as 80%. The development of fluidized-bed combustion plants with higher efficiencies has resulted in greater use of biomass and waste products in power and heat generation. In industrialized societies with a good electrical grid, the use of heat pumps is the most energy-efficient means of heating, so using biomass just for heating is not the best way of exploiting this resource. CHP plants can be very effective off-grid or where the biomass supply is close by. Since the cost of transport of biomass is expensive, because its energy density is typically quite low, small biopower plants that can utilize locally produced feedstock are often more appropriate.

Co-firing of biomass is the most cost-effective use of biomass for power generation, but in the longer term plants will need to incorporate carbon capture to meet emissions targets. Anaerobic digestion is most suitable for wet biomass but needs low-cost feedstock to be economic. There are only a few examples of gasification plants because of their cost and complexity, but they promise greater efficiency and lower costs. The use of ORC or Stirling engines (see Section 7.12.2) could become viable for small-scale applications or for distributed generation.

The production of biofuels for transport is expected to increase significantly, initially using first-generation crops, but increasingly from second-generation crops because of conflicts with food production and concerns over land-use change, as discussed in Section 8.5 (Environmental impact of biomass). Moreover, second-generation feedstocks have better FERs and so improve the reduction in GHG emissions. However, more R&D is required, in particular for lignocellulosic crops; these second-generation crops could possibly be commercial by ~2025, although the development of algae crops is expected to take longer.

As an example, rapeseed has been an important feedstock for biodiesel but it is a relatively expensive crop since it requires frequent rotation and considerable amounts of fossil-fuel based fertilizer, so is not economic without subsidies. But the use of a higher yielding crop such as jatropha, which can be grown in developing countries, may be a more cost-effective way of meeting demand, although as the feedstock would be imported it would not provide energy security.

Many countries hope that biomass will make a significant contribution to reducing their GHG emissions. An associated economic benefit is that biocrops give rural employment as well as energy security. The expanding areas are biomass pellet boilers, large CHP plants, and co-firing, with trade, e.g. in wood chips, vegetable oils, and residues, currently 1 EJ and growing quickly. The current rate of biomass expansion, though, is insufficient to meet many countries' targets, and policy incentives and supply chain improvements will be needed. The effect of weather variability would be reduced by enlarging the biomass markets and reserve stocks.

Biomass costs can be reduced by economies of scale, and its competitiveness is affected significantly by fossil fuel costs. GHG reduction is currently generally larger for co-firing biomass than from creating biofuels for transport. As batteries improve, electric cars may well prove a more effective way to reduce emissions from transport, coupled with improvements in efficiencies, though biofuels for hybrid (electric + internal combustion engine) cars may be useful. A niche market for biofuels may be in aviation and shipping.

Policies need to be sufficiently long-term to encourage investment, and both feed-in tariffs and renewable energy quotas have been used. Access to the grid for biomass generators is important and costs may be reduced by standardization of feedstocks. Sustainability is crucial and more use of residues and waste, which have the least environmental impact and give the largest GHG reduction, should be encouraged. Long-term, perennial lignocellulosic crops and algae have considerable potential.

8.7 Biomass outlook

Solid biomass will remain an important source of energy (~50 EJ in 2010), particularly in developing countries, and modern biomass for heat and power and for biofuels is expected to increase by ~1.5 times between 2010 and 2030, from ~12 EJ to ~18 EJ and ~3 EJ to ~4.5 EJ, respectively. Biomass has the potential to provide more energy, with the IEA estimating that 80 EJ for heat and power and 65 EJ for transport biofuels could be produced annually by 2050. This would be ~15% of global primary energy needs, with possibly a larger percentage in developing countries where improvements in the efficiency of biomass use could have a significant impact. The IEA estimate that the contribution to electricity generation by 2050 from biomass and waste could be ~2000 TWh y^{-1}, equivalent to ~200 GWe continuous output, which we will take as the accessible potential by 2050.

However, whether such an expansion in biomass would be beneficial in reducing carbon emissions is very dependent on the type of feedstock used and its location. The burning of tropical forests can cause a massive release of carbon, and land clearance can also adversely affect the land rights and land tenure of the poor in developing countries, as well as biodiversity. Policies on biomass need to consider their possible effect on food production and a whole life-cycle analysis should be carried out to evaluate their net carbon emissions.

In 2007 the IPCC concluded that sustainable use of forests would give the largest carbon mitigation benefit, and there is a considerable potential (~50–150 EJ y^{-1}) in forestry and agricultural waste. The use of only marginal land for energy crops such as cassava, jatropha, jojoba, and pongamia, which can be easily processed, should be encouraged. An interesting process that is under consideration is the use of biochar (see p. 61), which is a product of the pyrolysis of biomass, on a small scale as a way of sequestering carbon and at the same time improving the fertility of the soil. An increased use of black liquor, which is a by-product of the wood-pulping process, could also be beneficial.

Genetic engineering of bioenergy crops may also increase yield and allow more harvests per year. With more research and development, lignocellulosic crops could make an important contribution, and in the longer term the development of algae with their high yield could make a significant impact.

Although, like wind and solar power, biomass is affected by the weather, biomass can be stored. It can provide energy security and help rural economies, and biofuels can reduce the dependence on fossil fuels, and may be most beneficial in areas where electric power is unsuitable, such as in aviation and shipping. It can provide distributed heat and power generation, and its use locally avoids transport costs.

The introduction of tax incentives and in particular a percentage share of energy production or use, as in the EU biofuel directive, has helped the development of biomass technologies, but the land clearances that have occurred in producing the biofuel feedstock have had a very negative impact on their carbon mitigation. While biomass can bring employment opportunities and an important source of energy, future policies on biomass must incorporate LCA and evaluations of food production, biodiversity, and land rights. Biomass has the potential to contribute 10–20% of global energy needs by 2050, but it needs to be utilized very carefully.

 SUMMARY

- Plants store solar energy via photosynthesis as carbohydrates and as oils, with energy densities of ~16 MJ kg^{-1} and ~38 MJ kg^{-1}, respectively.

- The plant or biomass yield in the field is ~10 t ha^{-1} y^{-1}, which is equivalent to ~5 kW ha^{-1} or 0.5 MW km^{-2}.

- Biomass provides ~10% of global power usage (~1.5 TW in 2010) mainly for cooking and heating in developing countries.

- On a small scale, anaerobic digestion is widely used. On a larger scale, combustion of biomass is mainly utilized to provide both heat and electricity, with the potential in Europe to provide 10–20% of demand.

- The production of liquid biofuels from biomass is increasing rapidly, with annual global output now exceeding 100 billion litres, mainly the production of bioethanol through fermentation of sugar-containing plants in North and South America. The manufacture of biodiesel is also increasing fast.

- There is considerable environmental concern about carbon emissions and changes in land use from the deforestation and land clearances associated with the expansion of biofuel production.

- There is a large biomass resource (50–$150\,EJ\,y^{-1}$) in forestry and agricultural residues and wastes, whose use carries little environmental risk.

- The deployment of crops that will tolerate marginal land and require little water, such as jatropha, jojoba, cassava, and the lignocellulosic crops, switchgrass and miscanthus, should be encouraged.

- The development of bioethanol production from lignocellulosic plants could lead to sustainable energy crops with a good FER. In the long term, oil extraction from algae could be a very significant source of biomass with a very high yield per hectare.

- Biomass is a significant source of energy that could supply 10–20% of global energy demand by 2050, with an accessible potential for electricity generation of $\sim200\,GWe$. But a large area of land is required to produce a significant quantity of power (currently $\sim0.5\,MW\,km^{-2}$) and biomass must be used with care to avoid adverse impacts on the environment, food production, land use, and climate.

- The availability of water will become an increasing concern as demand is already (2012) greater than the sustainable supply and is expected to be over 50% greater by 2030.

FURTHER READING

Boyle, G. (ed.) (2004). *Renewable energy*. Oxford University Press, Oxford. Useful overview of bioenergy.

Cushion, E., Whiteman, A., and Dieterle, G. 2009. *Bioenergy development*. The World Bank, Washington, DC. Available at *siteresources.worldbank.org/INTARD/Resources/Bioenergy. pdf*. Very good review of the environmental issues surrounding the use of biomass.

Shepherd, W. and Shepherd, D.W. (2003). *Energy studies*. Imperial College Press, London. Useful overview of biomass.

Twidell, J. and Weir, T. (2006). *Renewable energy resources*. Taylor & Francis, London. Provides good detail about all of the biomass processes.

WEB LINKS

www.aebiom.org/IMG/pdf/Brochure_BiogasRoadmap_WEB.pdf Biogas Roadmap for Europe.

bioenergy.ornl.gov/papers/misc/energy_conv.html Conversion factors.

www.energy.ca.gov/2009publications/CEC-200-2009-017/CEC-200-2009-017-SF.pdf
Comparative costs of California central station electricity generation.

www.eere.energy.gov/biomass/ Source of information on biomass technologies.

www.fao.org/index_en.htm Food and Agriculture Organization of the UN.

www.iea.org/publications/freepublications/publication/biofuels_roadmap.pdf 2011 Biofuels for Transport Roadmap.

www.ieabioenergy.com 2007 IEA Bioenergy.

www.nrel.gov Information on renewables in the USA.

www.ren21.net 2011 status report on renewables.

projekt.sik.se/traditionalgrains/review/default.htm Stefan Wirsenius, Global use of agricultural biomass for food and non-food purposes.

www.worldenergy.org/publications/3040.asp World Energy Council 2010 Survey of Energy Resources (WEC2010).

www.fao.org/docrep/013/i1756e/i1756e00.htm 2010 FAO Forestry Paper 162.

wtert.gr/Pdfs/anaerobic_digestion_Ostrem_Thesis.pdf G. Guidotti, Biogas from excreta (AD04).

www.dti.gov.uk/files/file22065.pdf Biomass potential in the UK (UK2020).

siteresources.worldbank.org/INTARD/Resources/Bioenergy.pdf (WorldBank2009).

LIST OF MAIN SYMBOLS

AD	anaerobic digestion	LCA	life-cycle analysis
MSW	municipal solid waste	FER	fossil fuel energy replacement ratio
OSW	organic solid waste		
CCGT	combined cycle gas turbine plant	FAME	fatty acid methyl esters (biodiesel)

EXERCISES

8.1 Estimate the land area required to grow willow that would provide 1 GW of power in a region where annual solar radiation is $1500 \, \text{kWh} \, \text{m}^{-2}$.

8.2 A household disposes of 300 kg of domestic waste a year, 90% of which is carbohydrates and 10% of which is inert, in an anaerobic digester. The family uses the methane for cooking on an open fire. Estimate the number of litres of water that could be boiled annually. (Specific heat of water is $4.2 \, \text{kJ} \, \text{kg}^{-1} \, ^\circ\text{C}^{-1}$.)

8.3 Estimate the annual reduction in CO_2 emissions in tonnes if the average number of miles travelled by cars in the USA per litre of petrol were to increase by 30%. Take petrol to be 100% octane, density $0.73 \, \text{kg} \, \text{litre}^{-1}$ (see Example 8.4).

8.4 Estimate the heat of combustion of oleic acid, whose composition is $C_{18}H_{34}O_2$.

8.5 India has a land area of 2.97×10^6 square kilometres, 57% of which is cropland. If 5% of India's cropland were dedicated to producing jatropha plants, estimate the annual production of biodiesel. Compare your estimate with the global use of oil for transport and comment.

8.6 Discuss whether Europe should increase its biomass supply for heat and power or for liquid biofuel production.

8.7 Why is the USA promoting bioethanol production by enzymatic hydrolysis of cellulosic feedstock?

8.8 If all cars in the USA ran with petrol blended with 20% bioethanol produced by enzymatic hydrolysis of cellulosic feedstock, what would be the annual reduction in CO_2 production in tonnes?

8.9 Estimate what size of forest would need to be planted to absorb the carbon dioxide produced by a 3 GWe coal-fired power station. Is this a practical way to combat greenhouse gas emissions?

8.10 Discuss the extent to which energy derived from the following sources is carbon-neutral: (a) short rotation coppice willow; (b) jatropha; (c) corn; (d) sugar cane; (e) cassava; (f) palm oil.

8.11 (a) Estimate the annual reduction in CO_2 emissions if a 1 GWe coal-fired power station were replaced by a MSW plant. (b) How much waste would be required per year by the MSW plant?

8.12 A 1 GWe coal-fired power station is converted to use biomass as fuel. (a) Estimate the area of land required for energy crops. (b) What is the reduction in CO_2 and in C emissions per year? (c) What area would be required for an annual reduction of 1 Gt of carbon?

8.13 Calculate the amount of carbon emitted when 1 litre of biodiesel is burned. Take the composition of biodiesel to be 100% cetane (hexadecane), $C_{16}H_{34}$.

8.14 Estimate the number of miles per gallon (imperial) of petrol when a car travels at (a) 55 mph and (b) 70 mph. Assume that the resistance to motion is predominantly air resistance, the drag coefficient is 0.4, and the cross-sectional area of the car is $3\,m^2$.

8.15 In parts of South America the yield of sugar from sugarcane is 1600 tonnes per km^2 per year. The sugar is fermented into ethanol via the reaction

$$C_6H_{12}O_6 \rightarrow 2C_2H_5OH + 2CO_2$$

(a) Calculate the area of sugarcane required to produce sufficient ethanol to displace 2×10^{10} litres of petrol per year. (b) What would be the resulting reduction in carbon emissions, measured in tonnes of carbon dioxide per year? (Assume petrol is pure octane.)

8.16 Should the Brazilian bioethanol from sugarcane programme be adopted globally?

8.17 A strain of algae is developed that can produce $900\,GJ\,ha^{-1}\,y^{-1}$. What area is required to displace 400 million tonnes of conventional diesel?

8.18 What is the best way to increase the contribution of biomass to global energy production?

8.19 Discuss whether the CO_2 associated with converting land to biomass production, and the resulting loss of land and water for food production, make biomass an unsustainable energy resource.

9 Energy from fission

➜ **Introduction**

Nuclear power is associated in some people's minds with nuclear weapons and nuclear waste, but we will see that it is an abundant source of low-carbon energy that could play a big role in combating global warming. There are two forms of nuclear energy—one from controlling fission, the reaction used in the first 'atomic' bombs; the other from controlling fusion, the energy source in stars. Fusion power is now only at the prototype stage, but it holds the promise of almost unlimited power.

Commercial nuclear power plants are fission reactors, most of which use uranium for fuel, an element that occurs in many parts of the world, with Canada and Australia currently the main producers. Compared with the amounts of coal, oil, or gas required to fuel a conventional power station, remarkably small amounts of uranium are needed for a nuclear reactor—roughly 1 tonne of uranium will deliver an amount of energy equivalent to 20 000 tonnes of coal.

Uranium (U) is roughly as common as tin or zinc. It is in many rocks and in the sea. The average concentration in the Earth's crust is 2.8 ppm. Granite contains about 4 ppm U, while the sea has ~0.003 ppm, which corresponds to 4600 Mt. At $130 per kg U or less, the known reserves are 4.7 Mt, equivalent to 85 years operation at present consumption (IAEA), with an estimated additional 35 Mt recoverable that would come from unconventional sources such as phosphate rocks (U-reserves). High grade ore containing ~2% U is the cheapest to mine.

In this chapter we describe nuclear power from fission, and in Chapter 10 the progress on obtaining power from fusion. We will see that both come from converting part of the mass of nuclei into energy. In fission we find that the process is initiated by neutrons that then yield more neutrons, giving rise to the possibility of a chain reaction that has to be controlled safely in a nuclear reactor. A history of nuclear fission and fusion is given in Box 9.1.

We explain how a chain reaction is brought about, how it is controlled, what the power output from a reactor is, and what the fission waste products are. The chapter ends with a discussion of the worldwide implementation of nuclear power, its economics, safety, environmental impact, and the effect of public opinion on its use.

9.1 Binding energy and stability of nuclei

In order to understand why energy is released in the fission of heavy nuclei or in the fusion of light nuclei, we need to consider the relationship between the mass and the stability of nuclei. A nucleus consists of protons and neutrons (collectively referred to as nucleons) bound together by a short-range attractive force. Its mass is less than the sum of the masses of its constituent nucleons—the size of the difference ΔM gives the total binding energy B_E through Einstein's relation $B_E = \Delta M c^2$. This energy B_E is the energy that would be required to pull apart the nucleus into its constituent nucleons and determines whether a nucleus is stable or unstable. To illustrate this we will consider the element carbon, whose nucleus has six protons.

There are two stable isotopes: ^{12}C, which is the most abundant, and ^{13}C. The isotope ^{14}C is unstable and decays to ^{14}N, in which a neutron changes into a proton, with the emission of an electron and an anti-neutrino:

$$^{14}\text{C} \rightarrow {}^{14}\text{N} + e^- + \overline{\nu}$$

The half-life of this β-decay is 5730 years and the relative amount of ^{14}C to ^{12}C is used in radiocarbon dating of archaeological objects. The reason ^{14}C is unstable is that ^{14}N is more tightly bound with an equal number of protons and neutrons than ^{14}C with eight neutrons and six protons. The lighter isotope ^{11}C is also unstable and decays to ^{11}B. ^{11}C is less bound than ^{11}B since it has more electrostatic repulsion with six rather than five protons. In this β-decay a proton changes to a neutron and a positron and a neutrino are emitted.

Beta decay can leave a nucleus in an excited state that, being less bound, can decay to the ground state of the nucleus with the emission of a γ-ray (or γ-rays), just as an excited atom can decay with the emission of a photon (or photons). An excited state of a nucleus can also decay by transferring its energy to a bound electron, causing the electron to be emitted. This latter process is called internal conversion.

The balance between minimizing the electrostatic repulsion and trying to equalize the number of neutrons and protons to make a nucleus more bound determines the ratio of neutrons to protons and results in each element having at most only a few stable isotopes, as shown in Fig. 9.1. We will now look at how the nuclear binding energy varies across the Periodic Table.

The nuclear force between nucleons is short-range and attractive, unless the separation between the nucleons is very small ($\lesssim 1$ fm), when it becomes repulsive. This is like the force

Box 9.1 History of nuclear fission and fusion

In the mid nineteenth century scientists were baffled by what was powering the Sun: any chemical process gave the age of the Sun as only a few thousand years. In the late 1850s Kelvin and Helmholtz suggested that the source was gravitational potential energy, which gave the age of the Sun as 20 million years. Though this was much more realistic, it disagreed with Darwin's estimate in the *Origin of Species* (1859) for the age of the Earth. This estimate was 300 million years based on erosion of the 'Weald' in the south of England, which Darwin took as the timescale for evolution. The problem was not resolved until 1920 when Eddington proposed the fusion of hydrogen to helium as the source of energy. However, the mechanism of fusion was not fully understood until quantum tunnelling was discovered in the late 1920s.

The exceedingly hot hydrogen plasma in the core of the Sun, where the fusion reactions occur, is contained by the large gravitational forces arising from the enormous mass of the Sun. Since the 1930s numerous attempts have been made to contain a hot plasma using magnetic fields, with the eventual aim of producing fusion. These experiments have now reached that goal, but significant technical advances are still required, and it is still at least a few decades from being a commercial proposition.

Fission was observed (but not immediately recognized) shortly after the discovery of the neutron by Chadwick in 1932. Over the next few years, many different nuclei were bombarded with neutrons. In 1934 Fermi and collaborators found that several radioactive nuclei were produced following neutron bombardment of uranium. Initially some of these were thought to be transuranic elements (i.e. elements with atomic numbers greater than 92), but Noddack speculated (correctly) that these nuclei might be isotopes of known lighter elements arising from the splitting apart ('fission') of uranium nuclei by the neutrons.

However, it was not until 1938 that the radiochemists Hahn and Strassman established that some of the nuclei produced by neutron bombardment of uranium were actually radioactive isotopes of barium, rather than isotopes of the chemically similar radium, as they had thought initially. This possibility had been dismissed earlier as physically impossible, as only α (helium nucleus) and β (electron or positron) decays were known at that time. But early in 1939 Meitner and Frisch explained the phenomenon by noting that a uranium nucleus would behave rather like a liquid drop and that the capture of a neutron could cause the nucleus to oscillate and divide into two smaller nuclei.

In the fission process two to three neutrons are released, together with a considerable amount of energy, giving rise to the possibility of an explosive chain reaction through further neutron-induced fission. Fermi and Szilard (1942) demonstrated that this chain reaction could be controlled in a nuclear reactor but it was not until 1956 that the first prototype nuclear power station was built to utilize the energy released by fission reactions.

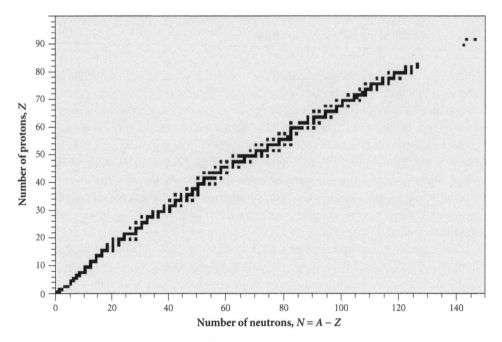

Fig. 9.1 Chart of stable nuclei.

between molecules in a water droplet. Nucleons are therefore on average the same distance apart and interact primarily with their nearest neighbours. So we expect the total binding energy of a nucleus to be approximately proportional to the number of nucleons A, or the binding energy per nucleon, $B_E / A \equiv b(A)$, to be approximately constant. A plot of $b(A)$ vs A is shown in Fig. 9.2 and, while it is roughly constant above $A \sim 12$, it has a maximum near iron (Fe; $A \sim 60$).

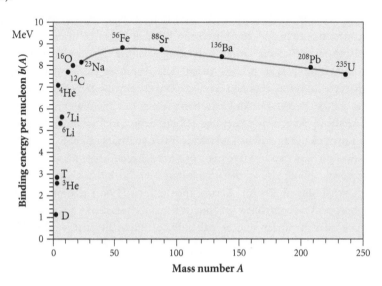

Fig. 9.2 Binding energy per nucleon $b(A)$ as a function of A.

The reason for this maximum is that nuclei would be most tightly bound if the number of neutrons N equalled the number of protons Z, were it not for the electrostatic repulsion between the protons, which increases with increasing Z. This causes heavy nuclei such as uranium to have more neutrons than protons (see Fig. 9.1), and produces the fall in $b(A)$ seen above $A \sim 60$. Below $A \sim 60$ the effect of the increase in the relative number of nucleons on the surface of the nucleus, which are less tightly bound as they have fewer neighbouring nucleons than those within, outweighs the reduced electrostatic repulsion from fewer protons, and $b(A)$ is lower.

We can now see why the fission of a heavy nucleus or the fusion of two light nuclei releases energy. Consider uranium with mass number A_1 splitting into two nuclei with mass numbers A_2 and A_3 and giving off two neutrons: the two nuclei will both be neutron-rich compared to stable nuclei of these mass numbers and will undergo β-decays until stable nuclei are reached. In these decays more energy is emitted. Each of the emitted neutrons will end up being absorbed by a nucleus. There is then a release of energy equal to the binding energy of the neutron, which is approximately $b(A)$. We can calculate the total energy release E_R by looking at the $b(A)$ values.

The nucleons in the lighter nuclei are more tightly bound than in the uranium nucleus and the total energy release E_R is given approximately by

$$E_R = A_2[b(A_2) - b(A_1)] + A_3[b(A_3) - b(A_1)] \tag{9.1}$$

EXAMPLE 9.1

Calculate the energy release due to the fission of ^{235}U, given that the resulting two lighter stable nuclei have mass numbers of 140 and 93.

The value of $b(A)$ for uranium ($A = 235$) from Fig. 9.2 is ~7.6 MeV, while for the two lighter stable nuclei with mass numbers of $A \sim 140$ and $A \sim 93$ it is ~8.35 and ~8.7 MeV, respectively. Substituting these values in eqn (9.1) gives

$$E_R = [140(8.35 - 7.6) + 93(8.7 - 7.6)] \approx 210 \text{ MeV}$$

About 10 MeV of this energy is taken away by neutrinos in the β-decays. Neutrinos interact extremely weakly with matter and hence escape. Therefore, ~200 MeV of energy will be deposited in the surrounding material if a uranium nucleus splits into mass 93 and 140 nuclei.

Energy is also released when deuterium ^{2}H and tritium ^{3}H nuclei fuse to form helium ^{4}He with the release of a neutron:

$$^2\text{H} + {}^3\text{H} \rightarrow {}^4\text{He} + \text{n}$$

The values of $b(A)$ for ^{2}H, ^{3}H, and ^{4}He are 1.1, 2.6, and 7.1 MeV, respectively, so approximately $(4 \times 7.1 - 2 \times 1.1 - 3 \times 2.6) \approx 18$ MeV is released in this fusion process.

We will now look at fission and how this process is harnessed to produce power, before describing in Chapter 10 the progress that has been made in obtaining fusion power.

9.2 **Fission**

Although energy is released by the fission of uranium, the natural occurrence of this process (called spontaneous fission) is very rare. This is because uranium is stable with respect to small deformations from its equilibrium shape, as illustrated in Fig. 9.3.

Such deformations increase the surface area of the nucleus and there is a consequent loss in binding energy, which is not offset by the decrease in electrostatic repulsion arising from the increased separation of charge. The result is a barrier of a form indicated in Fig. 9.3. Classically such a nucleus would be stable, as a ball would be in a dip in the top of a mound, but quantum mechanically decay by fission can take place through a process called quantum tunnelling.

This is the same process that occurs in α-decay. For $Z = 92$ the barrier is sufficiently low at $\sim 6\,\mathrm{MeV}$ that the process is just detectable for $^{238}\mathrm{U}$, with a half-life for spontaneous fission of $\sim 10^{16}\,\mathrm{y}$; its half-life for α-decay is comparatively much shorter at $4.5 \times 10^9\,\mathrm{y}$ so fission is very rare.

9.2.1 **Neutron-induced fission**

The probability of fission of uranium is increased enormously when a uranium nucleus captures a neutron. For a $^{235}\mathrm{U}$ nucleus this capture produces an excited $^{236}\mathrm{U}^*$ nucleus, $\mathrm{n} + {}^{235}\mathrm{U} \rightarrow {}^{236}\mathrm{U}^*$, whose energy is above the height of the fission barrier and can therefore fission promptly. However, for $\mathrm{n} + {}^{238}\mathrm{U} \rightarrow {}^{239}\mathrm{U}^*$, the excited $^{239}\mathrm{U}^*$ nucleus is $\sim 1\,\mathrm{MeV}$ below the top of the barrier, whose height is $\sim 6\,\mathrm{MeV}$. This difference in excitation arises because a neutron in $^{236}\mathrm{U}$ is slightly more strongly bound than in $^{239}\mathrm{U}$. In $^{236}\mathrm{U}$, which has 92 protons, all the neutrons are paired off, while in $^{239}\mathrm{U}$ there is one unpaired neutron and this is less tightly bound than a paired-off neutron.

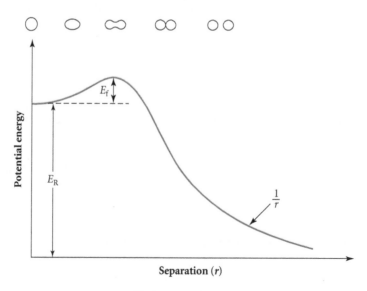

Fig. 9.3 Fission barrier.

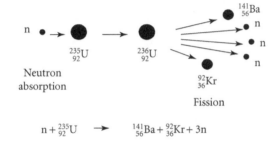

$$n + {}^{235}_{92}U \quad \longrightarrow \quad {}^{141}_{56}Ba + {}^{92}_{36}Kr + 3n$$

Fig. 9.4 Neutron-induced fission of ^{235}U producing ^{141}Ba and ^{92}Kr.

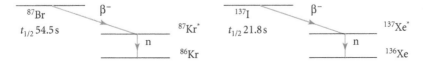

Fig. 9.5 Examples of beta-delayed neutron emitters.

Figure 9.4 shows an example of neutron-induced fission of ^{235}U. Typically two or three neutrons are released promptly. This is because the barium and krypton isotopes that are initially produced are very neutron-rich compared with the stable isotopes of these elements, the most abundant being ^{138}Ba and ^{84}Kr, respectively (see Fig. 9.1). These very neutron-rich nuclei are produced in states that emit neutrons promptly, leaving neutron-rich isotopes of barium and krypton that β-decay to stable isotopes.

Some of these isotopes decay to excited states of nuclei which are unstable to neutron emission (see Fig. 9.5). This process is called beta-delayed neutron emission because the neutron is emitted only after a β-decay, and happens about 0.6% of the time in the neutron-induced fission of uranium. We will see (in Section 9.4.3) that these β-delayed neutrons are extremely important for the control of the chain reaction in a nuclear reactor.

9.2.2 Energy release in fission

We have estimated the energy release in the fission of uranium from the $b(A)$ values in Fig. 9.2 to be about 200 MeV ≡ 3.2×10^{-11} J. This is about 50 million times more than that released in a chemical combustion reaction such as

$$C + O_2 \rightarrow CO_2 + 4.1 \text{ eV} \equiv 6.6 \times 10^{-19} \text{ J}$$

A carbon atom is much lighter than a uranium atom, so 1 tonne of ^{235}U is equivalent as an energy source to ~3.1 million tonnes of coal, as the specific energy from burning coal is ~80% that from burning carbon. Only 0.7% of natural uranium is ^{235}U, 99.3% being ^{238}U, so if only ^{235}U is used for fission,

1 tonne of uranium is actually equivalent to about 20 000 tonnes of coal.

The energy from fission can be divided into prompt release, and delayed release following the β-decay of the neutron-rich nuclei produced in the fission. The distribution of energy released is shown in Table 9.1.

Table 9.1 Energy release in fission of ^{235}U

	Energy released per fission (MeV)
Prompt release	
Fission products	168
Neutrons	5
γ-rays + internal conversion e$^-$	7
Delayed release	
β particles	8
γ-rays + internal conversion e$^-$	7
Antineutrinos	12

Source: Lilley2001.

In a reactor the antineutrinos escape because their interaction with matter is exceedingly weak. The neutrons that do not induce fission are captured by nuclei, producing nuclei in excited states which decay by γ emission. These γ-rays give another ~5 MeV, making the total energy absorbed as heat per fission close to 200 MeV.

EXAMPLE 9.2

How much energy is released when 1 kg of uranium enriched to 3% in ^{235}U is consumed in a nuclear reactor?

One mole of uranium corresponds to 238 g, so the number of uranium nuclei N_U in 1 kg is given by

$$N_U = 1000 \times 6 \times 10^{23}/238 = 25.2 \times 10^{23}$$

The energy release per fission of ^{235}U is 200 MeV. Since 3% of the uranium nuclei are ^{235}U, the total energy release E_T will be

$$E_T = 0.03 \times 25.2 \times 10^{23} \times 200 \times 10^6 \times 1.6 \times 10^{-19} \, \text{J} = 2420 \text{ GJ}$$

On average ~2.4 neutrons are emitted in the neutron-induced fission of ^{235}U, with a broad range of energies about a mean energy of ~2 MeV (see Fig. 9.6). The release of more than one neutron in the neutron-induced fission of ^{235}U, as in the reaction illustrated in Fig. 9.4, opens up the possibility of a chain reaction, which will occur if on average at least one of the neutrons released induces fission of other nuclei. Typically the fission is asymmetric, as shown in Fig. 9.4. It is important to note that not all of the neutrons are emitted promptly: 0.65% are β-delayed neutrons with a mean delay time of 13 seconds.

We now consider the conditions for a chain reaction to occur in uranium by looking at the relative probabilities for different neutron reactions on ^{235}U and ^{238}U.

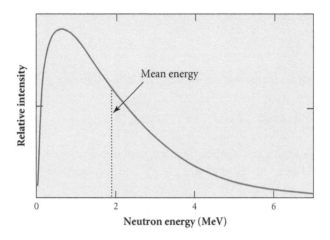

Fig. 9.6 Fission neutron energies.

9.2.3 **Chain reactions**

Whether a chain reaction actually occurs depends on the probability of neutron-induced fission relative to neutron loss. The dominant cause of loss is neutron capture, followed by γ emission. The main reactions that neutrons with energies from a fraction of an eV (thermal neutrons) to several MeV undergo with uranium are scattering (both elastic and inelastic), capture, and induced fission.

The probability that one of these reactions occurs can be described in terms of a cross section σ. This can be visualized as the effective cross-sectional area within which a target nucleus and an incident neutron will interact and give rise to a particular reaction. Its units are barns $(b) \equiv 10^{-28}\,m^2$. The cross-sectional area of a uranium nucleus is ~2 b. The value of σ may be much larger than the cross-sectional area of a nucleus. This is a consequence of the wave-like properties of a neutron. The cross section for any interaction is the total cross section σ_t, and that for absorption equals the sum of the capture and neutron-induced fission cross sections, i.e. $\sigma_a = \sigma_c + \sigma_f$.

Consider neutrons moving at a speed v through uranium where the number of ^{235}U nuclei per unit volume is n. If the cross section for neutron-induced fission is σ_f then in one second a neutron will sweep out a volume $\sigma_f v$ (see Fig. 9.7).

The number of ^{235}U nuclei enclosed within the volume swept out by this neutron per second is therefore $n_f \sigma_f v$, so if there are n neutrons per unit volume within the uranium then the fission reaction rate R_f per unit volume is given by

$$R_f = n_f \sigma_f v n \equiv \Sigma_f \phi \tag{9.2}$$

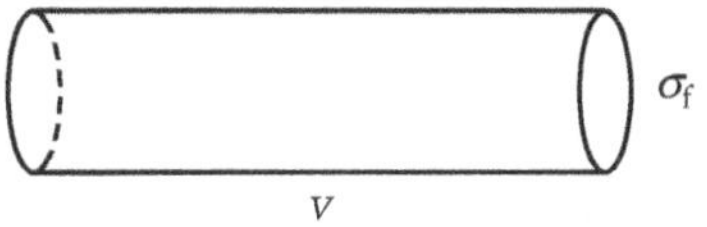

Fig. 9.7 Volume $\sigma_f v$ swept out by neutron in one second.

where

$$\Sigma_f = n_f \sigma_f \tag{9.3}$$

is called the macroscopic cross section for neutron-induced fission of ^{235}U. Its units are m^{-1}. The product

$$nv \equiv \phi \tag{9.4}$$

is called the neutron flux. The neutron flux ϕ is very important as it determines the power output in a nuclear reactor. Its units are $m^{-2}s^{-1}$ and a typical value for ϕ in a reactor is $10^{17}\,m^{-2}\,s^{-1}$.

EXAMPLE 9.3

Calculate the macroscopic cross section for fission in uranium that is enriched to 2% in ^{235}U. What is the energy produced per unit volume when the neutron flux is $10^{17}\,m^{-2}\,s^{-1}$? The density of uranium is $18\,900\,kg\,m^{-3}$ and the cross section for neutron-induced fission of ^{235}U is $579\,b$.

One mole of uranium corresponds to $238\,g$ so the number density of uranium nuclei n is given by

$$n = (18900/0.238) \times 6 \times 10^{23} = 4.77 \times 10^{28}$$

The amount of ^{235}U is 2% and the fission cross section is $579\,b$ so the macroscopic cross section Σ_f from eqn (9.3) is

$$\Sigma_f = 0.02 \times 4.77 \times 10^{28} \times 579 \times 10^{-28} = 55\,m^{-1}$$

The reaction rate per unit volume R_f from eqn (9.2) is

$$R_f = \Sigma_f \phi = 55 \times 10^{17}\,m^{-3}s^{-1}$$

So the energy generated per second per unit volume, P_f, is R_f multiplied by the energy release per fission E_f, i.e

$$P_f = R_f E_f = 55 \times 10^{17} \times 200 \times 10^6 \times 1.6 \times 10^{-19} = 176\,MWm^{-3}$$

For a medium containing a mixture of nuclei with number densities n_i and total cross sections σ_t^i, the macroscopic total cross section is given by

$$\Sigma_t = \sum_i n_i \sigma_t^i \tag{9.5}$$

The average number of nuclei with number density n_i that a neutron interacts with after it has travelled a distance x is $n_i \sigma_t^i x$, so the total number of nuclei that a neutron interacts with is on average $\Sigma_t x$. Putting $\Sigma_t x = 1$ determines the mean free path λ between interactions in the medium as

$$\lambda = 1/\Sigma_t \qquad\qquad (9.6)$$

The macroscopic cross sections determine the relative probabilities for a neutron giving rise to a particular reaction with different nuclei. For example, the ratio of Σ_f^{235} to Σ_f^{238} gives the relative probability that a neutron induces fission of ^{235}U rather than of ^{238}U.

We will now consider natural uranium, consisting of 99.28% ^{238}U and 0.72% ^{235}U, and look at what happens to the neutrons produced in fission, spontaneous or induced. The average number of neutrons emitted per fission, called v, is ~2.4, and their energies range between ~0 and 10 MeV with a mean energy of ~2 MeV. In natural uranium these neutrons most likely scatter off ^{238}U and it is only when they have energies less than ~5 eV that neutron-induced fission of ^{235}U is more likely than capture by ^{238}U.

When the neutrons have energies between ~5 and ~100 eV, sharp peaks (called resonances) occur in the cross section for the capture of a neutron by ^{238}U, which correspond to excited states in ^{239}U being formed. At these peaks the probability of capture is close to 1 as the cross section for capture is much larger than that for scattering: $\sigma_c \gg \sigma_s$. In this region the neutrons are only losing energy very slowly through elastic scattering off ^{238}U, and the chance that their energy falls under a peak in the ^{238}U capture cross section is high, with the result that very few (<1%) reach energies less than 5 eV and induce fission of ^{235}U.

Only a small percentage of the high-energy neutrons induce fission of ^{238}U (~8%) and of ^{235}U (~2%). So the average number of neutrons left after each fission is the percentage that induce fission (~10%) times the average number emitted per fission (~2.4), i.e. only about 0.24 neutrons, and there is therefore no chain reaction in natural uranium.

We can conclude that in order to produce a self-sustaining chain reaction in uranium either the isotopic abundance of ^{235}U must be increased from its naturally occurring percentage of 0.72%, by a process called enrichment, so that the percentage of induced fission by fast neutrons of ^{235}U is larger, or the probability of capture by ^{238}U must be reduced.

Enrichment allows a chain reaction to be maintained with fast neutrons, where fast neutrons are those with energies greater than 100 keV. Alternatively, if nuclei with a low atomic number, called moderators, are added to the reactor, then the change in energy of a neutron following an elastic collision with a moderator nucleus can be sufficiently large that the chance of capture by ^{238}U in one of the resonance peaks is significantly reduced. Once below the resonance region the probability of induced fission is large, and a chain reaction can be maintained with neutrons with energies ~0.05 eV. Neutrons with these energies are called thermal neutrons since they are at the same temperature as the uranium fuel.

Commercial nuclear reactors operate using thermal neutrons (obtained by the use of a moderator), and most use fuel enriched in ^{235}U to a few per cent. This enables a chain reaction to occur. As the time between a neutron from one fission inducing another fission is typically only ~0.1 ms, the chain reaction would be uncontrollable if it were not for the influence of the small fraction of beta-delayed neutrons that are also released. Their mean delay time of ~10 s allows sufficient time for the chain reaction to be controlled mechanically using control rods. Control rods contain nuclei with very high neutron absorption cross sections (e.g. ^{10}B, ^{113}Cd) and the total number of neutrons inside the reactor can be controlled by continually adjusting how far the rods are inserted into the reactor core.

We now look at the layout of a typical thermal reactor before looking more closely at the fuel, moderator, design, control, and safety requirements of a reactor.

9.3 Thermal reactors

The fissile material (the fuel) is generally in the form of fuel rods, which allows for easy refuelling. The rods are immersed in a chemically inert fluid such as water, CO_2, or He, which is heated by the fuel. Shown in Fig. 9.8 is a schematic diagram of a pressurized water reactor (PWR). The PWR is the most common reactor and illustrates the main features of a nuclear reactor. The controls rods are located above the core and the energy released by fission heats the fuel rods, which in turn heat the fluid. The hot fluid is pumped through the primary loop and releases its heat through the heat exchanger, and the steam produced drives a turbine.

The fuel is surrounded by the moderator and the first aspect of the design of a reactor is determining the conditions required of the fuel and moderator for a chain reaction to occur. Whether the neutrons emitted following neutron-induced fission (called the next generation

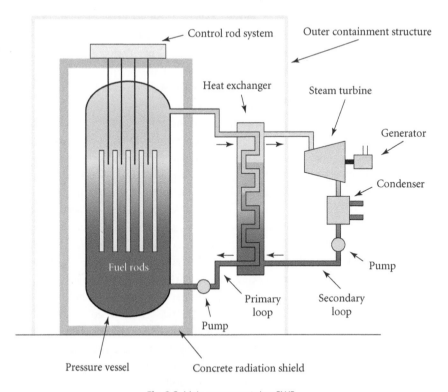

Fig. 9.8 Main components in a PWR.

of neutrons) actually lead to a chain reaction in a piece of fissile material depends on the ratio of the number of neutrons producing fission in one generation to the number producing fission in the previous generation. This ratio is called the multiplication constant k, and if $k \geq 1$ a chain reaction is possible.

The multiplication constant k is determined by the probability of neutron-induced fission relative to neutron loss, via other neutron reactions or by escaping from the reactor. The macroscopic cross sections Σ determine the relative reaction rates (eqn (9.2)), and the size of the core determines the escape probability. For an infinite core no neutrons can escape and the multiplication constant is then called k_∞, which is greater than k. A chain reaction is possible if $k_\infty > 1$, and the size of core for which $k=1$ is called the critical size. We will first consider the factors affecting k_∞.

9.3.1 Four-factors formula for the multiplication constant k_∞

In a core of uranium that is sufficiently large that neutron loss is negligible, there are four factors that determine k_∞:

- fast fission factor ε
- resonance escape probability p
- thermal utilization factor f
- number of neutrons produced per thermal neutron absorbed η.

Of the average initial number of fast neutrons emitted per fission, ν, the loss through capture is offset by the gain through fast neutron-induced fission of ^{238}U. For ^{235}U, $\nu=2.42$. The ratio of the number below the ^{238}U threshold of ~1 MeV to the initial number is called the fast fission factor ε. For low enrichment, the fast fission factor ε is close to unity.

Below the ^{238}U threshold a fraction escape resonance capture by ^{238}U and manage to slow down to thermal energies. The resonance escape probability p depends strongly on the choice of material for the moderator.

The fraction f of thermal neutrons that are absorbed by U, rather than by the moderator or the fuel can, is called the thermal utilization factor f. It is given by the ratio of the macroscopic cross section for absorption in the fuel (F) to that for absorption in the core (C):

$$f = \Sigma_a(F)/\Sigma_a(C) \tag{9.7}$$

The ratio η of the number of fission neutrons produced to the number of thermal neutrons absorbed is less than the average number ν of neutrons emitted, since some of the neutrons are captured by ^{238}U and ^{235}U. This ratio is given by

$$\eta = \nu\Sigma_f(F)/\Sigma_a(F) \tag{9.8}$$

Finally, combining all the above factors yields the four-factors formula:

$$k_\infty = \varepsilon p f \eta \tag{9.9}$$

We will now consider how the choice of moderator affects the resonance escape probability p.

9.3.2 **Moderators**

The average energy of the neutrons produced by the fission reaction needs to be reduced by many orders of magnitude in order to reach thermal energies, and the means of achieving this is called moderation. In a thermal reactor the fuel is generally in the form of fuel rods which are surrounded by moderating material. This arrangement of fuel and moderator gives a non-uniform core.

When a neutron elastically scatters off a nucleus of mass M it will transfer some energy ΔE, the amount depending on the angle of scatter and on the mass M: the larger the mass M, the smaller the loss ΔE (Exercise 9.6). The average relative energy loss $\Delta E/E$ is called the logarithmic decrement ξ and is given for moderating nuclei with mass number A by

$$\Delta E/E \equiv \xi \approx 6/(3A+1) \tag{9.10}$$

where the approximation is reasonable for $A \geq 12$. We can now see why neutrons slowing down in natural uranium by scattering off the uranium nuclei have only a small chance of not being captured by one of the ^{238}U resonances. Each peak in the capture probability occurring between ~5 and 100 eV extends over a small energy range, which is given by the width of the resonance. The peaks correspond to the formation of excited states in ^{239}U that are unstable, and the widths (Γ) are related to the mean lifetimes (τ) of the states through $\Gamma = h/(2\pi\tau)$, which is an example of the uncertainty principle. If the peak cross section is at an energy E_0 then the cross section is reduced by a factor of 2 at $E_0 \pm \Gamma/2$ and by a factor of 5 at $E_0 \pm \Gamma$.

One of the strong capture resonances is at a neutron energy of 6.7 eV, with a total width of 0.024 eV. The average neutron energy loss ΔE for a neutron with energy E when scattering from ^{238}U is given by ξE, so, for a 6.7 eV neutron, $\Delta E \approx 0.056$ eV by eqn (9.10). There is therefore a good chance that while losing energy through elastic scattering from U nuclei a neutron has an energy within Γ of the peak energy. As the peak cross section is over 100 times the elastic scattering cross section, the neutron is likely to be captured. We therefore need to choose a material with low mass nuclei as a moderator, which will have a larger $\Delta E/E$, to reduce the chance of resonant capture.

To be an efficient moderator, both ξ and the probability of scattering (which is proportional to Σ_s) should be large. The moderator should also have a small total absorption cross section Σ_a, and the moderating ratio $\xi\Sigma_s/\Sigma_a$ (a useful figure of merit for a moderator) should be as large as possible.

An approximate expression for the resonance escape probability p for a reactor core, where elastic scattering is predominantly by the moderator and loss is mainly by absorption by ^{238}U in the uranium fuel, is given by

$$p \cong \exp[-2.4(n_f/n_s\sigma_s)^{\frac{1}{2}}/\xi] \tag{9.11}$$

where n_f is the number density of fuel nuclei, n_s is the number density of moderating nuclei, σ_s is the scattering cross section in barns, and ξ is the logarithmic decrement of the moderator. In this expression p is closer to unity for larger ξ and Σ_s and for smaller n_f as expected.

We will now calculate k_∞ for a reactor core containing a mixture of enriched uranium and carbon in the form of graphite, for which the nuclear cross sections, densities, and values for ξ are given in Table 9.2.

Table 9.2 Nuclear and material properties*

Material	Density (kg m^{-3})	n (10^{28} m^{-3})	σ_f (b)	σ_s (b)	σ_a (b)	ξ	Σ_s (m^{-1})	Σ_a (m^{-1})
Graphite (C)	1 600	8.23	–	4.7	0.0045	0.158	37.7	0.037
^{235}U	18 700	4.79	579	10	680		47.9	3229
^{238}U	18 900	4.79	–	8.3	2.72		39.8	13.0

*σ are at thermal energies: 0.025 eV

EXAMPLE 9.4

Calculate the multiplication constant k_∞ of a reactor core containing a mixture of uranium, enriched to 1.7% in ^{235}U, and graphite, with a ratio of $n_s/n_f = 500$, where n_s is the number density of the graphite moderator.

The logarithmic decrement of the mixture is essentially the same as that of graphite as $n_s \gg n_f$. Using the values for cross section and logarithmic decrement given in Table 9.2 in eqn (9.11) gives the resonance escape probability as

$$p = \exp[-2.4(n_f/n_s\sigma_s)^{\frac{1}{2}}/\xi] = \exp\{-2.4[1/(500 \times 4.7)]^{\frac{1}{2}}/0.158\} = 0.731$$

The thermal utilization factor f for this graphite-moderated enriched-uranium mixture is given by eqn (9.7) as

$$f = \Sigma_a(\text{fuel})/[\Sigma_a(\text{fuel}) + \Sigma_a(\text{graphite})]$$

where

$$\Sigma_a(\text{fuel}) = n_{235}\sigma_a^{235} + n_{238}\sigma_a^{238} = (0.017\sigma_a^{235} + 0.983\sigma_a^{238})n_f$$

Applying eqn (9.3) to absorption in graphite gives

$$\Sigma_a(\text{graphite}) = n_s\sigma_a(\text{graphite})$$

So the thermal utilization factor f is given by

$$f = (0.017\sigma_a^{235} + 0.983\sigma_a^{238})/[(0.017\sigma_a^{235} + 0.983\sigma_a^{238}) + [n_s/n_f]\sigma_a(\text{graphite})]$$

Substituting in values from Table 9.2 and using $n_s/n_f = 500$ gives $f = 0.864$.

The ratio η of the number of fission neutrons produced to thermal neutrons absorbed is, from eqn (9.8),

$$\eta = \nu\Sigma_f(\text{fuel})/\Sigma_a(\text{fuel})$$

where $\Sigma_f(\text{fuel}) = (0.017\sigma_f^{235})n_f$ and $\nu \approx 2.42$. Substituting, we obtain $\eta = 1.673$.

For this low enrichment and high n_s/n_f ratio, $\varepsilon \approx 1$, so

$$k_\infty = \varepsilon p f \eta = 1.057$$

This means that a chain reaction is possible and will be critical, i.e. $k=1$, if the size of the reactor is such that the fractional loss, $(k_\infty-1)/k_\infty=(1.057-1)/1.057=0.054$. This loss is the probability for neutrons to escape from the core.

9.3.3 Effect of finite core size

Neutrons are lost from the reactor core because a neutron will typically undergo many elastic scatterings before reacting. Thereby they diffuse away from their point of origin on average a distance that we will call δ. Those neutrons, within δ of the outer surface of the core, will therefore have a good chance of escaping. The fraction of neutrons that are within δ of the outer surface reduces with increasing core size as the surface-to-volume ratio decreases. The size of a core a_c required for a reactor to be just critical ($k=1$) is given by

$$a_c=\frac{\pi\delta}{\sqrt{3(k_\infty-1)}} \tag{9.12}$$

where

$$\delta=1/\sqrt{\Sigma_a\Sigma_s} \tag{9.13}$$

(see Exercises 9.9 and 9.10). As the probability of absorption or scattering increases, i.e. Σ_a or Σ_s increases, δ decreases since the neutron is likely to be absorbed in a shorter distance from where it was created. For larger k_∞ the loss can be larger and hence a_c can be smaller. When the core is just critical, i.e. $k=1$, then the probability that neutrons do not diffuse out of the core equals $1/k_\infty$. The critical size of the core can be decreased by surrounding the core with a reflector—a material that reflects neutrons (such as graphite).

EXAMPLE 9.5

Find the critical radius of a uniform core of uranium, 1.7% enriched in ^{235}U, and graphite, with a ratio of moderator (M) to fuel (F) nuclei of 500.

We need to evaluate δ, so we require $\Sigma_a(C)$ and $\Sigma_s(C)$, where C stands for the reactor core. As the reactor has a high ratio of moderator (M) to fuel (F), then

$$\Sigma_s(C)\approx\Sigma_s(M)$$

The thermal utilization factor from eqn (9.7) is

$$f=\Sigma_a(F)/\Sigma_a(C)$$

Putting $\Sigma_a(C)=\Sigma_a(M)+\Sigma_a(F)$, we have

$$\Sigma_a(C)=\Sigma_a(M)/(1-f)$$

From Example 9.4, the core has $k_\infty = 1.057$ and $f = 0.864$. Also, from Table 9.2, $\Sigma_s(M) = 37.7\,\text{m}^{-1}$ and $\Sigma_a(M) = 0.037\,\text{m}^{-1}$. Substituting into eqn (9.13) gives

$\delta = 0.31\,\text{m}$

Putting this value into eqn (9.12) gives the critical radius as $a_c = 2.4\,\text{m}$.

9.4 Thermal reactor designs

The first reactor was built in a squash court in Chicago in 1942 under the direction of Enrico Fermi, as part of the Manhattan Project, which developed the first fission bomb. It was a graphite-moderated pile containing natural uranium fuel in the form of rods embedded in graphite blocks. Although a uniform mixture of graphite and natural uranium cannot go critical, a non-uniform core containing fuel in the form of rods can go critical.

The development of graphite-moderated reactors using natural uranium fuel was the approach adopted in the UK for the first generation of nuclear power stations, owing to the difficulty of producing uranium enriched in ^{235}U. These reactors were gas-cooled, using high pressure carbon dioxide as the primary fluid, and used uranium oxide fuel in a magnesium alloy container, hence their name: Magnox reactors. They operated at a relatively low temperature of 400 °C, which gave a low thermal efficiency of ~30%. In Canada, a heavy water moderated and cooled reactor (CANDU) was successfully developed. The use of non-enriched fuel, however, makes the core large, and the PWR developed in the USA, which uses enriched fuel and a more compact core, has been much more widely adopted. The techniques used to enrich uranium are described in Box 9.2.

9.4.1 Pressurized water reactor (PWR)

The PWR shown in Fig. 9.8 is the most widespread commercial reactor; out of a total of 435 nuclear reactors in 2010, 272 were PWRs. The PWR was initially developed for submarines since, unlike internal combustion engines, nuclear powered submarines do not need oxygen and can therefore remain underwater for much longer. The heat from the reactor produces steam to drive a turbine, and the relatively compact core proved a cost-effective design that could be scaled up to ~1 GW. The first prototype was operated in 1953.

In a PWR (see Fig. 9.8) water is circulated past the fuel rods in the primary loop at a high pressure of ~15 MPa to keep it in the liquid phase at a temperature of ~315 °C. It is passed through a heat exchanger, where the water in the secondary loop, which is at a lower pressure of ~5 MPa, is heated to produce steam to drive the turbine, after which the steam is condensed and the water is returned to the heat exchanger. The high neutron flux in the core activates the cooling water and makes it radioactive. This radioactivity is kept within the primary loop and inside the containment vessel. The thermal efficiency of a PWR is typically ~33%.

The water acts as a moderator as well as a coolant. It also absorbs neutrons. Should the pressure drop in the primary loop and the water start to boil, the creation of bubbles (voids) decreases the moderation and also the absorption. The effect on the moderation is more

Box 9.2 Enrichment

The percentage of ^{235}U in natural uranium is only 0.72%. As the isotopes ^{235}U and ^{238}U are chemically identical, enrichment is achieved by taking advantage of their slight difference in mass. Several techniques have been demonstrated in the laboratory but only two have been developed commercially: gaseous diffusion and the gas centrifuge process. Electromagnetic separation was used in the Manhattan Project but is not as economic: in this process, uranium ions are accelerated through a potential drop and then deflected in a magnetic field around a circular path whose radius depends on the mass of the ion. A process still under development is laser isotope separation. This method uses the isotopic shift of an atomic or molecular energy level to enable laser beams to selectively excite and then ionize ^{235}U, or molecules containing ^{235}U, which can then be collected.

Enrichment reduces the degree of mixing of the ^{235}U and ^{238}U isotopes. This produces a more ordered state with lower entropy, and requires work. In a gaseous separation plant there is a feed and two outputs—the product and tail streams. The amount of separation required depends on the mass and enrichment of all three streams and is measured in separative work units or SWUs. For example, to produce 1 kg of 3% ^{235}U using natural uranium as feed requires 3.8 SWU if the tail enrichment is 0.25% or 5.0 SWU if it is 0.15%. SWUs have dimensions of mass, and the amount of work required depends on the number of SWUs and the process used.

About 120 000 SWU are required to enrich the annual fuel loading for a typical 1 GW pressurized water reactor (PWR). The gaseous diffusion process uses about 2500 kWh per SWU, while gas centrifuge plants only require about 50 kWh per SWU. This energy, if produced from hydrocarbon fuels, accounts for the main greenhouse gas impact of nuclear power. However, if gas centrifuge plants are used, it is equivalent to only 0.1% of the CO_2 produced from a coal-fired plant of the same power.

Gaseous diffusion

In this process uranium hexafluoride gas is passed under pressure down the centre of a porous tube, as shown in Fig. 9.9(a). On average the molecules containing ^{235}U atoms are moving slightly faster, since they are lighter, and thus make more collisions per second with the porous membrane than those containing ^{238}U. As a result, the gas passing through the membrane gets enriched in ^{235}U. The amount of enrichment at each stage is very small and a cascade of over a thousand stages is required to produce a concentration of ~3% ^{235}U. Each stage needs a compressor and a cooler to remove the compressive heating of the gas, and these consume huge amounts of energy.

Gas centrifuge

In the gas centrifuge process uranium hexafluoride gas is injected near the axis of a very high speed centrifuge (see Fig. 9.9(b)). The heavier molecules tend to be closer to the wall than the lighter molecules as a result of the larger centrifugal force, and an axial flow causes more of the lighter molecules to leave through the top than the bottom tube.

Although the capacity of a single centrifuge is much smaller than that of a single diffusion stage, the isotopic separation is much higher. As a result, many centrifuges are arranged in parallel, but only some 20 stages are required to produce a concentration of ~3% ^{235}U. The main advantage of the gas centrifuge process is that it requires only about 2% of the energy per SWU used in a gaseous diffusion plant.

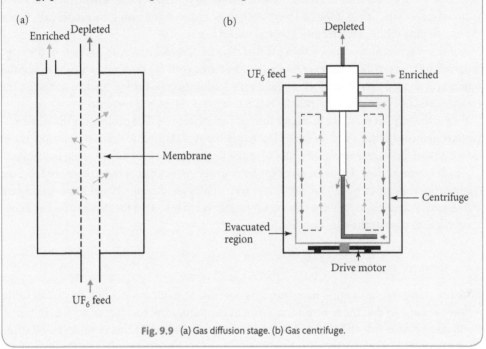

Fig. 9.9 (a) Gas diffusion stage. (b) Gas centrifuge.

significant, the chain reaction stops, and the reactor is no longer critical. The moderation is also decreased if the core temperature rises, as this increases the Doppler broadening of the ^{238}U resonances (see Section 9.4.4), which decreases the resonance escape probability p. Hence there is a negative temperature coefficient of reactivity, which tends to stabilize the power output, since an increase in power causes the temperature to rise and the reactivity to fall, and vice versa.

Over a long period the high neutron flux causes embrittlement of the reactor vessel as the metal becomes less ductile, and this limits the lifetime of the reactor. Corrosion in the steam-generating tubes must also be monitored. A loss-of-coolant accident (LOCA), in which the water in the primary loop is lost, requires additional emergency cooling to be available. While the core would no longer be critical, the heat generated from the decay of the fission products in the fuel rods could cause the core to melt. This possibility has been termed the China syndrome, referring to the (mistaken) idea that the molten core would bore through the Earth all the way from the USA to China!

9.4.2 Core design

Putting the fuel in the form of rods increases the resonance escape probability p significantly. Part of this increase arises because some of the fast fission neutrons, which have a good chance of escaping the fuel rods, are thermalized by the moderator before interacting with another fuel rod, and hence escape capture. However, the fuel rods are generally sufficiently close that this is not so significant. The greater effect arises because the flux of neutrons with energies in the resonance region is significantly reduced within a fuel rod because of the very strong absorption of these neutrons within the surface layer (<0.01 cm) of the rod. This means that most of the fuel is shielded from these neutrons, so the chance that a neutron is captured is considerably lower than if the fuel were uniformly mixed throughout the moderator; hence the escape probability is higher and a typical value for p is ~0.9.

A flux reduction also occurs within the fuel rods for thermal neutrons because of the large absorption cross section. However, since the mean free path for thermal neutrons in natural uranium is about 2.5 cm, the effect on the thermal utilization factor f is much less than on the resonance escape probability p, with the result that the product fp is increased.

Fuel in the form of rods also increases the fast fission factor ε. The probability that the emitted fast neutrons induce fission in ^{238}U is larger because, typically, the fast neutrons travel through pure fuel a distance of the order of the radius of the fuel rod before being moderated.

Besides these nuclear physics reasons for having fuel in the form of rods, there are excellent engineering reasons as well. The use of fuel rods allows good cooling, and hence good heat transfer, as well as easy refuelling. The control rods, which maintain the neutron flux at a constant value, also make the core non-uniform.

9.4.3 Reactor control

Whether using fast or thermal neutrons, it is essential to maintain the multiplication factor k close to unity, so that the neutron flux remains constant. In particular, as the fuel is 'burnt' k will decrease, so that the neutron absorption must be lowered. This is achieved by using electromagnetically operated control rods, which control the neutron flux in the core of the reactor. These contain nuclei with a high cross section for neutron absorption, such as ^{10}B and ^{113}Cd, and can be lowered into the reactor core to reduce the flux or raised out of the core to increase the flux. The reactivity ρ, defined by $\rho \equiv (k-1)/k$, is generally used when describing the time dependence of the neutron population.

Control can also be maintained by altering the absorption of the coolant, by adding a chemical containing a nucleus with a large neutron absorption cross section (chemical shim). If water is the coolant (as in a PWR) then boric acid can be used. Its effect on k is to alter the thermal utilization factor f. Its speed of response is slower than that of control rods, but it reduces the number of rods required.

Another method of control is to use burnable poisons. These are substances containing nuclei with very large thermal absorption cross sections, such as gadolina (Gd_2O_3) or erbia (Er_2O_3), which are included in some of the fuel rods. Initially they absorb sufficient neutrons that k is close to unity. As the fuel is consumed, the consequent decrease in k is partly offset by the reduction in the amount of burnable poison nuclei ^AX through the absorption reaction $n + {}^AX \rightarrow {}^{A+1}X$. This method has been used to extend the time between refuelling for a nuclear submarine reactor to over 10 years.

In a chain reaction there is a short period of time between the release of a neutron from a fission and it initiating another fission and more neutrons. This period, called the generation

Table 9.3 Half-lives and yields of beta-delayed neutron emitters following n + ^{235}U

$t_{1/2}$ (s)	Yield (%)
55.7	0.0215
22.7	0.1424
6.22	0.1274
2.30	0.2568
0.61	0.0748
0.23	0.0273

Source: Lamarsh2001.

time τ_g, for neutrons released promptly following fission, is given by λ_a/u, where λ_a is the absorption mean free path ($=1/\Sigma_a$) and u is the neutron speed. A typical value of τ_g for prompt thermal neutrons is $\sim 10^{-4}$ s. If the number of neutrons per unit volume in the reactor core is n, then the build-up of neutrons will be governed by the rate equation

$$dn/dt = (k-1)n/\tau_g \tag{9.17}$$

The -1 term arises because one neutron is absorbed in the chain reaction to produce the next generation, which has k neutrons, and this occurs on average every τ_g seconds. The population of neutrons therefore grows exponentially with a time constant

$$\tau = \tau_g/(k-1) \tag{9.18}$$

Hence, if k increases to 1.001, the neutron population increases by a factor of 20 in 0.3 seconds, which would make mechanical control exceedingly difficult. Fortunately, this is not what actually happens in practice because a small percentage of neutrons, $\beta \sim 0.65\%$ for ^{235}U, are emitted following the β-decay of neutron-rich fission fragments with a mean delay time $\tau_d \sim 13$ s ($\tau = t_{1/2}/\ln 2$). The lifetimes and fractions of β-delayed neutron emitters are listed in Table 9.3.

For k near unity, it turns out (see Exercise 9.15*) that τ_g effectively depends on the delayed neutrons alone and equals the average neutron lifetime τ_a, given by

$$\tau_a = \beta(\tau_d+\tau_p)+(1-\beta)\tau_p = \beta\tau_d+\tau_p \tag{9.19}$$

where τ_p is the generation lifetime for prompt neutrons (~ 0.1 ms). The time constant for $k=1.001$ then becomes ~ 85 s and mechanical control of the reactor is therefore possible.

When a reactor is started up, k is deliberately made sufficiently greater than unity that the flux increases to its operating value ϕ_0 within a reasonable time.

EXAMPLE 9.6

Estimate what multiplication factor k is required for a thermal ^{235}U fuelled reactor to have a time constant of 300 s. How long would it take for the reactor output to change from 1 W to 1 GW?

Using Table 9.3, the mean half-life of the β-delayed neutrons is 9.0 s, corresponding to a mean lifetime of $\tau_d = t_{1/2}/\ln 2 = 13.0$ s. Taking τ_p as 0.1 ms, the average neutron lifetime τ_a is given by eqn (9.19) as 84.6 ms. The generation time $\tau_g = \tau_a$, so the time constant τ is, from eqn (9.18), given by

$$\tau = \tau_a/(k-1) \quad \text{so } (k-1) = \tau_a/\tau = 0.0846/300 = 0.000282$$

Hence, $k = 1.000282$.

The power output depends on the flux of neutrons and therefore on their number density. This density changes as $\exp(t/\tau)$, so the time t is given by

$$\exp(t/\tau) = 10^9 \quad \text{so} \quad t = 9\tau \ln(10) \approx 6220 \text{ s} = 1.73 \text{ h}$$

9.4.4 Reactor stability

It is very important that the reactivity should decrease if the temperature of the core rises, so that the core temperature will be stable to small fluctuations in reactivity. In a thermal reactor this comes about through a broadening of the ^{238}U capture resonances with increasing temperature. The ^{238}U nuclei are vibrating and have a mean square speed that is proportional to the temperature of the uranium atoms. The neutron energy required to form an excited state of ^{239}U depends on the relative velocity of the neutron and ^{238}U nucleus and so has a spread Δ that increases with increasing temperature. This has the effect of increasing the neutron absorption rate. A simple example shows how this comes about.

Consider a capture resonance with a rectangular energy profile of width Δ and whose capture cross section is four times the scattering cross section. Let the neutron on average lose Δ at each scattering; then typically every fast neutron will fall once under this resonance as it loses energy, and the chance that it will be captured is $(1 - 1/5) = 4/5$. Now broaden the resonance to 2Δ. The total capture probability remains the same so the magnitude of the resonance is halved to twice the scattering cross section. A neutron now on average falls twice under the resonance and its chance of being captured is $(1 - (1/3)^2) = 8/9$. Increasing the core temperature therefore decreases the resonance escape probability and hence decreases k.

9.4.5 Power output of a thermal reactor

The distribution of energy released in neutron-induced fission of ^{235}U is shown in Table 9.1. In a reactor the antineutrinos escape because their interaction with matter is exceedingly weak. The neutrons that do not induce fission are captured through (n, γ) interactions and the γ-rays give another ~5 MeV, making the total energy absorbed as heat per fission close to 200 MeV. Typically >95% of the uranium in the core is ^{238}U, so nearly all of the captured neutrons produce ^{239}U, which β-decays first to ^{239}Np and then to the long-lived fissile nucleus ^{239}Pu. A nucleus that can be converted to a fissile nucleus by neutron capture is called fertile. In a few percent enriched uranium PWR, about 30–60% of the power output comes from the neutron-induced fission of ^{239}Pu produced through conversion of the ^{238}U.

The conversion ratio C is defined as the ratio of the amount of fissile material produced from fertile nuclei to the amount consumed. There is a contribution from fast neutrons resonantly captured by ^{238}U, C_f, as they slow down, and another from capture of thermalized neutrons, C_{th}. By comparison with eqn (9.2), the rate of thermal neutron capture per unit volume is given by $n_8 \sigma_{c8} \phi$, while the consumption rate of ^{235}U per unit volume is $n_5 \sigma_{a5} \phi$, where the subscripts '5' and '8' refer to ^{235}U and ^{238}U, respectively. The thermal neutron contribution C_{th} is therefore $n_8 \sigma_{c8} / n_5 \sigma_{a5}$. As there are η fast neutrons per neutron absorbed and a factor ε increase through fast neutron-induced fission, the fast neutron contribution C_f is $\eta \varepsilon (1-p)$, where p is the resonance escape probability. The initial conversion ratio C is therefore given by

$$C = \eta \varepsilon (1-p) + \frac{n_8 \sigma_{c8}}{n_5 \sigma_{a5}} \tag{9.20}$$

N fissile nuclei will therefore produce CN fissile nuclei from fertile nuclei and these in turn will produce $C^2 N$ fissile nuclei, and so on. The total increase is by a factor of $(C + C^2 + C^3 + \cdots) = C/(1-C)$ when $C < 1$. When $C > 1$, C is then called the breeding ratio and fissile material continues to be produced while fertile material remains, greatly increasing the amount of fuel (see Box 9.4, Breeders).

Conversion is particularly important for reactors using natural uranium as a fuel, for then $C_f \approx 0.25$ and, taking values from Table 9.2, $C_{th} = (99.3 \times 2.72)/(0.7 \times 680) = 0.57$, giving a conversion ratio $C = 0.82$. The percentage of fissile nuclei produced by conversion is therefore $0.7 C/(1-C) = 3.2\%$, so the maximum amount of uranium that could be consumed is 3.9% rather than 0.7% (neglecting changes in C caused by changes in the fuel composition and assuming that ^{239}Pu acts like ^{235}U).

For a reactor using fuel enriched to a few per cent, the conversion ratio is smaller; for example, for 4% enrichment and assuming C_f is the same as for natural uranium, then $C = 0.35$ and 6.2% is the maximum amount of uranium that could be consumed, so about 54% of the power comes from ^{239}Pu. The time interval between refuellings of a reactor is determined by the burn-up of the fuel, which is measured in GWday per tonne of fuel; modern reactors are designed to give 60 GWday per tonne burn-up.

In a reactor, the heat per fission of 200 MeV translates to an energy output of 0.95 GWday per kg of ^{235}U, or about 1 GWday per kg of fissile nuclei. To produce 1 GW for a year requires 384 kg of fissile nuclei. However, this is thermal power, not electrical power; the thermal efficiency of a typical PWR is ~33% so about 1150 kg of fissile nuclei per GWey is actually needed.

The power generated in a reactor core depends on the neutron flux, and the higher the flux the higher the power output. For a core of volume V containing N_{235} ^{235}U nuclei, the power P is given by the fission reaction rate multiplied by the energy release per fission, E_f. The reaction rate is the fission rate per unit volume, R_f, multiplied by the volume V. The rate R_f is given by eqn (8.2) as $\Sigma_f \phi$, so the initial power P is

$$P = V \Sigma_f \phi E_f \equiv N_{235} \sigma_f \phi E_f \tag{9.21}$$

where ϕ is the neutron flux in the reactor and $E_f=200\,\text{MeV}$ is the energy released per fission. Equation (9.3) has been used to equate $V\Sigma_f$ with $N_{235}\sigma_f$. The initial power P can also be expressed in terms of the mass M_{235} of ^{235}U (in tonnes) and flux ϕ (in units of $10^{18}\,\text{m}^{-2}\,\text{s}^{-1}$) as

$$P=4.75\,M_{235}\,\phi\ \text{GW} \tag{9.22}$$

EXAMPLE 9.7

A thermal reactor has a core volume of $14\,\text{m}^3$, a macroscopic fission cross section of $50\,\text{m}^{-1}$, and a neutron flux of $2\times10^{17}\,\text{m}^{-2}\,\text{s}^{-1}$. Calculate the initial power output and the initial mass of ^{235}U in the core. The fuel is 5% enriched initially and is replaced when the percentage of fissile material is reduced to 1.35%. Estimate the conversion ratio and the fuel burn-up in GWday per tonne of fuel.

The thermal reactor has: $\phi=2\times10^{17}$ neutrons $\text{s}^{-1}\,\text{m}^{-2}$, $\Sigma_f=50\,\text{m}^{-1}$, and $V=14\,\text{m}^3$, so from eqn (9.21) the initial power output P is

$$P=14\times50\times2\times10^{17}\times200\times10^6\times1.6\times10^{-19}\ \text{W}=4.48\,\text{GW}$$

From Table 9.2, σ_f^{235} is $579\,\text{b}$,

$$V\Sigma_f\equiv N_{235}\sigma_f,\quad \text{so } N_{235}=1.21\times10^{28}$$

As 1 mole of ^{235}U is 235 g, N_{235} nuclei correspond to a mass m equal to

$$m=[N_{235}/(6\times10^{23})](0.235)=4.74\ \text{tonnes of }^{235}U$$

The conversion factor is given by $C=0.25+(95\times2.72)/(5\times680)=0.326$, assuming $C_f=0.25$. The total percentage of fissile nuclei consumed is $(5-1.35)/(1-0.326)=5.4\%$, i.e. $54\,\text{kg}$ per tonne. At $0.95\,\text{GWday}$ per kg, the burn-up is $51\,\text{GWday}$ per tonne.

9.4.6 Fission products

During the operation of a reactor there is a build-up of fission products within the fuel rods. Some are very long-lived actinides arising through successive neutron capture reactions on uranium. When the amount of fissile material in a fuel rod is insufficient to maintain criticality, the rod is removed and the remaining fissile material is extracted chemically (called fuel reprocessing). It is reutilized in new fuel and the waste products are separated for storage. The presence of these actinides means that the waste must be stored safely for many thousands of years (see Section 9.10).

Some of the fission products are volatile fission fragments, and it is these that could be most easily released in the event of a major accident. The fission rate F is related to the thermal power output P in GW, assuming 200 MeV per fission, by

$$F = 3.13 \times 10^{19}\,P \text{ fissions per second} \tag{9.22}$$

If a fraction b_x, the fission yield, of fissions produce a nucleus x with a decay constant λ_x, then the decay rate R_x of the nucleus x, after a time t such that $\lambda_x t \gg 1$, equals the production rate $b_x F$, i.e. the activity R_x is given by

$$R_x = b_x F \qquad (9.23)$$

(This expression neglects any contribution from a fission product decaying to the nucleus x.)

EXAMPLE 9.8

Estimate the activity of ^{133}Xe when it is in equilibrium in a 1 GWe reactor. The fraction of fissions that produce ^{133}Xe is 0.0677 and the half-life of ^{133}Xe is 5.27 d. How long will it take after the reactor is shut down for the activity to decay to a rate of $10^6\,\text{s}^{-1}$ ($\equiv 1\,\text{MBq}$; see Box 9.3).

An electrical power of 1 GW corresponds to about 3 GW thermal power. Using eqn (9.22), the fission rate $F = 9.39 \times 10^{19}\,\text{s}^{-1}$. From eqn (9.23) the equilibrium activity R_x is

$$R_x = 0.0677 \times 9.39 \times 10^{19} = 6.4 \times 10^{18} = 6.4 \times 10^6 \text{ TBq}$$

Assuming production ceases on shutdown, the activity R_t after a time t is

$$R_t = R_x \exp(-t/\tau)$$

where $\tau = t_{1/2}/\ln 2$. So the time t is given by

$$t = \tau \ln(R_x/R_t) = (5.27/\ln 2)\,\ln(6.4 \times 10^{18}/10^6) = 224 \text{ days}$$

Table 9.4 lists three of the main noble gas and iodine fission products at the end of a 1 GWe PWR plant fuel cycle.

Table 9.4 Data on the equilibrium amount of three fission products from a 1 GWe reactor

Nuclide	Half-life	Fission yield (b_x)	Activity (TBq)
^{88}K	2.79 h	0.0364	2.52×10^6
^{133}Xe	5.27 d	0.0677	6.29×10^6
^{135}I	6.7 h	0.0639	5.55×10^6

Source: Lamarsh2001.

The activities in Table 9.4 are very large and in the case of a core meltdown some 20% of the ^{135}I in the fuel rods could be released into the reactor containment building and subsequently leak out into the atmosphere. The radioactive gas can then be dispersed by wind. Close to the reactor building the direct dose from the radioactive nuclides within the reactor building is dominant while farther away the internal thyroid dose from iodine isotopes is generally the largest. These considerations are clearly important in determining where to site a reactor and how much radiation shielding is required.

Box 9.3 Radiation

Radiation affects tissues as it causes ionization, which breaks molecules apart and gives rise to free radicals, which can damage cells. The scale of the effect depends on the energy deposited per unit mass of tissue, the dose D, and the type of radiation. Charged particles such as α particles cause relatively more damage than γ-rays or electrons depositing the same energy, since their energy loss per unit distance travelled (the linear energy transfer, LET) is higher. Likewise, neutrons transfer their energy in matter to nuclei, so their LET is also high. This difference in LET is taken into account by using a weighting factor w to give the equivalent radiation dose, $H = wD$. In SI units H is measured in sieverts (Sv) and the dose D in grays (Gy), with $1\,\text{Gy} \equiv 1\,\text{J}\,\text{kg}^{-1}$. An older unit for H still in use is the rem $\equiv 10^{-2}\,\text{Sv}$. The weighting factors for different types of radiation are given in Table 9.5.

Table 9.5 Radiation weighting factors

Radiation	Weighting factor w
X-rays, γ-rays, electrons	1
Neutrons	20*
α particles, fission fragments, heavy nuclei	20

*maximum value
Source: ICRP.

The dose received at a distance r in air from a radioactive source is proportional to $1/r^2$ and to the decay rate or activity of the source. The unit of activity is the becquerel (Bq), which is one disintegration per second. An older unit still used is the curie (Ci), where $1\,\text{Ci} \equiv 3.7 \times 10^{10}\,\text{Bq}$.

At a distance of r (m) from a source with activity A (MBq) emitting γ-rays of energy E (MeV), the flux F of γ-rays is

$$F = A/4\pi r^2 \, (10^6\,\text{m}^{-2}\text{s}^{-1})$$

An approximate expression for the equivalent dose rate D_{rate} is

$$D_{\text{rate}}(\mu\text{Svh}^{-1}) \approx A(\text{MBq}) \times E(\text{MeV})/6r^2(\text{m}^2)$$
$$\approx 2F(10^6\,\text{m}^{-2}\text{s}^{-1}) \times E(\text{MeV}) \tag{9.24}$$

when E is in the range ~0.1 to ~5 MeV.

The equivalent dose gives a measure of the biological effect when the whole body is irradiated uniformly. If only certain organs receive a dose, the effect on the person is less, and this is taken into account in the effective dose. An example is the effective dose following radioactive iodine inhalation. Iodine concentrates in the thyroid and if the thyroid receives an equivalent dose of 20 μSv then the effective dose is about 1 μSv.

Environmental radiation

There is naturally occurring radiation in the environment that comes from radioactive minerals and from cosmic rays. Uranium and thorium and their decay products, in particular radon, and the potassium isotope ^{40}K contribute most. Typical values for the annual effective dose in the UK from various sources are shown in Table 9.6.

Table 9.6 Typical radiation doses in the UK

Source	Average annual effective dose (μSv)
Cosmic rays	260
Food	300
Environmental*	1650
Medical	370
Miscellaneous	0.4
Fall-out	5
Occupational	8

*Higher in granite areas
Source: Lilley2001.

The cosmic ray contribution increases with altitude and is about three times higher at 2000 m. In an airliner at 10 000 m it is ~150 times larger than at sea level. The amount of radon in the air is higher in granite areas such as Cornwall, UK, where the total annual dose is about 7800 μSv.

For people working with radioactive materials, such as in the nuclear power industry, there is the possibility of increased radiation exposure. The principal long-term risk from exposure to significant amounts of radiation is an increased risk of cancer. The limits set on the annual radiation dose and on the amounts of any radioactive isotopes that might be ingested or inhaled are such that receiving these amounts would not cause the worker a significant risk in comparison with other occupational risks. The current whole-body annual dose limit for radiation workers in the UK is 20 mSv.

9.4.7 Radiation shielding

Shielding is required around the reactor core to reduce the radiation dose to safe levels. (See the discussion of radiation in Box 9.3.) Generally it is only necessary to shield against γ-rays and neutrons since α and β particles have very short ranges in matter. γ-rays of a few MeV passing through a concrete wall interact via the photoelectric effect, Compton scattering, or pair-production (the threshold is $2m_ec^2 \equiv 1.02$ MeV). While the intensity of an incident mono-energetic beam of γ-rays drops exponentially like $e^{-\mu x}$, where μ is the linear attenuation coefficient of concrete and x is the thickness of the wall, the emergent beam will contain γ-rays of lower energy. These arise primarily through Compton scattering, in which a γ-ray scatters off and transfers energy to an electron. (μ equals the total macroscopic absorption cross section

$\Sigma_t = n\sigma_t$.) The radiation dose is therefore not reduced by as much as $\exp(-\mu x)$. This effect is parameterized by using a build-up factor $B(\mu x)$, so that the dose is given by an effective flux F_{eff} of γ-rays of the incident energy given by

$$F_{\text{eff}} = B(\mu x)\exp-(\mu x) \qquad (9.25)$$

For example, for a beam of 2 MeV γ-rays passing through 2 m of water, for which $\mu x \sim 10$, $B(\mu x)$ is ~ 10, so the build-up is significant.

EXAMPLE 9.9

A beam of 2 MeV γ-rays of intensity $10^8\,\text{m}^{-2}\,\text{s}^{-1}$ is incident on a 0.05 m-thick lead shield. Calculate the attenuated flux and the effective flux behind the shield. Using eqn (9.24), estimate the dose rate behind the lead shield. The value of $\mu(2\,\text{MeV})$ for lead is $51.8\,\text{m}^{-1}$. The build-up factors $B(\mu x)$ of lead for 2 MeV γ-rays are $B(2) = 1.76$ and $B(4) = 2.41$.

The value of $\mu x = 51.8 \times 0.05 = 2.59$. The attenuated flux F_a is therefore

$$F_a = F\exp(-\mu x) = 10^8 \exp(-2.59) = 7.5 \times 10^6\,\text{m}^{-2}\,\text{s}^{-1}$$

Linearly interpolating between the values of $B(\mu x)$ for $\mu x = 2$ and 4 gives $B(2.59) = 1.95$. So the effective flux F_{eff} from eqn (9.25) is

$$F_{\text{eff}} = B(2.59)\,F_a = 1.46 \times 10^7\,\text{m}^{-2}\,\text{s}^{-1}$$

Substituting this effective flux into eqn (9.24) gives

$$D_{\text{rate}} \approx 2 \times F_{\text{eff}}(10^6\,\text{m}^{-2}\,\text{s}^{-1}) \times E(\text{MeV}) = 2 \times 1.46 \times 10^1 \times 2 \approx 58\ \mu\text{Sv h}^{-1}$$

For neutrons, absorption cross sections are much higher at low energies ($E_n \lesssim 1\,\text{keV}$) than at high energies ($E_n \gtrsim 100\,\text{keV}$), so good moderation is required of the fast fission neutrons. Water is a very effective neutron shield and boron can be added to the water to improve the absorption of low energy neutrons. Concrete is also effective because of its large hydrogen atom density (1/4 that of water). It can be cast into the required shape and is strong so it provides most of the shielding of a reactor.

9.5 Fast reactors

As mentioned above, a chain reaction can be maintained predominantly by fast neutrons ($E_n > 100\,\text{keV}$) if the fuel is enriched. Making a fast reactor is, however, technically more challenging than making a thermal reactor since it has a much higher energy density in the reactor core, and there are relatively few fast reactors in operation. However, when the supply of uranium eventually becomes more limited, fast reactors are likely to become particularly attractive for conserving uranium stocks since the conversion of fissile material is generally much higher than in a conventional fission reactor. For conversion, the core is surrounded by fertile material and the emitted neutrons convert it to fissile material. For example, when fuelled by ^{235}U some of the fission neutrons are absorbed by ^{238}U and produce ^{239}Pu, which is fissile, via the reactions

$$\text{n} + {}^{238}\text{U} \rightarrow {}^{239}\text{U} \rightarrow {}^{239}\text{Np} \rightarrow {}^{239}\text{Pu} \tag{9.26}$$

where the last two reactions occur via β-decay.

Conversion can be sufficient to yield more fissile material than is used in the core. When this happens the reactor is called a breeder reactor and the utilization of uranium can be increased from ~2% to ~50%, an increase of a factor of ~25 in the size of the nuclear energy reserve. The details of breeding are described in Box 9.4. The proven conventional fossil and uranium (for thermal reactors) fuel reserves are ~4×10^{22} J and ~3×10^{21} J, respectively. Breeder reactors could therefore provide an enormous amount of energy, without any significant greenhouse gas emissions.

Box 9.4 Breeders

In a reactor core η neutrons are emitted on average for every neutron absorbed by the fissile fuel. When fuelled by ^{235}U some of the fission neutrons are absorbed by ^{238}U and produce ^{239}Pu via the reactions in eqn (9.26). The ^{239}Pu can be used as fuel and, in a thermal reactor fuelled by ^{235}U, this can increase the percentage of uranium used to produce energy to ~2%. This production of fissile material from non-fissile nuclei is called conversion, and is quantified by the conversion or breeding ratio C, which is the average number of fissile nuclei produced per fissile nucleus consumed, as discussed in Section 9.4.5.

The breeding ratio C is related to η by $C < (\eta - 1)$, since one neutron must be absorbed by a fissile nucleus to maintain the chain reaction, i.e. to keep the core critical. When $C > 1$, more fissile material is produced from fertile material than is consumed and the reactor is then called a breeder reactor. This increases the percentage of uranium used to produce power up to an economic limit of around 50%, representing a considerable increase in the potential amount of energy available compared with conventional nuclear fission reactors.

For breeding to be possible, $\eta > 2$. In practice, however, because of neutron leakage and neutron absorption by elements in the core other than the fuel, η must be greater than 2.2. For the fissile elements ^{233}U, ^{235}U, and ^{239}Pu, only ^{233}U can be used to breed with thermal neutrons, but for fast neutrons (>100 keV) all three fuels can be used. ^{233}U can be bred from ^{232}Th; the process is

$$\text{n} + {}^{232}\text{Th} \rightarrow {}^{233}\text{Th} \rightarrow {}^{233}\text{Pa} \rightarrow {}^{233}\text{U}$$

where the last two reactions occur via β-decay. Thorium is about three times as abundant as uranium. India has large reserves of thorium and has a nuclear power programme designed to utilize them. They are estimated at ~800 000 tonnes, which is a huge energy source since 1 tonne of thorium corresponds to 2.6 GWy.

A measure of how good a reactor is at breeding is the time it takes for the amount of fissile material in the reactor to double; this is called the doubling time t_D. The breeding gain is $(C-1)$, as one fissile nucleus is consumed for every C fissile nuclei produced. Consider a reactor operating at a constant power P_0 (GW) with an initial mass of fissile material M_i. If m kilograms of fissile material are consumed per day per gigawatt of output then the gain in fissile material per day will be $(C-1)mP_0 \, \text{kg} \, \text{d}^{-1}$ and the doubling time t_D equals $M_i/[(C-1)mP_0]$.

This estimate assumes that all the bred fuel is left in the reactor for the whole time. The reactor only needs an amount of fissile fuel M_i to produce P_0 for a certain neutron flux, and the excess could be removed and used in another reactor. As reactors are refuelled once or twice a year this mode of operation could be approached, in which case t_D becomes

$$t_D = M_i \ln 2/[(C-1)mP_0] \tag{9.27}$$

(see Exercise 9.20). For example, consider a breeder reactor containing ^{235}U and ^{238}U which produces 1 GW, equivalent to 0.3 GWe if the thermal efficiency is 30%. This consumes about 1 kg d^{-1} of ^{235}U. For an initial 1000 kg of ^{235}U and a breeding ratio of 1.1, the doubling time would be 27 y without removal of the bred fuel, and 19 y with removal.

In a fast reactor, to ensure that fission is maintained predominantly by fast neutrons, no moderating material (or very little) is used and the fuel is enriched. As a result the core is very compact. For a reasonable power output the energy density in the core must be high and this requires a coolant with excellent heat transfer properties. Also, the coolant must not moderate the neutrons, thereby favouring coolants not containing low atomic number nuclei. Sodium has been used in many designs. It has a high boiling point of 882 °C at 1 atmosphere, so the thermodynamic efficiency is high without requiring high pressures and a large pressure vessel. However, sodium is chemically very reactive and is also activated through neutron absorption, producing ^{24}Na, which has a 15 h half-life, and so the coolant loop must be heavily shielded. As a result the design of a reliable, economic sodium-cooled fast reactor has proved difficult.

Figure 9.10 shows a schematic design of a pool-type sodium-cooled fast breeder reactor. The core consists of a central region with fuel rods surrounded by rods containing fertile material. Since the sodium in the primary coolant loop becomes radioactive, there is a secondary sodium loop that is used to generate the steam for the turbine. Although the reactor can be located partially underground to reduce the necessity for heavy shielding, access to all of the primary coolant components for repairs and maintenance is difficult because they are immersed in the sodium pool.

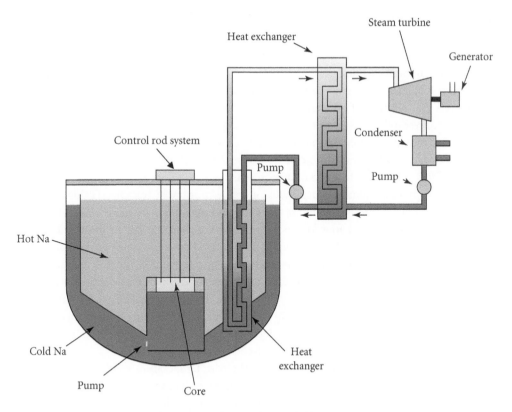

Fig. 9.10 A pool-type sodium cooled fast reactor.
Source: GIF.

9.6 **Present-day nuclear reactors**

Many reactor designs have been produced around the world, with different fuels, moderators, and coolants. The early prototype reactors, called generation I, were made during the 1950s and 1960s. Most of the reactors operating today are generation II reactors and were built during the 1970s and 1980s. The reactors operating in 2010 were PWRs (272), boiling water reactors (BWRs; 84), Russian graphite-moderated and water-cooled RBMK reactors (15), CANDU heavy water reactors (47), GCRs and AGRs (15), and fast reactors (2). Many light-water reactors in the USA were one-off designs, which led to duplication of effort and greater costs.

This prompted the development in the early 1990s of standardized advanced light-water reactor designs which are simpler, require fewer components and less piping, are easier to build, and include additional passive emergency cooling systems. (These are sometimes referred to as generation III or III+ designs.) Passive cooling depends on gravity or temperature differences, rather than pumps, and as a result it is expected to be much more reliable and closer to a fail-safe system. A few of these generation III reactors have now been built. The advantage of standardization in design is seen in France where, over nearly two decades,

34 0.9 GWe and 20 1.3 GWe nuclear plants have been built. They supply about ~75% of France's electricity.

By the end of the 1990s it was recognized that there was a need for an international effort to decide upon improved designs that addressed public concerns over safety and were more economic. In 2003, the Generation IV International Forum, representing 10 countries, announced the selection of six reactor technologies which would benefit from cooperative international research. These were chosen for being safe, clean, cost-effective, fuel-efficient, secure, and resistant to proliferation. Some of these systems use a closed fuel cycle, in which reprocessing is carried out on site and actinides (see Section 9.9) are recycled to minimize the amount of high-level waste that needs to be stored.

The very high temperature helium-cooled graphite-moderated reactor (HTGR), shown in Fig. 9.11, uses a once-through uranium fuel cycle and operates at a temperature of 1000 °C. This is considerably higher than water-cooled reactors, which are limited by pressure to a maximum of ~300 °C, and also higher than liquid metal cooled reactors, which are limited by corrosion to ~600 °C. A direct Brayton cycle gas turbine (see Section 3.10) could be used to give high thermal efficiency as shown in Fig. 9.11, or the reactor could be used for thermochemical hydrogen production. The high operating temperature allows the graphite to constantly anneal the damage caused by the fast neutrons and thereby avoid any build-up of stored energy. There is also a strong negative temperature coefficient of reactivity, and the chain reaction will stop if there is a loss of the helium coolant gas (loss-of-coolant accident or LOCA).

For a moderate power HTGR of 150 MWe, the core temperature would rise following a LOCA to 1600 °C—the rise eventually being limited by radiation cooling. The design of the fuel is aimed at preventing any significant loss of fission products at this temperature. The

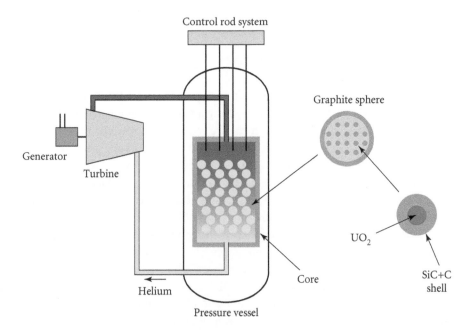

Fig. 9.11 Very high temperature gas cooled reactor.

fuel is in the form of small spheres of mixed fissile and fertile material contained in thin carbon and silicon carbide shells. As a result the reactor is called a pebble bed reactor. The relatively small power output makes the inventory of fission products low and would be further reduced by continuous refuelling. The uranium fuel would be quite highly enriched at ~8% and would not need to be reprocessed. The main safety concern is to avoid any air contact with the hot graphite, which could cause a fire. This form of fuel gives good fission product containment and surrounding the core with an outer containment building would provide additional safety, but would make the plant more expensive.

Another fast reactor design uses the natural convection of liquid Pb or liquid Pb-Bi to cool the core and is based on 40 years' experience of Pb-Bi cooling in Russian submarine reactors. There is also a sodium-cooled fast reactor design similar to the design discussed in Section 9.5.

An unusual technology forming part of the generation IV research is the molten salt reactor. The uranium fuel is dissolved in a sodium fluoride salt coolant which circulates through a graphite moderator. Fission products are removed continuously and the actinides are fully recycled, which reduces the high-level waste. The operating temperature is high enough for hydrogen production and there is passive cooling of the core, which improves safety. The other two systems selected are a supercritical water-cooled reactor and a very high temperature gas reactor.

Although mostly large reactors have been built up to now, some of the new designs are smaller. The latter better match the needs of less-developed countries. Smaller units also reduce the magnitude of the radioactivity inventory in a reactor. Prototypes of the generation IV reactors are planned for operation in 2020–2030. Nuclear power is one way to provide for low CO_2 energy and for greater national energy security by reducing the dependence on supplies of fossil fuel. The concern about the effect of fossil fuel emissions on global warming has renewed interest in nuclear power. Electricity generation using nuclear power has only slowly increased over the last two decades. The slow growth is due both to economic considerations and to public concern over safety, which was reinforced by the recent (2011) accident at Fukushima in Japan.

9.7 Safety of nuclear power

There are over 400 nuclear plants worldwide, most of them LWRs, operating at ~90% of maximum annual output. Operating experience has led to improved capacity as well as improved safety. However, there is public concern over the use of nuclear power, due in part to four serious accidents in commercial plants since the appearance of nuclear power in the 1950s. In 1952 there was a fire at a gas-cooled graphite-moderated reactor at Windscale in the UK. This was caused by the release of stored energy associated with the displacement of carbon atoms in the graphite by neutron irradiation (known as the Wigner effect), with a limited release of radiation into the atmosphere. In 1979 there was a LOCA in a PWR at Three Mile Island in Pennsylvania in the USA, caused by both mechanical and human failure, resulting in a 20% core meltdown but only a small release of radioactivity. In 1986 there was an uncontrolled reactor power increase in a water-cooled graphite-moderated RBMK reactor at Chernobyl in the Ukraine, causing a steam explosion and a huge release of a radioactivity.

The fourth accident occurred in 2011 at the Fukushima Daiichi site of six reactors, following an unprecedented massive earthquake (9.0 on the Moment scale) on 11 March off the east coast of Japan. Three reactors were operational at the time. A series of tsunamis breached the protective seawall around the reactors by several metres and flooded the site, which led to the loss of emergency power, and over the next few days several hydrogen explosions and fuel meltdowns occurred. There was a very large release of radioactivity, about 10 times less than that released at Chernobyl, which led to an evacuation of all people living within 20 km of the plant. On 16 December 2011 the reactors were declared stable, but it will take decades to clear up the contamination and decommission the reactors. The earthquake and tsunamis resulted in the loss of ~20 000 lives and massive damage in north-east Japan. Estimates of deaths from radiation-induced cancers are ~100–1000.

The outer containment building at the Three Mile Island PWR contained nearly the entire radioactivity released from the partial meltdown of the core. However, the costs of the clean-up were enormous, almost $1 billion, and the loss of public confidence in nuclear power, particularly in the USA, was considerable. The Chernobyl disaster was much worse since a significant fraction of the core inventory of fission products was released. The accident at Chernobyl was largely due to the unusual reactor conditions created for a safety test, coupled with procedures not being followed and an inherent weakness in the design of the RBMK reactor, in that certain parts of the reactor had a positive-feedback effect on the reactivity of the reactor when the amount of steam in the coolant water increased. (More steam reduced the average density and hence the neutron absorption of the water, which caused the reactivity to increase.)

About 50 people died as a direct result of the accident, most from high radiation doses. The radioactivity was carried great distances from the reactor. It is possible that, among the 600 000 people who received more significant radiation exposure, there might be 2/3 of a per cent increase in cancer mortality. There has been an increase in cases of thyroid cancer (most of which were curable) among children in the exposed areas of Ukraine, Belarus, and Russia. However, apart from this increase in thyroid cancer there has been no clear evidence from epidemiological studies for a radiation-induced increase in cancer deaths or mortality. Health, though, was significantly affected by the anxiety and the massive relocation caused by the accident. While the long-term effects of the Chernobyl disaster may not be as bad as had been feared, the effect on public confidence in nuclear power was significant and long-lasting.

Following the Chernobyl accident there has been increased international cooperation on safety and reactor design issues to reduce the chance of a similar accident. In the USA there have been about 3000 reactor years of operation, during which time there has been only one serious accident (at Three Mile Island). Based on probabilistic risk assessments, a similar frequency estimate of one core damage incident every 10 000 reactor years has been made. New reactor designs with both passive and active safety features will reduce this probability. There have also been improvements in operator training and in instilling the importance of safety at work. Being able to monitor the conditions in a reactor is also very important, since the operators can then be sure that conditions are satisfactory.

However, the Fukushima accident has again dented confidence in nuclear power. Although the increased risk of death by cancer, compared with that from other causes, in areas contaminated by radioactivity is very small (~0.1% as against 27%), the fear of ionizing radiation may be the greatest concern for the population in the affected areas (see Section 12.4). The accident has led to reviews of nuclear reactors around the world, and while these have reaffirmed that nuclear plants can be operated safely, the resulting recommendations on improving the defence of plants against natural disasters may well increase the capital costs of new reactors and affect the operation of some existing plants.

There is another aspect of reactor safety, and that is the ability of a reactor to withstand a terrorist attack. This has received particular attention following the 9/11 attacks. Fortunately, the strength of the construction in nuclear power plants would provide considerable protection from the effects of the crash of a hijacked aircraft or of a truck bomb. The spent fuel, which is initially stored on site, is in reinforced concrete pools and so is also protected. However, the security of nuclear power plants has been increased as a result of 9/11.

9.8 Economics of nuclear power

Nuclear reactor plants require large capital investment, and those operating today were originally state-owned or run by regulated utility monopolies. Under these arrangements the financial risks associated with capital expenditure could be more easily absorbed within a large state-protected framework (or borne by the consumers rather than by the suppliers, with a consequent reduction in the cost of capital), and projects with a long-term payback could be funded as part of the national infrastructure for long-term economic security. A competitive electricity supply market, however, tends to favour less capital-intensive energy sources and ones with shorter construction times. In the 1980s and 1990s in the USA, construction costs were higher than expected for nuclear plants and there were regulatory and political difficulties in obtaining site approval. During this period fossil fuel costs also decreased. All of these factors made nuclear power uncompetitive in the de-regulated economic power market in the USA at that time.

With fossil fuel prices now higher, nuclear power, with its relatively low operating costs, is close to being competitive with alternative fossil fuel sources for 'base-load' (high load factor) power. A group at MIT has recently (2009) updated its 2003 study of the economics and role of nuclear power in the USA. Table 9.7 shows their updated estimated relative costs for electricity produced by nuclear, coal, and gas power plants. The cost of borrowing capital is

Table 9.7 Relative costs of electricity in the USA

	Base case: 40-year plant life and capital for recovery, 85% lifetime capacity factor				
	Plant cost w/o interest ($/kW)	Fuel cost ($/MBtu)	Base case (¢/kWh)	w/ carbon charge $25/ tCO$_2$ (¢/kWh)	w/ same cost of capital (¢/kWh)
Nuclear	4000	0.67	8.4		6.6
Coal	2300	2.60	6.2	8.3	
Gas	850	7.00	6.5	7.4	

Source: MIT 2009.

higher for the nuclear base case than for coal or gas because of its poor track record for construction. Without this premium, nuclear would be competitive even without a carbon emissions tax; it is the high capital cost of a new plant and the high cost of borrowing that are the most significant factors. There is also regulatory risk for new designs, since it can take much longer to obtain approval, and delays are expensive. Furthermore, there is a technical risk with any first-of-a-kind project.

Another factor that can increase the chance of delays and hence costs is the experience of the construction workers—with relatively few reactors built during the last decade, skills can be lost and this makes delays more likely. An expansion would help drive down costs through the learning curve effect (see Section 12.3.3).

The Fukushima accident, besides its effect politically and on public opinion, is likely to result in an increase in the cost of nuclear plants due to the need for additional safety precautions. Such an increase would make it even harder for nuclear power to be competitive, without a significant carbon tax.

9.9 Environmental impact of nuclear power

The principal environmental advantage of nuclear power is the very small amount of associated CO_2 emissions, which makes it a very good energy source in light of global warming. The main environmental considerations are over the siting of nuclear reactors—in particular the seismology (as the Fukushima accident has demonstrated), meteorology, geology, risk of flooding, and population distribution in the area of the reactor. Meteorology is important since, in the event of an accident, the effect of the prevailing weather conditions on the dispersal of radioactivity has to be assessed. In addition the land requirements, the effect of thermal discharges to the environment, and the storage and disposal of waste have to be considered.

The waste generated in a nuclear reactor is classified into high-, intermediate-, and low-level waste. High-level waste is the main concern, and comprises ~25 tonnes of spent fuel per year for a 1 GWe PWR, which represents ~95% of the activity produced. It contains both transuranic and fission products: the transuranic waste contains long-lived isotopes of plutonium, americium, neptunium, and curium, which are all actinide elements, while after a few decades the fission products are mainly ^{90}Sr (and its daughter ^{90}Y) and ^{137}Cs, though ^{99}Te and ^{129}I are also quite abundant and potentially mobile.

The spent fuel is first stored for several years on site, to allow the intense short-lived activity to decay. On site, it is kept in storage pools to remove the heat and provide shielding of the radiation. The spent fuel can then be stored or reprocessed to recover the uranium and plutonium and the remainder immobilized by vitrification, in which the waste is incorporated into borosilicate glass. Another method is to incorporate the waste in natural stable mineral lattices. The spent fuel or vitrified waste can then be placed in a corrosion-resistant can and stored in an underground repository.

Caverns in dry stable rock formations or in salt deposits have been proposed for these repositories. Whether this high-level waste should be retrievable is a subject of debate. It would enable the waste to be reprocessed and so be a future source of fuel that might be more cost-effective than from mines. It could also be eventually disposed of as new technologies emerge in the future, such as accelerator-induced transmutation of the long-lived nuclides to shorter-lived ones. Sealing waste securely would protect it from terrorists but the main concern is whether the waste would remain well contained for thousands of years. In some sites there would be a risk in the long term that the waste could leak into the groundwater and contaminate drinking water. There is therefore a need to identify geologically stable sites. Deep boreholes have also been suggested as possible repositories, but the same issue arises. France plans to use a deep geological repository complex located in clay at Bure. The amount of waste generated during the early years of nuclear reactors, called legacy waste, is far greater than that generated by the current generation of reactors and also needs to be stored safely.

In the Oklo uranium mine in west Africa, a natural uranium reactor has been discovered. Two thousand million years ago the enrichment of the uranium was about 3%, because of the difference in the half-lives of ^{235}U (7×10^8 yr) and ^{238}U (5×10^9 yr). The presence of water in the ore provided the necessary moderation for the ore to become critical. The fission and transuranic products have remained within the ore, providing an example of long-term secure storage. However, the unavoidable question remains as to how certain we can be about the security of any proposed site.

9.10 Public opinion on nuclear power

The effect of the Fukushima accident in Japan in 2011 on public opinion has been significant. One year on from the accident, support for expanding nuclear power has decreased, but it is too early to gauge its long-term effect. A poll carried out about six months after the disaster found support had fallen in most countries with nuclear power, but had remained steady in the UK and USA, and was high in China and Pakistan. In Germany, where support has been ambivalent for some time, the government decided after Fukushima to close its nuclear programme by 2022, and Japan has indicated that it wishes to reduce its reliance on nuclear power.

An earlier survey in the USA, before Fukushima, had found that it was the relative cost of alternative energy sources that was the major factor in influencing people as to which source (nuclear, coal, oil, solar, or wind) should be developed. There was a similar finding in Europe, where the environment and cost were seen as important issues, with protection of the environment and the need to keep prices low the top priorities for energy policy. There was also a lack of trust among EU citizens as to whether the nuclear industry is open or provides accurate and sufficient information about radioactive waste.

There is also what is called the dread factor (see Section 12.4) associated with nuclear power, which causes the perception of risk to be much higher than the actual risk. Better communication with the public about the health risks may help, but there remains public concern about waste disposal and nuclear proliferation. Increased fossil fuel costs and energy security may be the most important factors that affect public opinion on nuclear power.

9.11 Outlook for nuclear power

The production of electricity from nuclear power rose steadily until 2004 and since then has been approximately constant at 2750 TWh per year. However, its share of the total world

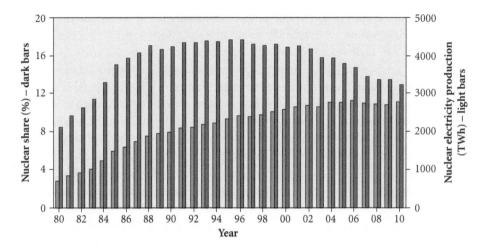

Fig. 9.12 Nuclear electricity production and share of total electricity production. *Source:* BP2011-EIA.

production has been decreasing since ~1996 as is seen in Fig. 9.12: 17.5% in 1996 and 13% in 2010. Table 9.8 gives some statistics on the actual and proposed number of reactors in countries that were in 2012 already generating a significant fraction of their electricity from nuclear power or proposed to increase their fraction.

In 2004, even before Fukushima, there were no plans to build any new reactors in the USA or in Europe—the proposed expansion was in China, India, Japan, Korea, and Russia. Germany had decided in 2000 to phase out its nuclear reactors and to promote the development of renewable energy sources. Although Germany changed its mind in 2009, the result of Fukushima was that Germany and Switzerland now plan to do without nuclear power and Italy has decided not to introduce it. However, in the UK, USA, and France the contribution from nuclear power is likely to be maintained, and the expansion in China, India, Korea, and Russia is still envisaged. The IEA were estimating prior to Fukushima that the installed nuclear reactor capacity could be 1200 GW (IEA2010). We will estimate the accessible potential by 2050 of nuclear power as ~1000 GWe of continuous output, which would contribute ~20% of the predicted global electricity demand.

The desire for both a secure energy supply and a reduction in global carbon emissions is a strong argument for nations to use nuclear power as a source of low-carbon electricity. However, the economic benefits are unattractive in the short term, except when oil and natural gas prices are high, and public support has been reduced as a result of Fukushima. More standardization and simplicity of design would improve the competitiveness of nuclear power, and the inclusion of both passive and active safety systems will reduce the likelihood of accidents. Public involvement in considering the option of nuclear power will be important in regaining public confidence.

The principal strategic considerations in favour of nuclear power are that it is a very compact fuel source, with a high energy density, that can be easily stockpiled. As there are diversified sources of supply of uranium, it is unlikely that the supply of fuel would be interrupted,

Table 9.8 Number of reactors and percentage of electricity generated in countries that use or propose to use nuclear power significantly

Country	Nuclear electricity generation 2010		Reactors operable (April 2012)		Reactors under construction (April 2012)	Reactors planned (April 2012)	Reactors proposed (April 2012)
	billion kWh	%	No.	MWe net	No.	No.	No.
Belgium	45.7	51.2	7	5943	0	0	0
Canada	85.5	15.1	17	12044	3	3	3
China	71.0	1.8	15	11881	26	51	120
France	410.1	74.1	58	63130	1	1	1
Germany	133.0	28.4	9	12003	0	0	0
India	20.5	2.9	20	4385	7	16	40
Japan	280.3	29.2	51	44642	2	10	5
Korea RO (South)	141.9	32.2	23	20787	3	6	0
Russia	159.4	17.1	33	24164	10	17	24
Spain	59.3	20.1	8	7448	0	0	0
Sweden	55.7	38.1	10	9399	0	0	0
Ukraine	83.95	48.1	15	13168	0	2	11
United Kingdom	56.9	15.7	17	10528	0	4	9
USA	807.1	19.6	104	101607	1	11	19
WORLD**	2630	13.8	435	372158	62	160	329

Source: WNA2011.

making it a very secure energy resource. However, the facilities used to enrich uranium for use in nuclear reactors could be used to provide highly enriched weapons-grade ^{235}U, so the development of nuclear power in any country has considerable political and security implications.

It will be more critical for any future reactor design to be very safe and economic than for it to be efficient. Public trust is essential and safe operation and waste disposal are paramount in any new reactor building programme. Reprocessing may not be required since fuel cost is not a major factor in reactor economics. The development of fast breeder reactors is less advanced than thermal reactors and the reserves of uranium are such that breeders are not essential in the short term.

Since any serious nuclear accident has global consequences, internationally approved designs and inspections would clearly separate the regulation from the management of nuclear plants and help build public confidence. France plans to use an advanced PWR for its next generation of reactors, which is a design based on existing PWRs that have had many reactor years of experience. There are advantages in the new proposed HTGR design, which is efficient, relatively small, and has good passive safety features, but there is much less operating experience with this type of reactor than with PWRs.

In some parts of the world the political and security issues arising from possible nuclear proliferation and terrorist attacks will be a significant additional concern over the development of nuclear power. However, without an expansion in nuclear power it will be much harder for the world to reduce its carbon emissions. How large a part it will play in providing low-carbon electricity and national energy security over the next 50 years is far from clear.

SUMMARY

- Nuclear power is a very compact source of low-carbon electricity, with 1 tonne of uranium equivalent in energy to 20 000 tonnes of coal.

- The known reserves of uranium are 4.7 Mt, equivalent to 85 years of operation at present consumption (IAEA), with an estimated additional 35 Mt recoverable. In PWRs, water is both the coolant and the moderator. The neutrons are moderated to thermal energies to allow the chain reaction to proceed with fuel only a few per cent enriched in ^{235}U.

- β-delayed neutrons allow sufficient time for mechanical control of the chain reaction through the insertion and withdrawal of control rods.

- The multiplication factor in the chain reaction is determined by the product $k_\infty = \varepsilon p f \eta$ and by the size of the reactor core.

- The initial power output in terms of the mass of ^{235}U (tonnes) and the flux ϕ in the core ($10^{18}\,\mathrm{m}^{-2}\,\mathrm{s}^{-1}$) is

$$P = 4.75\, M_{235}\, \phi \ \ \mathrm{GW}$$

- Fast reactors can utilize about 50% of the uranium fuel, rather than few percent in a thermal reactor, but are technically difficult and currently uneconomic.

- The safety of new nuclear reactors with passive safety features is predicted to be good, but public confidence in nuclear power has been affected by the Chernobyl and Fukushima accidents.

- Nuclear power is close to being competitive with fossil fuel plants for 'base-load' (high load factor) power, but its deployment is affected by its high capital cost.

- Underground depositories for nuclear waste are now proposed by several countries.

- Nuclear power provides good energy security and very low carbon emissions, which is particularly important in combating global warming. It is currently providing ~350 GWe and the accessible potential by 2050 is ~1000 GWe. But concerns about cost, safety, waste disposal, and nuclear proliferation make its future role uncertain.

FURTHER READING

Joskow, P.L. and Parsons, J.E. (2012). *The future of nuclear power after Fukushima*, CEEPR WP 2012-001. MIT Center for energy and environmental policy research, Cambridge, MA. Interesting report.

Lamarsh, J.R. and Baratta, A.J. (2001). *Introduction to nuclear engineering*. Prentice-Hall, Englewood Cliffs, NJ. Good detailed description of nuclear reactors.

Lilley, J. (2001). *Nuclear physics: principles and applications.* Wiley, New York. Clear discussion of the physics of nuclear reactors (Lilley2001).

Martin, R. (2012). *Superfuel: thorium, the green energy source of the future.* Palgrave Macmillan, New York. Description of an alternative source of nuclear fuel with less risk than uranium fuels.

Conversion coefficients for use in radiological protection against external radiation. *Annuals of the ICRP* **26** (3), 1 (ICRP).

WEB LINKS

www.iaea.org/ International Atomic Energy Agency (IAEA).

www.greenfacts.org/en/chernobyl/chernobyl-greenfacts.pdf A report on the legacy of the Chernobyl accident, 20 years afterwards.

bos.sagepub.com/content/67/5/27 Radiological and psychological consequences of the Fukushima accident.

www.globescan.com/news_archives/bbc2011_energy/ Poll about nuclear power, ~6 months after Fukushima.

www.gen-4.org/ GenIV International Forum on next-generation nuclear energy (GIF).

www.iea.org/publications/freepublications/publication/nuclear_roadmap.pdf (IEA2010).

web.mit.edu/nuclearpower/ (MIT2009).

www.bp.com/assets/bp_internet/globalbp/globalbp_uk_english/reports_and_publications/ statistical_energy_review_2011/STAGING/local_assets/spreadsheets/statistical_review_of_ world_energy_full_report_2011.xls Nuclear share (%) before 1990 from www.eia.gov (BP2011-EIA).

www.world-nuclear.org/info/reactors.html (WNA2011).

ieeexplore.ieee.org/stamp/stamp.jsp?tp=&arnumber=5462550 M. Ragheb and M. Khasawneh, Uranium fuel as a byproduct of phosphate fertilizer production (U-reserves).

LIST OF MAIN SYMBOLS

$b(A)$	binding energy per nucleon	k	multiplication constant
σ_x	cross section for reaction x	k_∞	multiplication constant for an infinite reactor
ϕ	neutron flux		
Σ_x	macroscopic cross section for reaction x	ε	fast fission factor
		p	resonance escape probability
λ	mean free path	f	thermal utilization factor
ν	average number of neutrons emitted per fission	η	number of neutrons produced per thermal neutron absorbed

ξ	logarithmic decrement	μ	linear attenuation coefficient
τ	mean lifetime	B	breeding ratio
Γ	width of excited state	GWe	gigawatts of electrical power
τ_{g}	generation time	n	number density

? EXERCISES

9.1 Calculate the energy released in the neutron-induced fission of ^{235}U into two nuclei with mass numbers 114 and 118 together with the emission of three neutrons.

9.2 How many tonnes of uranium fuel enriched to 3% in ^{235}U is equivalent to 100 000 barrels of oil?

9.3 Estimate the amount of natural uranium required annually to provide 10% of the primary global power consumption of 18 TW.

9.4 The global production of electricity by nuclear power in 2001 was about 2500 TWh. Estimate the annual consumption of uranium.

9.5* The isotope ^{157}Gd has a thermal neutron capture cross section of 255 000 b. Compare this value with (a) the cross-sectional area of the Gd nucleus and (b) the magnitude of $\pi\lambda^2$, where λ is the de Broglie wavelength of the thermal neutron. (A thermal neutron has a kinetic energy of 0.025 eV. $\lambda = h/p$, where p is the momentum of the neutron. The radius R of a nucleus with A nucleons $\approx 1.2 \times A^{1/3}$ fm.)

9.6* Derive the expression for the ratio r of the initial to the final neutron energy for a single scatter through an angle θ in the centre-of-mass frame:

$$r = E_1/E_0 = 1 - \alpha \sin^2(\theta/2)$$

where $\alpha = 4mM/(m+M)^2$.

9.7* Show that the relation for the logarithmic decrement $\xi \equiv \ln(E_1/E_0)$ is approximately equivalent to $\Delta E/E = -\xi$, where $\Delta E = E_0 - E_1$. The time Δt between scatterings is given by $\Delta t = 1/(u\Sigma_{\mathrm{s}})$, where Σ_{s} is the elastic scattering macroscopic cross section and u is the speed of the neutron. Show that the rate of loss of energy is approximately given by

$$\mathrm{d}E/\mathrm{d}t = -\xi u \Sigma_{\mathrm{s}} E$$

Assuming that Σ_{s} is independent of energy, show that the time t_{t} to thermalize a neutron of a few MeV is given by

$$t_{\mathrm{t}} \approx 2t/\xi$$

where t is the time between collisions at thermal energies, given by $t = 1/(u_{\mathrm{t}}\Sigma_{\mathrm{s}})$, with u_{t} the mean speed of thermal neutrons.

9.8* Neutrons scatter down from an energy interval A: $E \to E + \Delta E$ to B: $E - \Delta E \to E$, where $\Delta E = -\xi E$ is the average energy lost in scattering. The flux in A is ϕ and in B is $\phi' = \phi - (\partial\phi/\partial E)\Delta E$. By considering the rate of scattering and of absorption, show that

$$\Sigma_{\mathrm{s}}\phi = \Sigma_{\mathrm{a}}\phi' + \Sigma_{\mathrm{s}}\phi'$$

Hence show that

$$\phi_1 = \phi_h \exp\left(-\int_{E_1}^{E_h} \frac{\Sigma_a dE}{\xi E(\Sigma_s + \Sigma_a)}\right)$$

where ϕ_1 is the flux at E_1, and ϕ_h is the flux at E_h. The exponential term is the probability p that neutrons escape capture.

9.9* The diffusion equation for a single group of neutrons is

$$\nabla \cdot \boldsymbol{j} = -\partial n / \partial t$$

where $\boldsymbol{j} = -(\lambda/3)\nabla\phi$. The mean free path $\lambda = 1/\Sigma_s$. The production rate of neutrons per unit volume is the rate from induced fission minus the rate from absorption. So

$$\partial n / \partial t = (\nu\Sigma_f - \Sigma_a)\phi$$

For a spherical core the flux ϕ satisfies

$$(2/r)\partial\phi/\partial r + \partial^2\phi/\partial r^2 = -3\Sigma_s(\nu\Sigma_f - \Sigma_a)\phi \equiv -B^2\phi$$

Show that a solution to this equation is

$$\phi = (A/r)\sin Br$$

where A is a constant. For a spherical core of radius a, the flux ϕ is zero to a good approximation at $r=a$. In this single-group approximation, the neutron multiplication factor $k_\infty = \nu\Sigma_f/\Sigma_a$. Show that the smallest value of a that gives $\phi=0$ at $r=a$ is

$$a_c = \frac{\pi\delta}{\sqrt{3(k_\infty - 1)}}$$

where $\delta = 1/\sqrt{\Sigma_a\Sigma_s}$. Discuss the approximation $\phi=0$ at $r=a$.

9.10* Deduce that the critical size a_c of a reactor is of order $\delta/\sqrt{(k_\infty - 1)}$, where δ is the mean distance travelled by a neutron before it is absorbed, from equating the fraction of neutrons within δ of the surface of the core to the fractional loss of neutrons $(k_\infty - 1)/k_\infty$. Deduce that $\delta \sim 1/\sqrt{\Sigma_a\Sigma_s}$.

9.11 Consider a spherical reactor core containing a homogeneous mixture of graphite and uranium enriched to 2.0% in ^{235}U and with a ratio of graphite to uranium n_s/n_f of 600 (a) Find p, f, η, and k_∞ assuming $\varepsilon = 1$ (b) Use eqn (9.12) to calculate the critical radius a_c.

9.12* After the shutdown of a nuclear reactor, show that the number of ^{135}Xe nuclei n_p and of ^{135}I nuclei n_I satisfy

$$dn_p/dt = \lambda_I n_I - \lambda_p n_p \qquad dn_I/dt = -\lambda_I n_I$$

Show that the number of ^{135}Xe nuclei grows to a maximum ~ 11 h after shutdown. Neglect the equilibrium number of ^{135}Xe compared with the equilibrium number N_0 of ^{135}I present in the core prior to shutdown.

If the maximum amount of ^{135}Xe that can be compensated for by removing control rods is $0.2N_0$, how long is the reactor out of action after the shutdown? (For ^{135}I, $t_{1/2}=6.7$ h; for ^{135}Xe, $t_{1/2}=9.2$ h.)

9.13* A PWR plant releases radioactivity at rate R from a vent h metres above the ground while a wind of speed v is blowing. Downwind at a distance x from the plant the concentration χ of radioactivity is given by

$$\chi=(R/\pi v\sigma_y\sigma_z)\exp[-h^2/(2\sigma_z{}^2)]$$

where σ_y and σ_z, the horizontal and vertical dispersion coefficients, are functions of x.

Both σ_y and σ_z are very approximately proportional to x for $10^2<x<2\times10^3$ m in reasonably stable atmospheric conditions, and $\sigma_y=50$ m and $\sigma_z=20$ m at $x=10^3$ m.

The external γ-ray dose rate dH/dt received at x is given by

$$dH/dt=0.07\chi\langle E_\gamma\rangle\,\text{mSv s}^{-1}$$

where χ is in MBq m^{-3}, and $\langle E_\gamma\rangle$ is the average energy in MeV of the γ-rays emitted per disintegration.

The radioactive gas ^{135}Xe ($\langle E_\gamma\rangle=0.246$ MeV) is released for 10 minutes at a rate of 100 MBq s^{-1} from a vent 40 m above ground into a wind with $v=3$ m s^{-1}. (a) Calculate the total external γ-ray dose H received downwind at a distance of 10^3 m. (b)* At what distance would the dose be a maximum?

9.14 Neglecting the time for fast neutrons to thermalize, the generation time τ_g is given by $[u\Sigma_a(\text{core})]^{-1}$, where u is the mean speed of a thermal neutron. Estimate τ_g for a graphite-moderated core that has a thermal utilization factor f of 0.88.

9.15* (a) Consider a nuclear fission reactor in which it is assumed that neutrons are produced directly by fission and also indirectly by the β-decay of a single fission product. The concentration n of neutrons and the fission product C satisfy the equations

$$\frac{dn}{dt}=[k(1-\beta)-1]\frac{n}{\tau}+\lambda C \qquad \frac{dC}{dt}=\beta\frac{n}{\tau}-\lambda C$$

Explain the physical meaning of each of the four parameters k, β, τ, λ in the above equations and show that $k=1$ in the steady state.

(b) Consider a reactor transient in which the 'reactivity' $\rho=(k-1)/k$ is slightly positive. By considering solutions of the above equations of the form $n\propto\exp(mt)$ and $C\propto\exp(mt)$, show that m has two possible values, given by

$$m_1\approx-\left(\frac{\beta-\rho}{\tau}\right) \qquad m_2\approx\left(\frac{\lambda\rho}{\beta-\rho}\right)$$

Assume that $(\beta-\rho)/\tau\gg\lambda$ and $[(\beta-\rho)/\tau]^2\gg4\lambda\rho/\tau$. Typically, $\lambda=8\times10^{-2}$ s^{-1}, $\rho=2\times10^{-3}$, $\beta=6\times10^{-3}$, and $\tau=10^{-4}$ s.

9.16 The value of $\beta\tau_d$ for the delayed neutrons from the fissile nucleus ^{239}Pu is 0.0324. Calculate the time constant for control of a ^{239}Pu fuelled reactor when $k=1.0005$. Assume $\tau_g=1$ ms.

9.17 A PWR nuclear reactor is generating 1 GWe of electrical power. The reactor core contains 100 tonnes of uranium and the neutron flux density is 4×10^{17} neutrons s^{-1} m^{-2}. Estimate the enrichment in ^{235}U of the fuel.

9.18 (a) Estimate the ^{235}U enrichment required to give a burn-up of 60 GWday per tonne, if the fuel is replaced when the percentage of fissile material is reduced to 1.75%.

(b) * Show that the conversion ratio C_{th} can be expressed as $[(\eta_5/\eta_{fuel})-1]$, where η_5 is the value of η for pure ^{235}U and η_{fuel} that for the uranium fuel.

9.19 The fission yield of the nuclide ^{88}Kr ($t_{1/2}=2.79$ h) is 0.0364. Estimate the activity in the core from ^{88}Kr one day after a 2 GWe reactor is shut down. (Assume production of ^{88}Kr ceases on shut down.)

9.20 A beam of 3 MeV γ-rays of flux 10^9 m^{-2} s^{-1} is incident on a 1 m-thick water shield. The linear attenuation coefficient μ of water for 3 MeV γ-rays is 3.96 m^{-1}, and the build-up factor for the 1 m-thick shield is 3.55. Calculate (a) the effective γ-ray flux; (b) the effective dose rate D_{rate} (μSv h^{-1}).

9.21* For a breeder reactor show that if the excess bred fuel is removed and used in another reactor then the doubling time t_D is given by

$$t_D = M_i\ln2/[(C-1)mP_o]$$

A 2 GW breeder reactor contains ^{235}U surrounded by ^{232}Th. For an initial amount of ^{235}U of 1000 kg and a breeding gain of 0.05 to breed ^{233}U, calculate the doubling time with and without removal of the bred fuel.

9.22 (a) Estimate the volume of spent fuel generated annually by a 1 GWe PWR (without reprocessing). The fuel is uranium dioxide enriched to 3% in ^{235}U and the density of the fuel is 4×10^3 kg m^{-3}. (b) Reprocessing reduces the volume of waste by a factor of about 4. Estimate and comment on the volume of waste generated annually with reprocessing by PWRs with a total output of 0.9×10^6 GWh of electricity (~25% of the annual consumption in the USA).

9.23 Devise a demonstration (practical or computer) that illustrates how moderators increase the resonance escape probability in a reactor.

9.24 Discuss the relative merits of using carbon, heavy water (D$_2$O), or water as a moderator, given that the ratio of the average energy E_f of a neutron after impact to the energy E_i before impact is $E_f/E_i=(A^2+1)/(A+1)^2$, where A is the atomic mass number of the target nucleus.

9.25 Discuss the advantages of one of the proposed generation IV reactors compared with an existing PWR.

9.26 Summarize the main economic considerations that affect the choice of nuclear power compared with an alternative source of power.

9.27 Discuss *critically* the statement that the environmental risks associated with nuclear power are far less than those associated with global warming.

9.28 Does public opinion preclude the expansion of nuclear power?

9.29 Describe the options for the storage of nuclear waste; which one should be pursued?

9.30 Is nuclear power the only large-scale power source sufficiently developed that its widespread adoption can make a significant impact on averting global warming?

9.31 Is it better to build a single nuclear power plant or several smaller nuclear plants with the same total output?

10   Energy from fusion

→ **Introduction**

We have seen at the beginning of Chapter 9 that the increase in binding energy with mass number for light nuclei (Fig. 9.2) means that energy is released in fusion. The fusion of two light nuclei to form a heavier nucleus results in the release of energy, which is the source of energy in stars. In many stars, including our Sun, the energy results from a series of reactions that convert hydrogen into helium (the p–p chain), in which the initial step is the fusion of two hydrogen nuclei (protons) to form deuterium. This process is a weak interaction, as one of the protons changes to a neutron with the emission of a positron and a neutrino, and is the rate-determining step in the 'burning' of hydrogen.

In order for two protons to react they must have sufficient energy to overcome their mutual electrostatic (Coulomb) repulsion. The electrostatic potential energy of two protons is shown in Fig. 10.1(a) as a function of their separation. According to classical mechanics, protons require more kinetic energy than the Coulomb barrier (~700 keV) in order to fuse. However, quantum mechanical tunnelling enables the reaction to take place at the much lower proton energies that occur at the centre of the Sun (~1 keV). The protons have a Maxwellian energy distribution, $N(E) \propto E \exp[-E/(kT)]$, so the strong energy dependence means that the small high-energy tail of the distribution contributes most to the fusion reaction rate in the Sun (Fig. 10.1(b)).

The energy release in the fusion of hydrogen to form helium in the p–p chain is about 25 MeV. This is some 10^6 times larger than in a typical chemical reaction involving electrons. This is a similar yield to that in fission and again there is no release of greenhouse gases. However, the fusion of hydrogen is totally impractical for a fusion reactor since the reaction rate is too low, as it only involves the weak interaction. There are, though, other fusion reactions that occur via the strong interaction with a similar energy yield

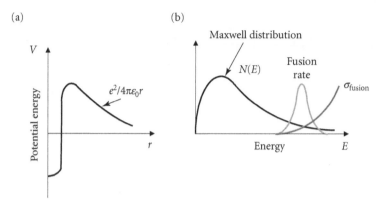

Fig. 10.1 (a) Coulomb barrier. (b) Fusion rate.

and with a reaction rate that can be made fast enough for a practical reactor. In particular, the fusion of deuterium and tritium to form helium and a neutron has the highest rate and releases 17.6 MeV. This has the potential to provide the world's energy needs for thousands of years. The energy density is enormous—only 100 kg of deuterium (D) and 150 kg of tritium (T) are required to produce about 1 GW of electricity for 1 year.

For the fusion of deuterium and tritium to occur at a sufficient rate the nuclei must have energies ~10 keV, corresponding to a temperature ~10^8 °C. At these temperatures deuterium and tritium are highly ionized and form a gas of charged particles called a plasma. No material can survive at such temperatures, so the plasma must be kept away from any containing walls. In the Sun the hydrogen plasma is so massive that it is contained by the huge gravitational force of attraction that balances the outward force resulting from the plasma pressure. In a terrestrial reactor the plasma must be contained by some other means. One method is to contain the plasma using magnetic fields, called magnetic confinement. Another is to compress the plasma by means of huge implosive forces, called inertial confinement. We will concentrate principally on magnetic confinement, as it is more developed.

10.1 Magnetic confinement

The earliest experiments were carried out in the USA in the late 1930s. The basic idea of using a toroidal plasma fusion reactor heated by radio-frequency (rf) waves was patented by Thompson and Blackman as early as 1946 in the UK. During the first decade of the Cold War, fusion research was conducted in secret. By the time it was declassified in 1958 the mirror, theta, and z-pinch machines and the stellarator had all been proposed. Progress in obtaining high-temperature confinement, however, was slow until a breakthrough was made in 1968 in the Soviet Union. This advance was made using a tokamak machine (Tamm, Sakharov), and since then most research has concentrated on this type of magnetic confinement. Although the plasma in a tokamak can be contained by the magnetic field, obtaining stability for significant periods has proved exceedingly difficult. It is only recently that the conditions necessary for a fusion reactor have been achieved, after over 50 years of research.

10.2 **D–T fusion reactor**

The fusion reaction of deuterium ($D \equiv {}^2H$) and tritium ($T \equiv {}^3H$) forms ${}^4He$ (α particle) plus a neutron:

$$D + T \rightarrow {}^4He + n + 17.6\,MeV \qquad (10.1)$$

(The nucleus of a deuterium atom is called a deuteron and that of a tritium atom a triton.) The 17.6 MeV of energy released is shared by the α particle (3.5 MeV) and the neutron (14.1 MeV). Deuterium is stable but tritium is unstable with a half-life of 12.3 years; however, tritium can be produced using the emitted neutrons. The energetic neutrons, which take ~80% of the fusion reaction energy, can be stopped in lithium external to the plasma containment vessel. The energy deposited would be used as a source of heat for a thermal power station and the neutrons would produce tritium in the lithium via the following two reactions:

$$n + {}^6Li \rightarrow T + \alpha + 4.78\,MeV \qquad (10.2)$$

$$n + {}^7Li \rightarrow T + \alpha + n - 2.87\,MeV \qquad (10.3)$$

The charged α particles from the D+T reaction provide a source of heat to maintain the plasma at a temperature of $10^8\,°C$, which must be kept away from the vessel walls. There is a loss of energy by the plasma through radiation, particle diffusion, and heat conduction to the walls. For steady-state operation this loss would be balanced by the heat generation in the bulk of the plasma from the α particles. There would also be some additional heating and fuel injection used to maintain optimum running conditions. The ejected particles would be diverted away from the main body of the plasma, and thereby exhaust the helium ('ash') and heat from the plasma.

A schematic design for a D–T fusion reactor is shown in Fig. 10.2. The fast neutrons (14.1 MeV) pass through the plasma containment wall and are stopped in a lithium blanket,

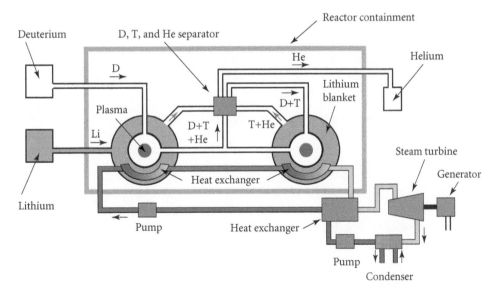

Fig. 10.2 Schematic layout of a fusion reactor.

where they produce tritium. The natural isotopic abundance of the lithium isotopes is 7.4% ^{6}Li and 92.6% ^{7}Li. Enriching the percentage of ^{6}Li enhances the tritium breeding ratio (TBR), which needs to be greater than unity for self-sufficiency. The ^{7}Li reaction produces a neutron which can make another T nucleus. Nuclei with significant (n, 2n) reactions, such as Be and Pb, could also be added to the lithium blanket to increase T production. Making the blanket material in the form of pellets aids the diffusion of the tritium, which is purged by helium gas.

The 14.1 MeV neutrons from the D+T reaction interact with the walls of the plasma chamber and other components. Their interaction can produce radioactive nuclei, i.e. activate the materials. A low-activation stainless steel has been developed in which Cr, W, and Ti replace Mo, Ni, and Nb, which allows the steel to be recycled after 50–100 years. Ceramic and fibre-composite materials are also being examined. The level of waste is significantly less than that from a fission reactor and there would not be a large nuclear waste disposal problem. Also, the tritium does not pose a serious radiation risk.

10.2.1 D–T fuel resources

The fuels for D–T fusion are deuterium and lithium. Deuterium is very plentiful as there are about 35 g m^{-3} of seawater (1 in 6500 hydrogen atoms are deuterium). The Earth's reserves of lithium are estimated to be greater than 10^6 tonnes. Less easily available are the amounts in the sea where there are more than 10^{11} tonnes, at a concentration of Li$^+$ of about 175 mg m^{-3}. The energy density of the fuel for a fusion reactor is enormous—only 100 kg of D and 150 kg of tritium, equivalent to ~300 kg of ^{6}Li, are required to produce ~1 GW of electricity for one year (Example 10.1). One tonne (~1200 litres) of oil has the same energy content as 50 mg of D plus 150 mg of ^{6}Li. The cost of deuterium is ~€1 per g and lithium is ~€20 per kg, so the contribution of the cost of the fuel to the cost of electricity is negligible at 0.001c per kWh. The reserves of lithium in the Earth alone would provide the world with 10 TWy of energy for ~1000 years.

EXAMPLE 10.1

What is the energy released when 100 kg of deuterium and 150 kg of tritium are consumed in 1 year in a fusion reactor? If the reactor's thermal plant is 35% efficient, find the average continuous electrical power output over the year.

100 kg of deuterium and 150 kg of tritium each correspond to 5×10^4 moles. The energy release E is therefore

$$E=5\times10^4\times6\times10^{23}\times17.6\times10^6=5.28\times10^{35}\text{ eV}$$
$$=5.28\times1.6\times10^{16}\text{J}=8.45\times10^{16}\text{J}$$

An amount of energy E in a thermal plant of efficiency ε can provide a continuous electrical power output P for a time t given by

$$\varepsilon E=Pt \quad \text{i.e. } P=\varepsilon E/t=0.35\times8.45\times10^{16}/(3.15\times10^7)=0.94\text{GW}$$

10.2.2 Characteristics of the D+T→⁴He+n reaction

The reaction rate for a flux ϕ of deuterons interacting with N tritons is $\phi N \sigma$ (eqn (9.2)). Hence the reactivity R, which is the reaction rate per unit volume, is given by

$$R = \phi n_T \sigma = n_D n_T \langle v\sigma \rangle \tag{10.4}$$

since the flux ϕ of deuterons equals $n_D v$, where n_D and n_T are the number of deuterons and tritons per unit volume, respectively. The average value $\langle v\sigma \rangle$ is the product of the relative velocity of deuterons and tritons and the fusion cross section σ, averaged over the velocity distributions of the particles. Figure 10.3 shows that $\langle v\sigma \rangle$ is of similar magnitude for the D+³He and D+D reactions

$$D + {}^3He \rightarrow {}^4He + p + 18.3\,MeV \tag{10.5}$$

$$D + D \rightarrow T + p + 4.0\,MeV \text{ or } \rightarrow {}^3He + n + 3.3\,MeV \tag{10.6}$$

but that over the temperature range corresponding to 10–20 keV, the D+T fusion reaction has the highest rate of reaction and the value of $\langle v\sigma \rangle$ for D+T is approximately proportional to T^2. This strong temperature dependence comes from the sharp increase with velocity of the probability of penetrating the Coulomb barrier.

For a D–T fusion reactor the fusion power P_{fusion} produced by the plasma is divided between the α particles, P_α (20%), and the neutrons, P_n (80%). The α particles are stopped within the plasma, so P_α heats the plasma. The neutrons escape the plasma and pass through the containment walls. Outside they are stopped in the lithium blanket, where tritium is produced, and thereby heat the fluid for the power turbine.

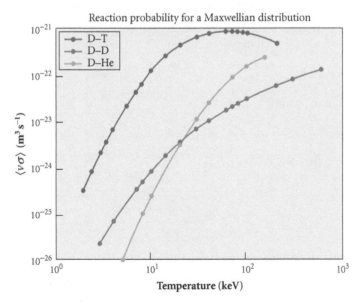

Fig. 10.3 Rate of fusion reactions as a function of temperature.
Source: ITER.

External power P_{ext} may be required in addition to P_α to compensate for losses from the plasma and to optimize the plasma conditions. The power loss P_{loss} arises from particles and heat that diffuse from the plasma centre to the walls of the container. In addition there is radiation (bremsstrahlung and synchrotron), together with line radiation from impurities emitted by the containing walls. The power loss P_{loss} is related to the energy containment time τ_E and the total plasma energy W by

$$P_{loss} = W/\tau_E \qquad (10.7)$$

where τ_E is the time for plasma energy to be lost to the walls when the plasma is in its operating state but with no energy input. (NB τ_E is not the same as the plasma duration (burn) time: P_α and P_{ext} heat the plasma during a burn so the duration time is longer.)

A quality factor Q is defined by

$$Q = P_{fusion}/P_{ext} \qquad (10.8)$$

Break-even is defined as $Q=1$ and ignition as $Q=\infty$. Ignition corresponds to $P_\alpha = P_{loss}$ and $P_{ext} = 0$. We show in Derivation 10.1, when deriving the Lawson criterion, that for a 50–50 mixture of D and T the requirement for ignition is

$$n\tau_E T \geq 3 \times 10^{21}\,\mathrm{m^{-3}\,s\,keV}\,(T\,\text{in keV}) \qquad (10.9)$$

Temperatures are often expressed in keV; i.e. T is equivalent to $kT/|e|$ so $10^8\,\mathrm{K} \sim 10^4\,\mathrm{eV}$ or $\sim 10\,\mathrm{keV}$.

For $T \sim 10\,\mathrm{keV}$ and $n \sim 10^{20}\,\mathrm{m^{-3}}$ the Lawson condition requires a confinement time $\tau_E \gtrsim 3\,\mathrm{s}$.

Derivation 10.1 Lawson criterion

For a D–T reactor the rate of change of the total plasma energy W is given by

$$dW/dt = P_\alpha + P_{ext} - P_{loss}$$

so under equilibrium conditions ($dW/dt = 0$)

$$P_{input} \equiv P_\alpha + P_{ext} = P_{loss}$$

where $P_{loss} = W/\tau_E$ (eqn (10.7)). Therefore

$$P_\alpha + P_{ext} = W/\tau_E \qquad (10.10)$$

Defining

$$f_\alpha \equiv P_\alpha/P_{fusion} = E_\alpha/E_{fusion} \qquad (10.11)$$

and, noting from eqn (10.8) that $P_{ext} \equiv P_{fusion}/Q$, eqn (10.10) is equivalent to

$$(f_\alpha + 1/Q)\,P_{fusion} = W/\tau_E \qquad (10.12)$$

The reaction rate per unit volume R, or reactivity, is given by eqn (10.4)

$$R = n_D n_T \langle v\sigma \rangle$$

Each reaction produces energy E_{fusion}, where

$$D + T \rightarrow \alpha + n + E_{fusion}$$

and

$$E_{fusion} = E_\alpha + E_n = 17.6 \, \text{MeV}$$

The power released is the reaction rate multiplied by the energy released per reaction, i.e.

$$P_{fusion} = RV \times E_{fusion} \tag{10.13}$$

where V is the volume of the plasma.

For a 50–50 mixture of D and T, $n_D = n_T = n/2$. Plasma neutrality means that the electron and ion densities are equal, so that $n_e = n$. Hence the total kinetic energy per unit volume W/V is given by

$$W/V = 2n\left(\tfrac{3}{2}kT\right) = 3nkT \tag{10.14}$$

assuming the electron and ions are in thermal equilibrium (i.e. at the same temperature). Using eqn (10.4) with $n_D = n_T = \tfrac{1}{2}n$ then, P_{fusion} (eqn (10.13)) becomes

$$P_{fusion} = \tfrac{1}{4}n^2 \langle v\sigma \rangle E_{fusion} V \tag{10.15}$$

Substituting eqns (10.14) and (10.15) into eqn (10.12) gives

$$\tfrac{1}{4}(f_\alpha + 1/Q)n^2 \langle v\sigma \rangle E_{fusion} = 3nkT/\tau_E$$

The break-even condition ($Q = 1$) can be written as

$$n\tau_E \geq 12kT/[(1+f_\alpha)\langle v\sigma \rangle E_{fusion}] \tag{10.16}$$

which is a function of T only.

In the operating temperature region of 10–20 keV the value of $\langle v\sigma \rangle$ is approximately proportional to T^2 (see Fig. 10.3), i.e.

$$\langle v\sigma \rangle \sim 1.2 \times 10^{-24} T^2 \, \text{m}^{-3} \, \text{s}^{-1} \tag{10.17}$$

where T is in keV so this condition can then be expressed as one on the triple product $n\tau_E T$. The requirement for ignition ($Q = \infty$) is

$$n\tau_E T \geq 3 \times 10^{21} \, \text{m}^{-3} \, \text{s} \, \text{keV} \quad (T \text{ in keV}) \tag{10.9}$$

This is known as the Lawson criterion. For $T \sim 10$ keV and $n \sim 10^{20}$ m^{-3}, $\tau_E \geq \sim 3$ s.

With no external heating, equilibrium requires $P_\alpha = P_{loss}$. However, as a result of diffusion, the loss will be greater than just that from bremsstrahlung due to electron–ion collisions. The temperature dependence of the reactivity R_b for bremsstrahlung interactions is $T^{1/2}$ which is weaker than that for fusion, which depends on T^2. This means that at sufficiently high temperatures $P_\alpha > P_b$. The operating temperature for D–T fusion plasmas is ~10 keV, which corresponds to a temperature of ~10^8 °C. At these temperatures deuterium and tritium are almost completely ionized and form a very hot plasma. This plasma has to be contained in the reactor, away from the walls.

10.3 Performance of tokamaks

Figure 10.4 shows the progress in obtaining better confinement since 1968, when the results from tokamak T3 were announced. The increase of $n\tau_E T$ in the following 40 years has been four orders of magnitude, with the Joint European Torus (JET) obtaining Q ~ 0.6 and τ_E ~ 1 s in 1997 with a D–T plasma.

Energy confinement in the TFR tokamak in France in the mid-1970s was 20 ms at $T = 1$ keV. We will assume that the plasma conditions were such that heat loss was mainly via particle diffusion and heat conduction. Then, as these are diffusive processes, the energy containment time τ_E is proportional to the square of the diameter (or to the cross-sectional area of the toroid). In TFR the area was 0.13 m² while in JET it was 8 m². This increase in area would then predict a confinement time of 1.2 s, which is close to that achieved by JET. However, to achieve plasma conditions where the energy transport is primarily diffusive and the plasma is stable has proved very difficult. To understand why, we need to consider the characteristics of plasmas.

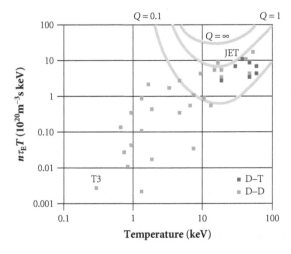

Fig. 10.4 Progress on increasing $n\tau_E T$ since the Russian tokamak T3 in 1968.
Source: Tok.

10.4 **Plasmas**

In the centre of the Sun, hydrogen is highly ionized and consists of a gas of charged particles, hydrogen ions, and electrons called a plasma. The long-range electrostatic interactions between the charged particles cause the plasma to exhibit long-range collective effects. (This can also occur when a gas is only partly ionized and contains neutral as well as charged particles.) The average kinetic energy of the particles is related to the temperature T of the gas by

$$\tfrac{1}{2}m\langle v^2 \rangle = \tfrac{3}{2}kT \qquad (10.18)$$

where m is the mass of a particle and v is its speed. A typical particle velocity is therefore $\sim (kT/m)^{1/2}$. For a plasma obtained from ionizing neutral atoms, there will be a roughly equal number of positive and negative charge carriers per unit volume.

Any time-independent external electric field is shielded from the interior of a plasma by a polarization layer of charge (known as a plasma sheath), the thickness of which is of order the Debye length,

$$\lambda_D = (\varepsilon_o kT/ne^2)^{1/2} \qquad (10.19)$$

and is typically a fraction of a millimetre for plasma machines. The typical distance that particles are separated in a plasma is of order

$$r_s \sim n^{-1/3} \qquad (10.20)$$

The relative kinetic energy of a pair of charged particle is about kT, so the separation r_c, when their electrostatic potential energy $U_c = e^2/(4\pi\varepsilon_0 r_c)$ equals their relative kinetic energy, is given by

$$r_c \sim e^2/(4\pi\varepsilon_o kT) \qquad (10.21)$$

If their typical distance apart is much greater than r_c, i.e. $r_s/r_c \gg 1$, then binary Coulomb interactions are rare. It is then a good approximation to treat such hot diffuse plasmas as weakly coupled.

The pressure p exerted by the plasma is given by

$$p = nkT \qquad (10.22)$$

EXAMPLE 10.2

A plasma has a temperature of $10\,\mathrm{keV}$ and a number density $n = 10^{20}\,\mathrm{m^{-3}}$, typical for fusion. Calculate the plasma pressure, the Debye length, and the ratio r_s/r_c.

Using eqn (10.22) for p and (10.19) for λ_D,

$$p = 10^{20} \times 10^4 \times 1.6 \times 10^{-19} = 1.6 \times 10^5\,\mathrm{Pa} = 1.6\,\mathrm{bar}$$

$$\lambda_D = (8.85 \times 10^{-12} \times 10^4 \times 1.6 \times 10^{-19} / [10^{20} (1.6 \times 10^{-19})^2])^{1/2} = 0.7 \times 10^{-4} = 0.07 \text{ mm}$$

The distances r_s and r_c are given by eqns (10.20) and (10.21) so

$$r_s / r_c = (4\pi\varepsilon_0 kT) n^{-1/3} / e^2 \equiv 4\pi\lambda_D^2 n^{2/3} = 4\pi (0.7 \times 10^{-4})^2 \, 10^{40/3} = 1.3 \times 10^6$$

Hence $r_s / r_c \gg 1$, so the plasma is weakly coupled.

In a magnetized weakly coupled plasma, the charged particles spiral quite freely along the magnetic field lines with a radius of gyration ρ. In a fusion plasma the dimensionless magnetization parameter

$$\delta \equiv \rho / L \tag{10.23}$$

where L is the typical distance over which the field is essentially constant, is very small. The magnetized plasma is therefore highly anisotropic, with motion parallel and perpendicular to the field being very different. Small perturbations can also set up an enormous variety of motions, both oscillatory and turbulent, some of which are unstable and lead to exponential growth. This has been a subject of intensive research for many years and not all forms of instability are yet understood.

For magnetic confinement, though, a necessary condition is that the tightly spiralling particles must remain within the magnetic fields generated externally and by their own motion. To understand how the magnetic field in a tokamak can confine a hot diffuse plasma, we need to consider first the motion of charged particles in magnetic and electric fields. We will consider later the stability of the confined plasma.

10.5 Charged-particle motion in *E* and *B* fields

The force experienced by a charged particle of mass m in an electric field E and a magnetic field B is the Lorentz force F, given by

$$F = qE + qv \times B \tag{10.24}$$

where q is the charge and v is the velocity of the particle. In a region where E is zero and B is uniform, then the particle experiences a constant force $v_\perp \times B$, where $v_\perp$ is the component of velocity perpendicular to B. This force causes the particle to gyrate about a magnetic field line at an angular frequency ω_c (the cyclotron frequency), in a helical path with a radius ρ (Larmor radius), where

$$\rho = mv_\perp / qB \tag{10.25}$$

and

$$\omega_c = qB / m \tag{10.26}$$

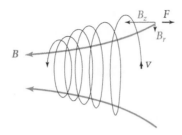

Fig. 10.5 Charged particle moving in a non-uniform axially symmetric magnetic field.

Since in a plasma at a temperature T the mean kinetic energy $\frac{1}{2}mv^2$ equals $\frac{3}{2}kT$, only one component of velocity is parallel to B, so $mv_\perp^2 = 2kT$. For a temperature of $10\,\text{keV}$ and a B field of 5 tesla (T), typical values in a fusion plasma are

Deuterons: $\rho = 4.1\,\text{mm}$ $\omega_c = 239 \times 10^6\,\text{rad s}^{-1} = 38\,\text{MHz}$

Electrons: $\rho = 67\,\mu\text{m}$ $\omega_c = 879 \times 10^9\,\text{rad s}^{-1} = 140\,\text{GHz}$

When a charged particle is moving in a non-uniform magnetic field it will spiral along the magnetic field lines provided that the field varies sufficiently slowly that B is essentially uniform over a distance of order ρ. A particle moving into a region of higher field (Fig. 10.5) experiences a decelerating force as there is a radial component of B.

The decelerating force is in the direction of varying field and is given by (see Exercise 10.5)

$$F_z = -\mu \partial B / \partial z \tag{10.27}$$

where

$$\mu \equiv \frac{1}{2} v_\perp \rho q \tag{10.28}$$

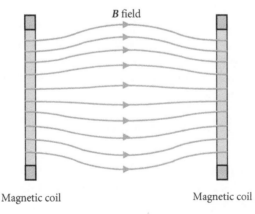

Fig. 10.6 A magnetic mirror configuration showing the magnetic field between the coils; the field extends beyond the coils.

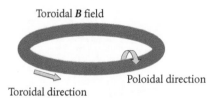

Fig. 10.7 Toroidal magnetic field.

is the magnetic moment associated with the spiralling motion of the charged particle. Since the particle does not experience a force in the azimuthal direction (since B is slowly varying), the angular momentum of the particle will be conserved, i.e. $mv_\perp\rho$ is constant, so as ρ decreases $v_\perp$ increases. Hence the magnetic moment μ is conserved; it is called an adiabatic invariant. The kinetic energy of the particle is also invariant as the magnetic force does no work on the particle, since it acts perpendicular to the motion of the particle.

The decelerating force causes the component of velocity v_z parallel to the field to decrease as the particle spirals along the field line, and it can reverse and so confine the particle's motion. A field configuration that confines particles in this way is called a magnetic mirror. Figure 10.6 shows the magnetic field configuration in a magnetic mirror, where the magnetic field is a minimum midway between the two coils ($z=0$) and a maximum close to the position of the coils ($z=\pm z_m$). Whether a particle is reflected or passes through a coil as it spirals along the field lines depends on the pitch angle α of the spiralling particle, given by

$$\tan\alpha = v_\perp(0)/v_z(0) \tag{10.29}$$

All particles with $\alpha > \alpha_m$, where

$$\sin^2\alpha_m = B(0)/B(z_m) \tag{10.30}$$

will be reflected and contained in the field between the coils (see Exercise 10.6).

Unfortunately, this magnetic mirror field configuration is not a good one for containing a plasma. Collisions occur within the plasma, which cause particles to change their pitch angle such that they are lost through the coils.

To avoid this loss we need to consider a field that has no ends, i.e. a toroidal field as illustrated in Fig. 10.7, which shows the toroidal and poloidal directions. However, the curvature and non-uniformity of such a field cause the spiralling motion of a charged particle to drift away from the toroidal magnetic field lines. This drift is explained in Derivation 10.2; it can be counteracted by adding a poloidal field.

Derivation 10.2 Drift in toroidal and poloidal fields

We can see why drift occurs in a toroidal field by considering a particle initially moving with speed v in a circle of radius ρ in a uniform magnetic field B. It then experiences a force F at right angles to B (see Fig. 10.8).

The particle, which starts on the left-hand side of Fig. 10.8, initially curves to the right. When it is moving in the z-direction the force F is in the opposite direction to the magnetic

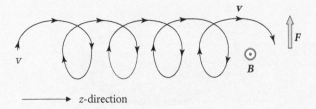

Fig. 10.8 The drift and track of a positively charged particle in a uniform *B* field when a force *F* perpendicular to the magnetic field is applied.

force $q\boldsymbol{v}\times\boldsymbol{B}$ so the curvature of its trajectory is less. When the particle has curved round and is travelling in the negative *z*-direction, the force *F* and the magnetic force are in the same direction and the curvature is greater. This causes a drift in the *z*-direction as shown.

The motion is equivalent to a circular motion plus a drift in the direction $\boldsymbol{F}\times\boldsymbol{B}$ with a drift velocity $v_{d\perp}$ such that the Lorentz force $qv_{d\perp}B$ balances the force *F*. The drift velocity produced by a constant force *F* is therefore

$$v_{d\perp} = F/qB$$

The magnetic field in a toroidal field (Fig. 10.9) varies in the radial direction, so a spiralling charged particle experiences a force, as in the magnetic mirror configuration, that is also in this direction, i.e. radially. The centre of the helical path is circular so the particle is on average accelerating radially. It therefore experiences a further radial force. The result is that ions and electrons will drift in the direction $\boldsymbol{F}\times\boldsymbol{B}$, i.e. vertically (Fig. 10.9(b)), but in opposite directions, which will set up a vertical *E* field and thus a radial $\boldsymbol{E}\times\boldsymbol{B}$ outward drift, leading to loss of confinement.

This drift can be counteracted by adding a poloidal component to the magnetic field producing a helical field, as illustrated in Fig. 10.10. The helical field lines connect the top and bottom of the torus, with the result that drifts cancel, no charge separation arises, and there is no $\boldsymbol{E}\times\boldsymbol{B}$ drift outwards.

The helical toroidal magnetic field can be produced by a series of twisting magnetic coils in a configuration called a stellarator, or by a tokamak. Fusion research with tokamaks is the most advanced, and we will now describe the main features of a tokamak.

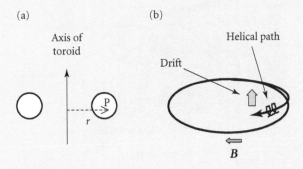

Fig. 10.9 (a) The drift of a charged particle in a toroidal magnetic field. (b) The magnetic field at a point P depends on $1/r$, where *r* is the distance from the axis of the toroid.

Helical magnetic field

Fig. 10.10 A helical field produced by adding a poloidal field to a toroidal field.

10.6 **Tokamaks**

In a tokamak, the toroidal magnetic field is produced by a series of toroidal field-coils (Fig. 10.11). For continuous operation in a fusion reactor, superconducting coils would be used to reduce the power requirements for generating the large fields of several tesla required. The vertical field-coils are used to position and shape the plasma. Negative feedback is used to maintain the plasma in position.

To create the poloidal field a current is induced in the plasma; the resulting field is helical, as shown in Figs. 10.10 and 10.11. The current is induced in the plasma by a large time-varying magnetic field that passes through the centre of the torus. The plasma acts like the secondary circuit in a transformer. If the flux linkage is Φ, then

$$V = -\mathrm{d}\Phi/\mathrm{d}t \quad \text{and} \quad I = V/R$$

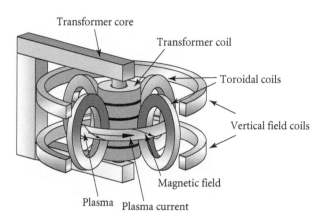

Transformer core

Transformer coil

Toroidal coils

Vertical field coils

Magnetic field

Plasma   Plasma current

Fig. 10.11 Principal components of a tokamak and the confining helical magnetic field.
Source: Tok.

where R is the resistance of the plasma. In JET at Culham in the UK, a voltage of 10 V was required to establish the current since initially R was relatively high. Once the plasma was fully ionized by ohmic heating then a voltage of 0.3 V was sufficient to drive a 3.5 MA current. JET had a flux swing of 34 V s and was able to sustain a current for about a minute ($[\Phi_{max} - \Phi_{min}] = -\int V \, dt$).

The need to apply an external field reduces the efficiency, or availability, of a fusion reactor. Fortunately a radial pressure gradient within the plasma causes a toroidal current to flow. As it is produced by the plasma itself, it is called a bootstrap current and reduces the amount of external current drive required to maintain the toroidal current.

In the early tokamak experiments the temperature of the plasma was raised by ohmic heating. However, the resistivity of the plasma decreases with increasing temperature (as $T^{-3/2}$), owing to the decrease in the cross section for collisions (see eqn (10.21)) and the time spent colliding. The ohmic heating from the current required to give the necessary poloidal field raised the temperature to only a few 10^7 °C. At this temperature the power losses due to radiation and transport balanced the power input by ohmic heating. So we need to raise the temperature of the plasma by other means.

One method of raising the temperature is to inject a beam of neutral deuterium atoms into the plasma. Charged deuterium ions are accelerated and then neutralized by passing them through a neutral deuterium gas. In the gas, charge exchange yields high-energy (~80 keV) neutral deuterium atoms that can cross magnetic field lines. Once inside the plasma they transfer energy to the plasma particles and also become ionized. (D or T would be used in a fusion reactor.)

Another technique to heat the plasma is to use high (radio) frequency electromagnetic waves, a technique called rf heating. The oscillating electric field in the rf wave resonates with the cyclotron motion of the plasma particles, either ions or electrons depending on the frequency. The frequencies are: 20–50 MHz for ion cyclotron resonance heating (ICRH) or 70–140 GHz for electron cyclotron resonance heating (ECRH). Ohmic, rf, and neutral beam heating are illustrated in Fig. 10.12. Both the neutral beam and rf can also be used to drive a current within the plasma. Using high speed pellets of frozen D, a high central plasma density can also be obtained.

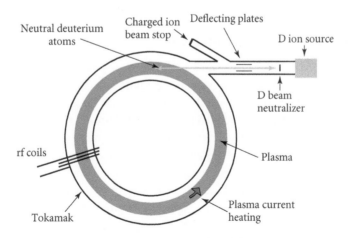

Fig. 10.12 Ohmic, rf, and neutral beam heating techniques.

10.7 Plasma confinement

Confining the energy within a hot plasma requires that the plasma is held in stable equilibrium. There is also the need to reduce the rate of loss of energy from the plasma to the containing walls. Instabilities can be caused by gradients in the pressure or in the current. The tokamak magnetic configuration can provide macroscopic stability under certain running conditions, i.e. stability over distances of order of the size of the tokamak. The rate of energy transport, however, depends in particular on turbulence occurring over smaller distances ($\sim \rho_{\text{Larmor}}$), which is only partially understood. We will first consider the conditions for equilibrium, then the stability of the equilibrium, before discussing the factors that limit the energy containment time.

10.7.1 Plasma equilibrium

In a hot plasma, collective effects among the plasma particles make the plasma act like a fluid over distances $\gg \rho_{\text{Larmor}}$, characterized by a temperature T, pressure p, and particle number density n. This is a useful approximation for low-frequency, large-scale phenomena. The current density j is given by the sum of the ion and electron current densities:

$$j = n_i e v_i - n_e e v_e = n e (\mathbf{v}_i - \mathbf{v}_e)$$

since $n_i = n_e = n$. The velocity v of the fluid is approximately equal to that of the ions, $\mathbf{v}_i$, because the fluid is restrained by the inertia of the ions ($m_i \gg m_e$). The relatively small difference in ion and electron velocities gives rise to the current. This single-fluid model for the motion of a plasma in a magnetic field is called magnetohydrodynamics (MHD).

At higher frequencies the relative motions of the electrons and ions have to be taken into account and a two-fluid model is used. For even higher frequencies the effect of the distribution of particle velocities must be considered. For example, certain departures from a Maxwellian velocity distribution can be unstable and give rise to an exponential growth of small-amplitude waves, although in tokamaks micro-instabilities arise from temperature and density gradients. We will consider the conditions for equilibrium only in the MHD approximation.

10.7.2 Plasma equilibrium in a tokamak

When a current is induced in the plasma in a tokamak, the current flows in a ring. A short section of the current can be approximated by a cylindrical section. The current gives rise to a very large poloidal magnetic field, which exerts an inward force on the plasma current—a pinch effect, as shown in Fig. 10.13. For equilibrium the plasma current must be large enough to generate sufficient magnetic force to balance the outward pressure of the plasma.

We can estimate the magnitude of this current I by finding the magnetic field B at the edge of the plasma, which has a radius R. The magnetic field B due to the plasma current is $\mu_0 I / 2\pi R$. Approximating the current to be all at a radius R, the force on a length l of plasma is IBl. The force acts inwards over an area $2\pi R l$ so the magnetic 'pressure' (i.e. force per unit area) p_B is

$$p_B \sim \mu_0 I^2 l / [(2\pi R)^2 l] = \mu_0 I^2 / (2\pi R)^2 = B^2 / \mu_0$$

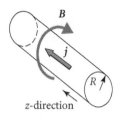

Fig. 10.13 z-pinch configuration.

The pressure of the plasma p in the cylindrical section is nkT or $NkTl/\pi R^2 l$, where N is the number of charged particles per unit length and l is the length of the plasma section. The magnitude of current required to balance p is therefore given by

$$I^2 \sim 4\pi NkT/\mu_0$$

The exact relationship is derived in Derivation 10.3 for the case of a cylindrical plasma, called a z-pinch, and is

$$I^2 = 16\pi NkT/\mu_0 \tag{10.31}$$

EXAMPLE 10.3

Estimate the current required to contain a tokamak plasma at a temperature of 10^8 K, a number density $n = 10^{20}$, and a cross-sectional area of 1 m^2.

The linear number density $N = 10^{20}$. For equilibrium we require a current

$$I = (16\pi 10^{20} \times 1.38 \times 10^{-23} \times 10^8 / 4\pi 10^{-7})^{1/2} = 2.35\,\text{MA}$$

This is the order of magnitude of the current required in a tokamak.

The exact relationship for the magnetic pressure in terms of the field B is $B^2/2\mu_0$. The ratio $\beta \equiv p/(B^2/2\mu_0)$ of the plasma pressure (which is equivalent to its kinetic energy density as $p = nkT$) to the magnetic pressure (which is the same as the magnetic energy density) is an important measure of the effect of the thermal motion on the confining ability of the magnetic field. This ratio, called the plasma beta, needs to be <1 for confinement, and typically $\beta \approx 0.1$.

Derivation 10.3 The z-pinch (Bennett pinch)

Consider a short section of plasma carrying a current I as shown in Fig. 10.13. Equilibrium requires balancing the force arising from the pressure gradient ∇p within the plasma with the magnetic force $j \times B$ arising from the plasma current density j in the magnetic field B. For equilibrium the resultant force on an element of the plasma is zero, i.e.

$$-\nabla p + j \times B = 0 \tag{10.32}$$

We can derive another relationship from Maxwell's equations assuming the plasma is sufficiently charge neutral that any electrical forces can be neglected. We will also assume, for low-frequency effects, that the displacement current $\partial D/\partial t$ can also be ignored. Then

$$\mu_0 j \approx \nabla \times B \tag{10.33}$$

This approximation, taking the conductivity as effectively infinite, is called ideal MHD.

Rather than considering the current flowing in a tokamak, we analyse the simpler z-pinch configuration shown in Fig. 10.13. The axial current density j gives rise to an azimuthal B field. The force per unit volume is $j \times B$. If the plasma pressure is p then

$$\mu_0 \nabla p = (\nabla \times B) \times B \quad (\text{since} \nabla \times B = \mu_0 j) \tag{10.34}$$

$$\mu_0 \, dp/dr = -(B/r)d(rB)/dr$$

$$\mu_0 \int r^2 (dp/dr)dr = -\int (rB)d(rB)$$

$$\mu_0 \left[r^2 p \right] - 2\mu_0 \int rp\,dr = -\frac{1}{2}\left[(rB)^2 \right] \tag{10.35}$$

Since $p(R)=0$ at the boundary of the axial plasma current, the first term of eqn (10.35) is zero. Since $p = (n_+ + n_-)kT$ and $n_+ = n_- = n$, we have

$$2\mu_0 \int rp\,dr = 2\mu_0 \, kT \int 2nr\,dr = \frac{1}{2}\left[(rB)^2 \right]_{r=R}$$

Defining the total number of particles per unit length $N = \int 2\pi nr\,dr$, we can write

$$2\mu_0 \, NkT = \frac{1}{2}\pi \left[(rB)^2 \right]_{r=R}$$

From $\nabla \times B = \mu_0 j$, we have $(1/r)d(rB)/dr = \mu_0 j$ and

$$[rB]_{r=R} = (\mu_0/2\pi)\int 2\pi jr\,dr = \mu_0 \, I/(2\pi)$$

Finally, we obtain the Bennett pinch relation, eqn 10.31:

$$I^2 = 16\pi NkT/\mu_0$$

10.7.3 Plasma stability

In a z-pinch configuration the plasma is not contained at the ends, but using a toroidal field, as in a tokamak, avoids this problem. However, while the z-pinch is in equilibrium it is not a stable equilibrium—it is unstable to sausage and kink instabilities, which are illustrated in Fig. 10.14.

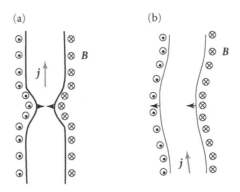

Fig. 10.14 (a) Sausage instability. (b) Kink instability.

In the sausage mode, Fig. 10.14(a), a small radial contraction of the plasma current causes the azimuthal magnetic field and the radially inward force to both increase, leading to further radial contraction. The volume of the plasma can remain constant by bulging out so there is no compensating force and the plasma column is highly unstable. Likewise in the kink mode, Fig. 10.14(b), the magnetic field increases on the inward side, thereby increasing the force and the magnitude of the kink. The sausage instability is also seen in the break-up of a thin stream of water falling from a tap. Surface tension (Rayleigh) waves grow in amplitude as the jet falls and lead to the formation of droplets.

In a tokamak the toroidal component of magnetic field tends to stabilize the plasma to both sausage and kink instabilities. How the sausage instability is stabilized is explained in Derivation 10.4. The tokamak safety factor q is defined as the ratio of toroidal to poloidal turns of the magnetic field and is close to unity near the centre of the plasma. The variation of q with distance from the plasma centre $(S=(r/q)\,dq/dr)$, called the magnetic shear, is an important parameter in determining the transport of energy and particles.

Derivation 10.4 Stabilization of the sausage instability in a tokamak

Consider a cylindrical section of plasma carrying a current I as in Fig. 10.13, but with a magnetic field B_z along the axis. The magnetic pressure p_B equals $B^2/2\mu_0$. The azimuthal component B_θ produces an inward pressure on the cylindrical plasma of magnitude $B_\theta^2/2\mu_0$. In the sausage instability p_B increases as the radius a of the plasma decreases. This is because $B_\theta=\mu_0 I/(2\pi a)$, which increases as a gets smaller.

The axial field B_z produces an outward pressure on the plasma. A contraction of the plasma's radius causes B_z to increase as currents are induced in the plasma opposing the change. In the MHD approximation the plasma acts like a perfect conductor. In this limit the induced currents exactly maintain the magnetic flux through a section perpendicular to the axis of the plasma. The flux Φ equals the field multiplied by the cross-sectional area, i.e. $\Phi=B_z\pi a^2$. The current I is constant so both $B_z a^2$ and $B_\theta a$ are constant when the plasma radius changes. So $\partial(B_z a^2)/\partial a$ and $\partial(B_\theta a)\partial a$ are zero, i.e.

$$\partial B_\theta/\partial a=-B_\theta/a \quad \text{and} \quad \partial B_z/\partial a=-2B_z/a \tag{10.36}$$

For stability we want the inward magnetic pressure p_B to increase if a increases. The pressure p_B can be expressed as

$$p_B = (B_\theta^2 - B_z^2)/2\mu_0 \qquad (10.37)$$

Differentiating p_B with respect to a and using eqn (10.36) gives

$$\partial p_B/\partial a = (B_\theta\, \partial B_\theta/\partial a - B_z\, \partial B_z/\partial a)/\mu_0 = (2B_z^2 - B_\theta^2)/a\mu_0 \qquad (10.38)$$

This expression is positive provided

$$B_z^2 > B_\theta^2/2 \qquad (10.39)$$

i.e. the plasma is stable against a sausage instability provided the inequality (10.39) is satisfied.

10.7.4 Energy confinement

The energy confinement time τ_E is determined by the time taken for particles and energy (both kinetic and radiation) to be transported from the hot plasma to the outside wall. A typical deuteron in a plasma at a temperature of $10\,\text{keV}$ has a velocity of $\sim\!1000\,\text{km s}^{-1}$ so in a pulse of a few seconds duration the deuteron travels a great distance. Plasma particles spiralling along a magnetic field will scatter and transfer energy to other particles over a distance of about ρ, the Larmor radius. This diffusion process, called classical transport, will have a diffusion coefficient of order $\rho^2/\tau_\perp$, where $\tau_\perp$ is the characteristic time between collisions. There is a larger contribution (called neoclassical transport) from trapped particles. Particles with only a small component of velocity in the direction of the helical $\boldsymbol{B}$ field in the tokamak can be reflected through the mirror effect as they move to a smaller radius where the field is greater as it depends on r^{-1} (see Fig. 10.15).

The characteristic size of these banana-shaped trapped orbits is greater than ρ and so they contribute more to diffusion losses. However, transport is observed in many operating configurations to be much faster where it is caused predominantly by micro- and macro-turbulence.

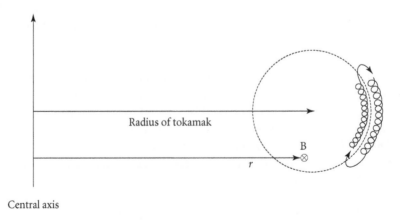

Fig. 10.15 Trapped particle in a banana orbit.

The control of macro-turbulence, involving disturbances over distances of the order 0.1 m to 1 m, has been achieved in tokamaks largely by choosing particular operating conditions. Micro-turbulence involves fluctuations over distances of order 0.1 mm to 10 mm (ρ_{Larmor}), due to density and temperature gradients. These fluctuations form unstable waves that transport energy across field lines from the hot centre to the cooler edge of the plasma and then to the containing wall. The process is not well understood and much research effort is being directed towards modelling turbulence.

A significant advance in improving confinement came with the discovery of an H-mode (high confinement mode) in the ASDEX tokamak in Germany. It was associated with the use of a divertor in which the magnetic field is altered to keep the main interactions of the plasma with any material away from the central plasma region.

10.8 Divertor tokamaks

Improved performance was obtained in Russia in the early tokamak experiments by introducing an artificial boundary, known as a limiter. It was used to define the radial extent of the plasma, as illustrated in Fig. 10.16. Diffusing particles, once they pass the last closed flux surface, rapidly spiral along the field lines and deposit their energy on the limiter. The limiter can be cleaner, have a higher melting point than the containing wall, and also have a lower Z. The latter is important as high Z impurity atoms in the plasma cause a high radiative loss through line radiation. Low Z atoms are fully ionized in the hot centre of the plasma and can only radiate by bremsstrahlung, which is much weaker than line radiation. However, it is difficult to keep the deposited power density low enough to avoid significant melting or sublimation of the limiter material.

By elongating the field vertically in a tokamak it is possible to form a cross-over, called a separatrix, in the magnetic field configuration near the wall. This configuration diverts the plasma away from the main central region onto divertor plates, as shown in Fig. 10.17.

The plasma is incident over a larger area which helps reduce the power density. In addition, gas is added near the plates to increase the radiative loss and, through charge exchange, neutralize the energetic charged particles. Both the radiation and the neutral particles spread the heat load. The temperature of the plasma must be sufficiently low, of order eVs, for the gas, such as nitrogen, to retain some of its atomic electrons, and thereby emit line radiation. If the temperature is too high, the gas atoms are nearly fully ionized and can only radiate through bremsstrahlung, a much weaker process than line radiation.

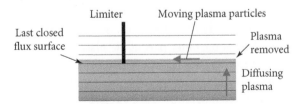

Fig. 10.16 Limiter in a tokamak.
Source: JETscience.

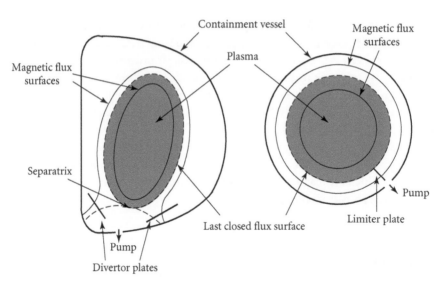

Fig. 10.17 Divertor and limiter tokamak configurations.

The divertor needs to be made from a low Z material so that any evaporated atoms are fully ionized in the hot plasma; carbon, with its very high sublimation temperature of 3825 °C, and beryllium have been used.

10.8.1 H-mode confinement

When sufficient plasma heating was supplied in the ASDEX divertor tokamak, a high confinement mode was found with τ_E approximately twice as large as in the normal configuration (low confinement L-mode). The H-mode develops as turbulence at the plasma edge is suppressed, which gives rise to a transport barrier. The pressure distribution in different modes is shown in Fig. 10.18. The L-mode can be affected by periodic disruptions, called sawtooth oscillations, which reduce the stored energy but expel impurities from the plasma core. These are associated with the formation of magnetic islands in the field distribution in which magnetic energy is converted to kinetic energy. (This process, called magnetic reconnection, is important in solar flares.) They can be stabilized by additional plasma heating.

The H-mode shows a sharp pressure drop near the edge of the plasma, and over an extended region transport is close to neoclassical (see Section 10.8.4). Above a critical pressure gradient MHD instabilities develop, giving edge localized modes (ELM). These modes reduce the edge pressure periodically and expel particles and energy from the plasma.

Improved confinement can also be achieved by generating a radial current distribution in the plasma to give negative magnetic shear; this occurs when the centre of the plasma has less current flowing than in the outer part of the plasma. Negative magnetic shear can reduce ion turbulent transport and set up an internal transport barrier (ITB), resulting in higher central temperatures. Feedback can also be used to control instabilities, for example by using ECRH to change the temperature and therefore the resistivity of the plasma, and hence the current flow.

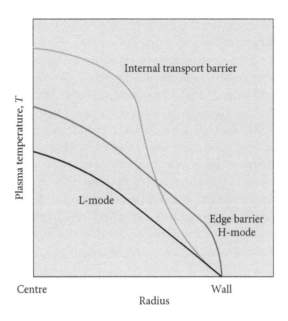

Fig. 10.18 Radial pressure distribution for different modes in a tokamak.
Source: Tok.

10.9 **Outlook for controlled fusion**

The successful culmination of the experiments at JET in 1997, when fusion with a 50–50 D–T plasma produced 16 MW of power corresponding to a Q ~ 0.6, made the prospect of a commercial fusion reactor by 2050 much more feasible. The H-mode provides a fusion plasma that can be confined and from which the output heat and particles can be extracted. However, considerable research is still required on the transport problem and on the materials to be used, particularly for the divertor, and on avoiding excessive heat loads from the ELM instabilities.

These problems will be addressed in the International Thermonuclear Experimental Reactor (ITER), which will be a tokamak designed to give Q ~ 10 and provide data upon which a demonstration reactor (DEMO) can be built. ITER will produce significant power under quasi-stationary conditions, investigate α-particle heating and particle wall interactions, and test the tritium breeder lithium blanket design. An international collaboration of China, the EU, Switzerland, India, Japan, South Korea, Russia, and the USA is involved. ITER is planned to have an inductive drive that will last for 8 minutes. It will also have auxiliary current drives that could extend operation up to 30 minutes. The fusion power output is planned to be 500 MW. In particular ITER will provide information on the lifetime of the torus wall components. It is currently under construction in Cadarache near Aix-en-Provence in southern France, with first plasma planned for 2020. The tokamak will be nearly 30 m tall and weigh 23 000 tonnes, with a plasma volume of 830 m³. ITER will use superconducting magnets to reduce energy consumption. The largest currently operating tokamaks, JET in the UK and JT-60 in Japan, both have plasma volumes of 100 m³.

The use of graphite in the divertor, although it has a high sublimation temperature, shows a high erosion rate—both from sputtering and also from chemical attack to form hydrocarbons.

Graphite also traps tritium and this loss may be serious. Using tungsten, with a melting point of 3695 °C, may be a remedy, as below that temperature it has a very low erosion rate and does not trap tritium. Carbon fibre composites and ceramics are also being considered.

In a fusion reactor neutron bombardment can deposit ~2 MW m^{-2} and cause transmutation of structural materials. This can give rise to activation and embrittlement due to α production. It can also cause lattice defects that can weaken the material. High temperatures, though, can cause partial annealing. Regular replacement of certain wall components will be necessary; the radioactivity within the torus will require remote operation, which has been pioneered at JET.

Controlled fusion offers an enormous source of power with essentially no greenhouse emissions, and the progress that has been achieved, as shown by the increase of four orders of magnitude in the triple product $n\tau T$ over the past 40 years, justifies a continued investment in fusion research. Research is also ongoing into inertial confinement fusion; this method is described in Box 10.1. Though a commercial reactor may still be several decades from fruition, the long-term benefit to future society cannot be overstated.

Box 10.1 Inertial confinement fusion (ICF)

Principle of ICF

Another approach to obtaining controlled fusion is to use lasers to heat up small spheres of deuterium and tritium to very high temperatures and pressures. By focusing the laser light so that the spheres are uniformly illuminated, it is predicted that the density can be raised sufficiently for fusion to take place. The process is illustrated in Fig. 10.19.

When pulses of laser light strike the surface of the D–T spheres a plasma is created. The hot plasma expands rapidly into the vacuum surrounding the spheres. This expansion causes a reaction pressure on the surface of the spheres (Newton's third law) that compresses them to very small radii. This compression raises the density by a large factor and initiates fusion. If the final radius is r_f then the plasma confinement time can be estimated as the time for the sphere's volume to double. The expansion is limited to that of the speed u_s of sound in the plasma so the confinement time τ_E is given by

$$\tau_E \sim r_f/4u_s \tag{10.40}$$

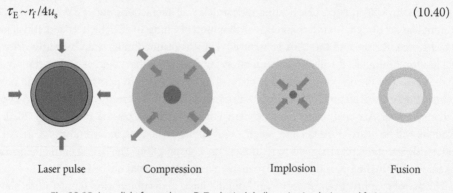

| Laser pulse | Compression | Implosion | Fusion |

Fig. 10.19 Laser light focused on a D–T spherical shell causing implosion and fusion.

The plasma is confined by its own inertia, hence the name inertial confinement fusion.

A particular challenge is the existence of a hydrodynamic instability known as the Rayleigh–Taylor instability. This instability occurs when two fluids in contact have density and pressure gradients of opposite sign. An example would be a film of water floating on a layer of oil. The pressure in the oil layer would be higher than in the film of water, as the weight of water acts on the oil. However, the oil has a lower density than water. The result is that the layers are unstable and the oil would rise on top of the water if the layers were disturbed.

A similar situation occurs in the implosion of the D–T spheres. The density of the sphere is higher than the plasma on its surface caused by the laser pulse. But the pressure in the plasma is much higher than in the sphere. As a result any imperfections in the spherical symmetry of the sphere or in the illumination will grow in size and can cause the sphere to break up before it has been fully compressed.

ICF reactors

In magnetic confinement fusion (MCF), the plasma has $n \sim 10^{20}$, so the confinement time τ_E required to satisfy the criterion for break-even (eqn (10.16)) is greater than ½ second. For ICF, the initial density of the D–T fuel is $\sim 200 \, \text{kg m}^{-3}$. The implosion of the shell raises the pressure and density enormously. At the very centre the density is expected to be $\sim 10^5 \, \text{kg m}^{-3}$. This density corresponds to $n \sim 10^{31}$. For break-even at a temperature of 10 keV the requirement on the product $n\tau_E$ is given by eqn (10.16) as

$$n\tau_E > 5 \times 10^{19} \, \text{m}^{-3} \, \text{s}$$

so the confinement time τ_E in ICF is correspondingly much smaller and only $\tau_E > 5 \times 10^{-12}$ second is required for break-even.

In ICF τ_E is given by eqn (10.40) so the radius r_f of the very dense core of the plasma, where ignition occurs, must be greater than

$$r_F \gtrsim 4\tau_E u_s$$

The speed of sound is about $\sim 500 \, \text{km s}^{-1}$, so the initial burn would then occur within a sphere of radius $\sim 10 \, \mu\text{m}$. This burn would heat up the remaining fuel to fusion.

EXAMPLE 10.4

Calculate how many D–T spheres would be required to fuse per second to provide energy for a 1 GWe power station. The radius of the D–T volume is 2 mm. Estimate the maximum Q.

The density of the fuel is $\sim 200 \, \text{kg m}^{-3}$ so the amount is 6.7 mg. A mole of D–T is 5 g so the total number of D–T pairs is 8.0×10^{20}. When these fuse they release 17.6 MeV per D–T fusion, 2.25 GJ in total. A 1 GWe power station requires ~ 3 GW thermal power, so a minimum of about 1.3 spheres per second would be required.

The energy stored in the compressed plasma at 10 keV is $6NkT \sim 7.68$ MJ as there are two electrons per D–T pair. The maximum Q is therefore ~290. In practice, Q would be much lower because of losses.

Outlook for ICF

Laser-induced inertial confinement fusion is being actively pursued in several laboratories around the world. At the National Ignition Facility (NIF), which became operational in 2009 at Lawrence Livermore National Laboratory in California, there are 192 neodymium glass lasers which can deliver 1.8×10^6 J pulses of energy at a power of 400×10^{12} W. This energy can be focused directly onto millimetre-size spheres at the centre of the 10 m target chamber or indirectly via an outer container called a hohlraum. In this indirect approach, the D–T sphere is enclosed in a gold container. The laser light vaporizes the gold into a plasma which radiates X-rays strongly. These X-rays are then absorbed in the surface of the D–T sphere and cause it to implode. This indirect method is predicted to give good uniformity. NIF is hoping to achieve ignition by the end of 2012 but progress has been slower than expected.

One of the most difficult tasks facing ICF is to design a device, called a driver, that can produce MJ pulses of energy of duration of about a nanosecond, at a repetition rate of 1–10 Hz. Good efficiency is also required and one alternative to lasers under consideration is to use particle accelerators. An interim report on ICF by the US National Academy of Sciences in 2012 concluded that it was unclear yet what the preferred driver for ICF was, but that NIF should play a major role in assessing the feasibility of ICF. Although there are still problems to solve in ICF, the potential benefit to mankind from harnessing fusion power argues for pursuing both MCF and ICF research.

SUMMARY

- Fusion has a very high energy density: 100 kg of deuterium and 150 kg of tritium would produce 1 GWe for one year. Tritium is produced by neutron reactions in lithium.

- Resources of lithium and deuterium are sufficient to provide the world with 10 TWy of energy annually for ~1000 years.

- Fusion requires plasma at a temperature of ~10^8 °C, equivalent to ~10 keV, and the plasma must be contained away from the reactor walls.

- The tokamak magnetic field configuration cancels the effects of magnetic drift of charged particles in a toroidal field and provides good macroscopic stability and containment.

- Divertor tokamaks have a magnetic field configuration that diverts particles diffusing outwards onto plates away from the main central region. With sufficient plasma heating, improved containment is possible (H-mode).

- Plasma heating is obtained initially by resistive heating and then by neutral current injection and rf heating.

- For a self-sustaining fusion plasma the particle density n, the temperature T, and the containment time τ_E must satisfy the Lawson criterion:

$$n\tau_E T \geq 3\times10^{21}\,\mathrm{m^{-3}\,s\,keV}\,(T\,\mathrm{in\,keV})$$

For $T \sim 10\,\mathrm{keV}$ and $n \sim 10^{20}\,\mathrm{m^{-3}}$, $\tau_E \geq {\sim}3\,\mathrm{s}$.

- The increase of $n\tau_E T$ over the last 40 years has been four orders of magnitude, with JET obtaining a ratio of fusion to input power of about 0.6 and a τ_E of ~1 s in 1997 with a D+T plasma.

- Inertial confinement fusion, in which pellets of D+T are heated by intense laser pulses, is also under development. In this method τ_E is much shorter than in magnetic confinement, but n is much larger. The inertia of the extremely compressed ions gives the confinement.

- ITER, which is a tokamak design, is under construction in the south of France. Data from this facility will be used to design a demonstration reactor (DEMO). It is hoped that the first commercial reactor will be operational by 2050.

- The importance of controlled fusion as a means of providing an almost unlimited supply of carbon-free energy, both for improving the standard of living worldwide and for tackling the consequences of climate change, cannot be overstated.

FURTHER READING

Boyd, T.J.M. and Sanderson, J.J. (2003). *The physics of plasmas*. Cambridge University Press, Cambridge. A good textbook on plasma physics.

Chen, F.F. (2006). *Introduction to plasma physics*. Springer, Berlin. A good introduction to plasma physics.

WEB LINKS

dx.doi.org/10.1051/epn:2005203 J. Lister and H. Weisen (2005), What will we learn from ITER? Europhysics News 36, March/April, p. 47 (ITER).

farside.ph.utexas.edu/teaching/plasma/lectures/Plasmahtml.html Good lecture notes on plasma physics by R. Fitzpatrick.

www-fusion-magnetique.cea.fr/gb/ Useful quantitative discussion of fusion.

www.iter.org/ ITER tokamak fusion reactor.

lasers.llnl.gov/programs/nic/icf/ National Ignition Facility, ICF research.

www.iop.org/Jet/fulltext/JETR99013.pdf J. Wesson, The science of JET. Interesting description of the science of JET (JETscience).

www.jet.efda.org Information about JET and article by B. Unterberg and U. Samm, *www.carolusmagnus. net/papers/2005/docs/unterberg_overview_tokamak_results.pdf* (Tok).

LIST OF MAIN SYMBOLS

τ_E	energy containment time	n	number of particles per unit volume
P_{fusion}	fusion power	p	plasma pressure
P_{ext}	external power	T	plasma temperature
Q	quality factor	ρ	Larmor radius
W	total plasma energy	ω_c	cyclotron frequency

EXERCISES

10.1 Calculate the amount of tritium required to fuel a 5 GWe power station for 3 years.

10.2 Using the conservation of momentum, explain why in the fusion of D+T the α particles receive 3.5 MeV and the neutrons 14.1 MeV.

10.3 Calculate the number density of a plasma at a temperature of $10^8\,^\circ$C and a pressure of 2 bar.

10.4 A plasma has a pressure of 2 bar and a number density $n=10^{20}\,\text{m}^{-3}$. Calculate the plasma temperature, the Debye length, and the ratio r_s/r_c.

10.5* A plasma consists of ions and electrons with charge $+e$ and $-e$, respectively, with the same temperature T and number density (number of charges per unit volume) n_0. Suppose that a positive point charge Q is introduced into the plasma (at the origin, $r=0$), which attracts electrons and repels ions, such that the number density of ions and electrons are Boltzmann distributions:

$$n_e=n_0e^{eV(r)/(kT)}, \qquad n_i=n_0e^{-eV(r)/(kT)}$$

where $V(r)$ is the electric potential distribution in the neighbourhood of the point charge.

Assuming that $|eV(r)/kT|\ll 1$, and that $V(r)$ satisfies Poisson's equation in spherical polar coordinates,

$$\frac{\varepsilon_0}{r^2}\frac{d}{dr}\left(r^2\frac{dV}{dr}\right)=-e(n_i-n_e)$$

verify that the function

$$V(r)=\frac{A}{r}\exp\left(-2^{1/2}r\big/\lambda_{\mathrm{D}}\right)$$

satisfies Poisson's equation, where $\lambda_{\mathrm{D}}=\sqrt{\dfrac{\varepsilon_0 kT}{n_0 e^2}}$ is the Debye length of the plasma.

10.6 What is the Larmor radius of a triton in a plasma at a temperature of 15 keV where the magnetic field is 5 T?

10.7 The energy U of the dipole in a magnetic field is given by $-\boldsymbol{\mu}\cdot\boldsymbol{B}$. The force acting in the z-direction on the dipole is given by $F_z=-\partial U/\partial z$. Deduce eqn (10.27) for the force acting parallel to the field. (Note the direction of the magnetic dipole moment.)

10.8 The kinetic energies of the charged particles in a magnetic mirror remain constant and their magnetic moments are conserved. Deduce the condition on $\sin^2\alpha_{\mathrm{m}}$ given by eqn (10.30).

10.9 Calculate the ion cyclotron resonance frequency for tritons in a plasma when the magnetic field is 4 T.

10.10 A current of 3 MA flows in a tokamak plasma which has $n=2\times10^{20}$. Estimate the temperature of the plasma.

10.11 Estimate the distance travelled by a triton in a 15 keV plasma contained for 5 seconds.

10.12 Describe the main features and principles of operation of a tokamak.

10.13 Compare and contrast inertial and magnetic confinement fusion.

10.14 Discuss *critically* the statement: 'Even though fusion power plants that can contribute significantly to base-load power will not be available until after 2050, and therefore will not contribute to reducing our carbon dioxide emissions before then, it is still very important to invest in their development'.

11 Electricity and energy storage

Introduction

In the earlier chapters we considered the ways in which various forms of primary energy are exploited to do work on a turbine. We now describe how the rotational energy of the turbine is converted into electrical energy and how electricity is transmitted to consumers. We also investigate various forms of energy storage, batteries, and fuel cells.

11.1 Generation of electricity

The basic principle of electricity generation is illustrated in Fig. 11.1. Consider a planar loop of conducting wire of area A, rotating at a constant angular velocity ω in a uniform magnetic field B. Suppose the loop consists of N turns and subtends an angle $\theta = \omega t$ to the magnetic field at some instant t. The magnetic flux intersecting the loop is given by

$$\varphi = NBA \cos\theta = NBA \cos\omega t \tag{11.1}$$

By Faraday's law of electromagnetic induction, the electromotive force (i.e. the work done on unit charge) is equal to the rate of change of magnetic flux. Thus

$$V = -\frac{d\varphi}{dt} = NBA\omega \sin\omega t \tag{11.2}$$

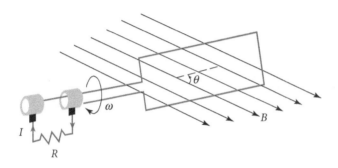

Fig. 11.1 Loop rotating in a uniform magnetic field.

If the electrical circuit is completed by connecting the ends of the wire across a resistance R, then an alternating electric current flows through the resistance, given by

$$I=\frac{V}{R}=I_0\sin\omega t \tag{11.3}$$

where

$$I_0=\frac{NBA\omega}{R} \tag{11.4}$$

In order to make electrical contact between the moving and stationary parts, slip rings and graphite brushes are inserted between the rotating conductor and the external circuit. It should be noted that the current induced in the wire is an effect that only exists while the conductor rotates in the magnetic field; thus $I\rightarrow0$ as $\omega\rightarrow0$.

EXAMPLE 11.1

A coil with 1000 turns of cross-sectional area $1m^2$ rotates at $50\,Hz$ in a uniform magnetic field of 0.5 T. What is the maximum current flowing through a load of $1000\,\Omega$?

From eqn (11.4), the maximum current is given by

$$I_0=\frac{NBA\omega}{R}=\frac{10^3\times0.5\times1\times2\times3.14\times50}{10^3}\approx157A$$

In a typical large generator, the magnetic field is produced by passing a direct current through coils mounted on a central rotating shaft connected to the turbine, called the rotor. There is a small loss of energy due to resistance heating of the coils but it is more economic than using permanent magnets. The rotor is surrounded by a set of stationary coils wound around an iron core, called the stator. The rotating magnetic field due to the rotor intersects the stationary coils in the stator and induces a current. The frequency of generation is determined by the angular velocity of the rotor. It is chosen to be high enough to avoid flickering of electric lights: 60 Hz in North America, South America, and parts of Japan, and 50 Hz in Europe.

The configuration of the windings in the stator is very complicated. The general principle is to maximize the emf (electromotive force ≡ voltage generated), but it is also necessary to minimize the flow of eddy currents, self-circulating currents which produce unwanted components of magnetic field and cause losses. Eddy currents are reduced by increasing the resistance of the paths through which eddy currents flow, i.e. by laminating the core using thin sections of steel alloy, thereby forcing the current to flow mainly through the laminations. The evolution of the design of rotors and stators has been something of a black art and details of particular machines tend to be commercially sensitive.

Electricity is usually generated as three-phase current rather than single-phase current; this is achieved by employing three independent sets of windings, 120° apart, around the generator (Fig. 11.2). The idea of using three phases was originated by Nikola Tesla, a pioneer in the early years of electricity generation and transmission. Unlike single-phase power, three-phase power never drops to zero and the power delivered to a resistive load is constant in time (Exercise 11.5). Another advantage of three-phase current is that it needs only ~75% of the material to conduct the same quantity of power as that in single-phase.

Heat is dissipated in the stator, which needs to be cooled in order to maintain it at a constant temperature; at higher temperatures the resistance of the wiring increases and the life of the insulation is shortened. For medium-sized machines, forced cooling using a rotor-mounted fan is sufficient but, for large machines, stators are cooled using de-ionized water or hydrogen (which is a better heat conductor than air).

Mechanical stability is also an important consideration, particularly in large machines. A plant outage to repair a large generator is a time-consuming and costly business, and usually means that less efficient generating plant has to be used. It is therefore normal practice to monitor mechanical vibrations for early signs of metal fatigue and cracks before a major incident occurs.

Apart from electricity generation by conventional turbines (i.e. steam, gas, or water turbines) described earlier in the book, special-purpose generators are needed for exploiting certain kinds of alternative energy. The case of wind turbine generators is described in Box 11.1.

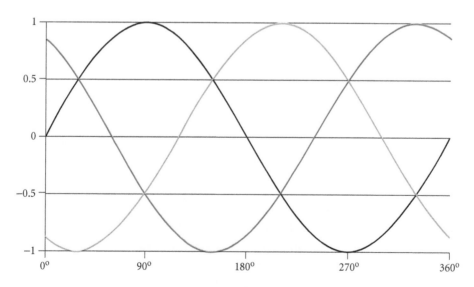

Fig. 11.2 Three-phase current.

Box 11.1 Wind turbine generators

Generators work by moving magnetic field lines through conductors and thereby inducing a current. The magnetic field can be provided by current flowing in coils, or by permanent magnets. The declining cost of rare earth magnets (in particular NdFeB magnets) has led to permanent magnet generators being built with capacities of several MW. NdFeB has a very high BH product (where B and H are the magnetic field density and the magnetic field strength, respectively), which means that only a small amount of magnetic material is required to produce a strong magnetic field. The magnets are attached to a cylindrical drum (the rotor) connected to the turbine shaft. Surrounding the rotor are conducting coils attached to the inside of a fixed cylinder (the stator). As the magnets rotate, an alternating current is induced in the coils.

In a variable speed turbine the alternating current is first rectified to direct current and then converted back to alternating current at a fixed frequency for connection to the grid. A torque T is required to turn the rotor, since the induced current produces a magnetic field that opposes the motion of the magnets (Lenz's law). The opposing tangential force per unit area (i.e. the shear stress), $\langle\tau\rangle$, depends on the surface current density K (amperes per metre) and on the strength of the magnetic field B (NB For a single conductor of length L carrying a current I normal to a field B, the force is BIL). The magnitude of the current density is limited by dissipation and the maximum B field is limited by magnetic saturation; as a result, $\langle\tau\rangle$ is typically between 25 and 100 kPa.

The size of generator required depends on the torque T produced by the wind turbine, which is given by the product of $\langle\tau\rangle$, the surface area of the rotor and the radius of the rotor, i.e.

$$T=2\pi rl\langle\tau\rangle r \quad \text{or} \quad T/V_{\mathrm{R}}=2\langle\tau\rangle$$

where V_{R} is the volume of the rotor. From Chapter 6, the power absorbed from the wind is given by

$$P=\tfrac{1}{2}C_{\mathrm{P}}\rho Au_0^3=T\Omega$$

where C_{P} is the power coefficient, ρ is the air density, u_0 is the wind speed upstream of the turbine, Ω is the angular velocity, and A the swept area of the turbine. Ω is related to the radius R of the turbine, the tip-speed ratio λ, and the wind speed u_0 by

$$\Omega=\lambda u_0/R$$

Hence

$$V_{\mathrm{R}}=C_{\mathrm{P}}\rho Au_0^3/(4\Omega\langle\tau\rangle)=\pi R^3 C_{\mathrm{P}}\rho u_0^2/(4\lambda\langle\tau\rangle)$$

For a 2 MW turbine operating with $C_{\mathrm{P}}=0.45$, a wind speed of 12 ms^{-1}, $R=35$ m, and a tip-speed ratio $\lambda=8$, the volume of the rotor with $<\tau>=80$ kPa is ~4.1 m^3 ($\rho_{\mathrm{air}}\sim1.2$ kgm^{-3}), i.e. ~2 m in diameter and 1.3 m long. We can see that the generators become very large for outputs of order MW but the simplicity of having no gearbox reduces maintenance costs (and noise) and improves reliability.

11.2 High voltage power transmission

To transmit large quantities of power over large distances the following issues need to be addressed:

- What is the optimum voltage for long-distance transmission?
- How can the voltage be increased and decreased?
- Is it better to transmit AC or DC?

To answer the first issue we consider the heating due to the resistance of a long-distance transmission line, known as ohmic (or Joule) heating. In principle, a superconducting cable would be the perfect solution, but no material has yet been found that is superconducting at ambient temperatures.

Consider a wire of resistivity ρ, cross-sectional area A, and length L. The total resistance of the wire is

$$R=\frac{\rho L}{A} \qquad (11.5)$$

Suppose the wire conducts a current I at a voltage V. The loss of power due to the resistance of the wire is

$$\Delta P=RI^2=\rho\frac{I}{A}LI$$

Putting $I=P/V$, we have

$$\frac{\Delta P}{P}=\rho\frac{I}{A}\frac{L}{V} \qquad (11.6)$$

In practice there is an upper limit to the current density I/A that can be conducted; otherwise the wire would get too hot. Hence, the fractional loss of power $\Delta P/P$ is proportional to L/V, and it follows from eqn (11.6) that the operating voltage should be as high as possible to minimize the power loss. The total loss of power for a national grid due to long-distance high voltage transmission and local distribution is typically about 5–10%.

EXAMPLE 11.2

A power plant transmits 100 MW of power along a transmission line of length 50 km with a resistance of $0.01\,\Omega\,\text{km}^{-1}$. Calculate the percentage loss of power if the line is at (a) 10 kV; (b) 400 kV.

The resistance of the complete line is $R=0.01\times50=0.5\,\Omega$. For a line at 10 kV, the current needed to transmit a power of 100 MW is $I=\frac{P}{V}=\frac{10^8}{10^4}=10^4\,\text{A}$. The fractional power loss is $\frac{RI^2}{P}=\frac{0.5\times10^8}{10^8}=50\%$. For a line at 400 kV, the current is $I=\frac{P}{V}=\frac{10^8}{4\times10^5}=250\,\text{A}$ and the fractional power loss is $\frac{RI^2}{P}=\frac{0.5\times250^2}{10^8}\approx0.03\%$.

High voltage overhead lines operate between about 110 kV and 1200 kV. In Europe the typical voltage for long-distance transmission is around 400 kV. The capital cost of overhead lines is typically about of 10% of that for underground lines with the same load capacity. Underground lines are used only where overhead lines are unacceptable, e.g. built-up areas and underwater crossings. Increasing or decreasing the voltage is straightforward for AC transmission, using transformers (see Section 11.3). However, in the case of DC transmission, electronic devices are required which are more expensive (see Section 11.4).

Power is normally transmitted as three-phase, each phase being conducted at a different height above the ground. The tops of the transmission towers are connected by a cable at earth potential, which helps to protect the transmission cables from lightning strikes. The height of the cables is set to ensure that the electric field at ground level is too low to endanger life.

The maximum voltage for transmission is determined by the electrical breakdown strength of air. The maximum electric field occurs at the surface of the cable and if it becomes too large the surrounding air becomes ionized, forming a corona discharge that conducts electric current to ground. The electrical breakdown strength of air is around $3 \times 10^6 \, \mathrm{Vm^{-1}}$ for dry air, but is lower in damp conditions.

Overhead transmission cables consist of twisted strands of conducting wire (Fig. 11.3). The outer strands are usually made of aluminium (owing to its low resistivity and low cost) and the inner strands are made of steel, for mechanical strength. The electric field at the outer surface varies inversely with the radius of the cable. For voltages over 110 kV, cables are usually configured in bundles, which reduces corona losses. Two-cable bundles are used for 220 kV lines and three- or four-cable bundles for 400 kV lines.

There are important differences between AC and DC high voltage transmission systems. In AC transmission, the inductance and capacitance of the line present added complications that give rise to a reactive current, which produces extra losses. Capacitors and other components are required to control the reactive power flow and maintain the stability of the system voltage. For long-distance power transmission, high voltage direct current (HVDC) transmission may be more economic than high voltage alternating current (HVAC) transmission, because of lower ohmic and corona losses and smaller construction costs, which offset the additional cost of converters at the ends of the line.

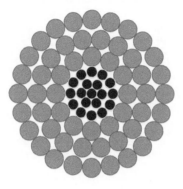

Fig. 11.3 Cross section of stranded conductor for high voltage power transmission.

Public concerns that overhead power lines are a health hazard to people living in close proximity have been expressed. Workers in electric substations in the former USSR were paid an extra allowance because of perceived health risks, though these have not been substantiated elsewhere.

11.3 **Transformers**

The electricity in power stations is typically generated at around 18–20 kV. The power is fed into a substation, where a series of step-up transformers increases the voltage to that of the transmission line. Conversely, a step-down transformer is used to reduce the voltage from the transmission line to that of the local distribution network. The principle of a transformer is shown in Fig. 11.4.

A transformer basically consists of two coils of wire wrapped around a common iron core. In the case of a step-up transformer, the number of turns N_2 in the secondary coil is greater than the number N_1 in the primary coil, and vice versa for a step-down transformer. The iron core has a high magnetic permeability to ensure that the bulk of the magnetic flux is concentrated in the iron core.

A time-varying voltage applied across the primary coil produces a rate of change of magnetic flux given by

$$V_1(t) = -N_1 \frac{d\varphi}{dt} \tag{11.7}$$

Assuming there is no flux leakage, the voltage across the secondary coil is similarly given by

$$V_2(t) = -N_2 \frac{d\varphi}{dt} \tag{11.8}$$

Eliminating $\dfrac{d\varphi}{dt}$ between eqns (11.7) and (11.8) yields

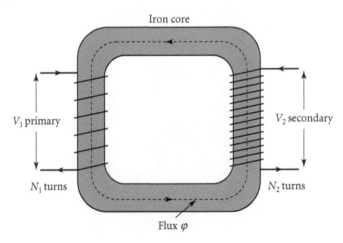

Fig. 11.4 Transformer.

$$V_2 = \frac{N_2}{N_1} V_1 \qquad (11.9)$$

Hence, for $V_2 \gg V_1$ we require $N_2/N_1 \gg 1$.

Large power transformers (over 50 MVA) are typically over 99% efficient. The main energy losses are due to the resistance of the windings; cooling systems are used to maintain transformers at a steady temperature.

EXAMPLE 11.3

Calculate the ratio of the number of turns in the secondary coil to the number in the primary coil of a transformer required to step up the voltage from 20 kV to 400 kV.

From eqn (11.9), the required ratio is $\dfrac{N_2}{N_1} = \dfrac{V_2}{V_1} = \dfrac{400}{20} = 20$

11.4 High voltage direct current transmission

Direct current is used for long overhead high voltage transmission lines and for underground or underwater cables, where reactive power losses for alternating current transmission would be high, such the English Channel or the Baltic Sea between Sweden and Germany. Direct current is produced by rectifying alternating current. The original devices for rectification (from AC to DC) and conversion (from DC to AC) were gas discharge devices called mercury arc rectifiers, but since the mid-1970s they have been displaced by solid state devices called thyristors. A thyristor consists of four semiconducting layers of alternate n-type and p-type materials. The on-phase and off-phase are controlled by a gate which can be biased positively or negatively. A thyristor has three basic states of operation: (a) a reverse blocking mode, (b) a forward blocking mode, and (c) a forward conducting mode. By arranging the thyristors in stacks, it is possible to achieve any desired voltage for a high voltage direct current system.

11.5 Comparison of AC and DC for high voltage transmission

The long-distance transmission of electricity began with direct current, around the end of the nineteenth century, but was soon displaced by alternating-current systems, which were much easier to step up and step down to different voltages (using transformers) and because alternating three-phase synchronous generators were technically superior to direct-current generators.

However, high voltage AC transmission has significant drawbacks. The inductance and capacitance of an AC transmission line generate losses which do not arise in DC transmission. In addition, capacitors and other components are required to control the reactive power flow and maintain the stability of the system voltage. The net effect is to limit the transmission distance and the transmission capacity of a high voltage AC transmission line. Also, alternating current is predominantly conducted in the outer layer of a cable, the skin effect, which

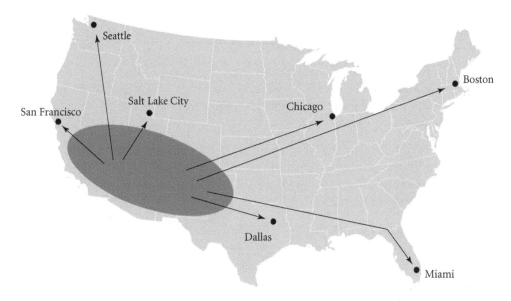

Fig. 11.5 Possible HVDC grid for transmitting solar energy across the USA.
Source: DOE2007.

increases the effective resistance of the line. Furthermore, it is not possible to make a direct connection between two AC systems operating at different frequencies, and even connections at the same frequency can lead to system instability.

For very long distance transmission overland, and for transmission undersea and under cities, HVDC transmission systems may be more economic than HVAC systems, despite the additional cost of converters at each end of the transmission line. There are numerous undersea HVDC links in Europe under the Baltic, the North Sea, and the Mediterranean. Opening in 2012, the Rio Madeira link from the Amazonas to the Sao Paolo area will be the longest overland HVDC line in the world, at over 2500 km. Figure 11.5 shows a possible HVDC grid for transmitting solar power generated in the south-west of the USA across the country.

11.6 **National electricity grids**

In most countries, generating plants are interconnected by a 'grid' of high voltage transmission lines and substations. National electricity grids are complex networks managed by a central control unit, which decides how much power each plant is allowed to generate, taking account of the cost of the electricity generation, plant availability, load distribution, and the need to minimize the energy losses along the transmission lines, subject to various constraints such as the maximum current rating for each line. It is also important to maintain stability of the grid, because unbalanced networks can produce excessive currents along certain lines and lead to blackouts. In order to prevent such events from occurring, transmission systems tend to be built with a degree of redundancy, with multiple pathways linking generating stations to large consumers. The loss of power along any line can then be overcome by diverting power through a different line. Nonetheless, when grid overloads do arise the results can be spectacular. In 2003, a massive power outage in parts of the United States and Canada affected about 50 million people and caused financial losses of around $6 billion. A blackout in New York City in 1977 was followed 9 months later by a 35% increase in the birth rate. In 2012 there was a massive blackout in India which left about 700 million without electricity.

11.7 Smart grids

The establishment of national electricity grids in the twentieth century provided most consumers with electricity on demand, but subject to the terms dictated by the grid controller, not the individual consumers. In some cases, this increased the capital cost of the system, e.g. peaks in electricity demand meant that fast generation equipment had to be kept on standby (e.g. gas turbines or pumped storage schemes), even though it was only required for a few hours a day.

The abundance of cheap digital communication technology in the twenty-first century has opened up the possibility of two-way communication between the electricity consumer and the electricity provider. The term 'electricity provider' is no longer restricted to bulk suppliers but will increasingly include small generators supplying electricity to the national grid who are able to take advantage of locally favourable conditions.

Smart meters, smart sockets, and smart devices have been available to savvy consumers for some years to time their electricity consumption to coincide with periods of the day when electricity prices are lower. A limited version of this principle has been available for a few decades in some countries like France, with a predominantly nuclear generation capacity, where cheaper electricity tariffs at night-time have been exploited to drive washing machines, dishwashers, etc., and for heat storage. A smart grid takes the idea a stage further by controlling the operation of tens of millions of smart domestic and industrial devices (each fitted with sensors and communication systems) from a central grid control. The potential economic benefits to countries with smart grids are immense, significantly reducing peak demand and improving system stability, maximizing the exploitation of renewable energy supplies (notably wind and solar) and thereby mitigating global warming, and providing cheap electricity for an electric-car society.

11.8 Energy storage

Electricity grids need the flexibility of being able to store energy over time intervals varying from a few seconds to about a day. Energy storage reduces generation costs during periods of peak demand and enables the grid controllers to cope with sudden changes in electricity demand or unexpected losses in generation capacity until alternative generating units can be brought into action. Unfortunately, storing large quantities of energy in the form of electrical energy is not a viable option for large-scale electricity supply purposes, but there are many other ways that energy can be stored and converted into electrical energy when needed. Energy storage is also an important factor in the case of alternative energy sources of generation such as solar, wave, and wind, because of their inherent variability; i.e. it is important to be able to store the energy and to supply it to the grid when there is demand.

We now give a brief description of some of the different forms of energy storage.

11.9 Pumped storage

Nuclear reactors need to produce a steady output in order to maintain reactor stability and to minimize ageing effects. For countries with nuclear plants and suitable high-level terrain, pumped storage can be an attractive option. At night-time, when electricity demand is lowest, water is pumped from a low-level reservoir to an upper reservoir using (predominantly) nuclear-generated electricity. During periods of peak demand, when electricity prices are at a maximum, water is discharged from the upper reservoir back to the lower reservoir. The same machine is used for pumping water to the upper reservoir and for generating electricity, i.e. it acts as a reversible pump-turbine. Pumped storage is essentially a form of hydropower, described earlier in Chapter 5. The operating principles are illustrated in Fig. 11.6.

A major advantage of pumped storage is that it can respond continuously to fluctuations in demand and to a sudden surge in demand, e.g. due to the loss of a generator elsewhere on the grid. The Dinorwig pumped storage plant in North Wales has a working volume of $6 \times 10^6\,\mathrm{m}^3$, a head of 600 m, and a total storage capacity of 7.8 GWh. There are six generating units, each of which can deliver 317 MW in 16 seconds from rest, or 10 seconds if they are already spinning in air.

The environmental impact of pumped storage is generally confined to mountainous areas (since large heads are required) and usually hidden from view. The capital cost of pumped storage tends to be high. In 2012, the global capacity of pumped storage was about 127 GW. In Japan, which has a large nuclear capacity, about 10% of electricity is generated using pumped storage. The potential for pumped storage is mainly in the developing world. A pumped storage facility on Okinawa pumps seawater and, if the measures taken to avoid corrosion due to salt in the seawater prove to be successful, this technology could considerably expand the

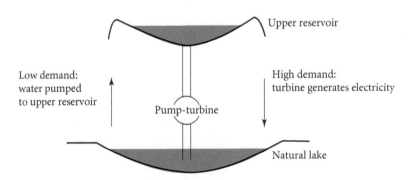

Fig. 11.6 Pumped storage.

global number of possible sites by including some by the sea. Another possibility for expansion of pumped storage is to obtain a suitable head between a surface reservoir and an underground storage cavern such as a disused mine.

An estimate of the energy density of pumped storage can be obtained from the expression (eqn (5.1)) for the power P of a dam with a head h, $P = \eta \rho g h Q$, where η is the efficiency. Since $P = E/t$ and $Q = V/t$, where V is the volume of water that flows for a time t, then the energy density $U \equiv E/V$ is given by $U = \eta \rho g h_{av}$, where h_{av} is the average head. For the Dinorwig pumped storage facility, $h_{av} \approx 500$ m and $\eta \approx 0.75$, so $U \approx 1$ kWh m^{-3} of water.

EXAMPLE 11.4 Pumped storage

It is proposed to build a pumped storage plant with a head of 500 m. How large a working volume is needed if the plant is required to generate 100 MW for 3 h a day? Assume the efficiency of generation is 90% and the density of water is 10^3 kg m^{-3}.

From eqn (5.1) the volume throughput required to generate 100 MW is given by

$$Q = \frac{P}{\eta \rho g h} = \frac{10^8}{0.9 \times 10^3 \times 10 \times 5 \times 10^2} \approx 22.2 \text{ m}^3 \text{ s}^{-1}$$

The working volume needed to deliver 100 MW for 3 h is

$$V = Qt \approx 22.2 \times 3 \times 60 \times 60 \approx 240\,000 \text{ m}^3$$

11.10 Compressed air energy storage

Another method of storing energy using cheap off-peak electricity is to pump air at high pressure into a large chamber until needed. Storing compressed air in large air-tight pressure vessels is uneconomic, but underground caverns or disused mines are feasible provided they are leak-tight. Aquifers (see Chapter 2) and salt caverns are particularly suitable, salt being self-sealing under pressure, though there is public concern over the risk of an explosion.

The first large compressed air storage facility, at Huntdorf in Germany, uses an old salt cavern with a capacity of 300 000 m^3 and air compressed to around 70 bar. The plant can generate 300 MW for 2 h. A compressed air energy storage in McIntosh, Alabama generates 110 MW, and a plant under construction near Cleveland, Ohio will generate 2700 MW.

The energy stored by compressing n moles of air isothermally (approximated when the heat of compression is dissipated with heat exchangers) from V_i to V_f is

$$E = -\int_{V_i}^{V_f} p\,dV = -\int_{V_i}^{V_f} \frac{nRT}{V}\,dV = nRT\ln\left(\frac{V_i}{V_f}\right) = p_f V_f \ln\left(\frac{V_i}{V_f}\right) \tag{11.10}$$

assuming air acts like an ideal gas. The advantages are quick start-up (about 10 minutes) and long storage capability. The air is usually mixed with natural gas and burned in a combustor, after which it is expanded through a turbine and generates electricity. The option of expanding air alone (~adiabatically) is not feasible because the air temperature at discharge would be very low, causing material and seal degradation. To improve efficiency, the heat of compression could be stored, e.g. in a spray of water, and then used for the expansion.

EXAMPLE 11.5   Compressed air energy storage

A compressed air energy storage chamber has a capacity of $300\,000\,\mathrm{m}^3$ and compresses air from 1 bar to 70 bar. Assuming that the compression process is at constant temperature, estimate the energy stored in the chamber. Also estimate the electrical power output if the air is discharged over a period of 2 h, assuming the overall efficiency of the plant is 24%.

From eqn (11.10), the energy stored is $E=3\times10^5\times70\times10^5\times\ln(70)\approx8.9\times10^6$ MJ. The average power output over a 2 h period is $P=\dfrac{0.24\times8.9\times10^6}{2\times60\times60}\approx300$ MW.

11.11 **Thermal storage**

Thermal storage has been used for a long time in buildings where solar heat absorbed by material within the building during the day is emitted during the night. This idea can be extended to seasonal thermal stores where heat is pumped into an underground store, e.g. into the soil itself or a water tank when cooling a house in the summer, and then extracted for heating during the winter. Another scheme makes a large amount of ice that is stored underground when energy is cheap and the temperature is low. The store is compact since ice has a large latent heat of fusion; the ice is then used to cool the building during the summer.

Phase-change materials have also been used to both regulate and store heat. One method is to fill a cavity wall with a wax-impregnated insulation chosen such that the wax melts at a suitable room temperature, e.g. 22 °C. In the morning, the Sun's radiation heats up the wall to 22 °C and the wax then melts with the wall remaining at 22 °C. The amount of wax is designed so that not all of it melts by sundown, after which the wax will re-solidify, giving up its latent heat to the house.

Thermal storage is also used in concentrated solar thermal power (CSTP) plants where the solar heat is stored in a suitable material (e.g. molten salt) during the day and the stored heat is then used overnight to generate steam for a steam turbine generator. The thermal material can store energy either as sensible heat or latent heat, the latter stores being more compact. Such stores may offer a cost-effective way of storing energy (see Example 11.6).

EXAMPLE 11.6

The heat source for a 1 kWe Stirling engine generator is a thermal store containing molten magnesium chloride at 714 °C. The efficiency of the generator is 40%. Calculate the mass of magnesium chloride required to run the generator for 10 h and compare with the mass of lead–acid batteries that would give the same output. The latent heat of fusion of magnesium chloride is 452 kJ kg^{-1}, and the energy density of the battery is 30 Wh kg^{-1}.

The required heat input H equals the electrical energy E, which is given by the power multiplied by the time, divided by the efficiency, so

$$H = 1000 \times 10 \times 60 \times 60 / 0.4 = 90 \text{ MJ}$$

The mass of salt M_{salt} equals the heat divided by the latent heat, so

$$M_{salt} = 90 / 0.452 = 199 \text{ kg}$$

The mass M_{lead} of the equivalent energy store using lead–acid batteries is E, which equals $10\,000$ Wh, divided by the gravimetric energy density of lead–acid of 30 Wh kg^{-1}, so

$$M_{salt} = 10\,000 / 30 = 333 \text{ kg}$$

An interesting method of energy storage is to liquefy air and store it in well-insulated cryogenic containers at atmospheric pressure. When energy is required, the liquid air is first compressed to high pressure, then vaporized and heated to the ambient air temperature. The high-pressure superheated air (since the gas is above its critical point) passes through a turbine generator, in the same way as superheated steam is used in a thermal power station (see Chapter 3). When waste heat is used to heat the air, then about 70% of the energy used to liquefy the air can be recovered. The challenge with this method is to achieve a sufficient rate of liquefaction of air.

In principle it is possible to store electrical energy by using an engine to pump heat from one reservoir to another and to then recover it by running the system in reverse. The difficulty with such a process is keeping the overall efficiency high, i.e. keeping the process as reversible as possible, since there are inefficiencies in the conversion of electrical to shaft work, shaft work to compressive energy, and compressive energy to thermal energy (arising from the temperature difference required for adequate heat flow), and back again. However, it is not only the efficiency that matters but also the size and cost of the store. For a thermal store using gravel as its storage medium (specific heat capacity 800 J C^{-1} kg^{-1}, density 2400 kg m^{-3}), the cost of the storage material is very low and the energy density for a temperature difference of 200 °C, assuming an efficiency of 40%, is considerably higher at ~40 kWh m^{-3} than that of pumped storage, which is typically ~1 kWh m^{-3} (see Section 11.9).

11.12 Flywheels

Another means of storing mechanical energy is in the form of rotational kinetic energy. The idea is not new:

- Grid controllers use the 'spinning reserve' of rotors to make minor adjustments to power supply and frequency.
- Flywheel-powered buses were used in Switzerland in the 1950s.
- The flywheel in a car provides kinetic energy to keep the engine turning between piston strokes.

In recent years, flywheels for energy storage have been developed for niche markets, e.g. providing power for testing switchgear equipment, which would otherwise cause large disturbances to the local distribution network due to sudden drops in current.

Conventional flywheels for energy storage are metallic with mechanical bearings and rotate at up to around 4000 rpm. By using strong and light materials such as plastics, epoxies, and carbon composites, together with magnetic bearings in vacuum to minimize friction, up to 100 000 rpm can be achieved.

The kinetic energy of a flywheel with a moment of inertia I and angular velocity ω is given by

$$E = \frac{1}{2} I \omega^2 \tag{11.11}$$

The moment of inertia is of the form

$$I = kmr^2 \tag{11.12}$$

where m is the total mass of the flywheel situated and r is the outer radius. Hence

$$E = \frac{1}{2} kmr^2 \omega^2 \tag{11.13}$$

$k = 1$ for a thin ring and $k = \frac{1}{2}$ for a uniform disc. Since the kinetic energy varies as the square of the radius but linearly with the mass, it is more effective to rotate flywheels faster than to make them heavier. For dynamic equilibrium, the centrifugal force is balanced by the tensile stress; the maximum angular velocity is determined by the maximum tensile stress σ_{max} that the material can withstand without breaking (see Exercise 11.14).

Typically, the energy storage capacity of flywheels made from epoxies or plastics is about $0.5 \, \text{MJ kg}^{-1}$, i.e. about 10 times that of steel. For a 100-tonne flywheel, the storage capacity is about 15 MWh. The storage capacities of modern flywheels are comparable with batteries but flywheels can be energized and de-energized much more rapidly than batteries. The typical efficiency of a flywheel is about 80%. The major problems are safety and cost. Flywheel explosions due to material failure or bearing failure can be catastrophic, so strong containment vessels are essential, adding to the capital cost. The cost of flywheels is typically an order of magnitude greater than that of batteries, but they are particularly suited where high power is repeatedly required for a short time.

EXAMPLE 11.7 Flywheel energy storage

A 1-tonne flywheel is a uniform circular disc of radius 1 m and rotates at 4000 rpm. Calculate the kinetic energy of the flywheel.

Putting $k = \frac{1}{2}$ in eqn (11.13), we have

$$E = \frac{1}{2} \times \frac{1}{2} \times 10^3 \times 1 \times \left(\frac{2 \times 3.14 \times 4 \times 10^3}{60} \right)^2 \approx 43.8 \, \text{MJ}$$

11.13 Superconducting magnetic energy storage

Superconducting magnetic energy storage (SMES) is the storage of energy in the magnetic field due to the flow of direct current in a superconducting material. The energy stored in the magnetic field is released by discharging the current in the coil. Since superconductors have no resistance, the current and the associated magnetic field do not decay with time once a direct current has been induced to flow in a superconducting coil.

Essentially, a SMES system consists of three components: a superconducting coil, a cooling system, and a power conditioning system (which converts AC to DC and vice versa). The overall efficiency of SMES is typically 95% after allowing for losses in AC/DC conversion and cryogenic cooling of the superconducting material.

The advantages of SMES are that there are no moving parts, power is available almost immediately, and a very high power output can be delivered for short periods. The disadvantages are that superconductors operate only at low temperatures, the capital cost is high, and the energy content is fairly small. The magnetic energy stored per unit volume is

$$E = \frac{B^2}{2\mu_o} \approx 4 \times 10^5 \, B^2 \, \text{Jm}^{-3}$$

Thus, a magnetic field of 4 T has an energy density of about 6.4 MJ m^{-3}.

The main applications of SMES are for improving power quality for electricity supply utilities, e.g. smoothing fluctuations due to intermittent loads on transmission lines, and for manufacturing processes where an ultra-smooth power supply is important.

11.14 Batteries

Batteries are electrochemical devices for storing energy in a form that can be readily converted into electrical energy. The chemicals are stored within the device from manufacture, unlike a fuel cell in which the chemicals are renewed continuously (see Section 11.16). Batteries are either non-rechargeable (primary) or rechargeable (secondary), but only rechargeable batteries are of interest for large-scale energy storage.

11.14.1 Basic principles

A battery contains galvanic (also called voltaic) cells, each of which has two or more half-cells: one for oxidation and one for reduction. Each half-cell contains an electrolyte solution and an electrode. A simple example is the copper–zinc galvanic or Daniell cell. A zinc electrode is immersed in a zinc sulfate solution and a copper electrode in a copper sulfate solution, with the solutions separated by a porous barrier which allows ions to pass through, as shown in Fig. 11.7.

When a metal is placed in a solution, some metallic atoms at the surface of the electrodes go into solution as positive ions, thereby leaving the metal negatively charged. The ions are attracted by the negatively charged metal and a double layer of charges is formed, separated

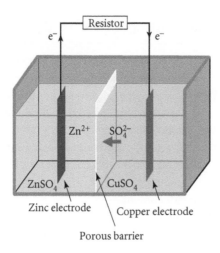

Fig. 11.7 Daniell cell.

by about an ion diameter, with a potential difference across the layer of order 1V. Only a very small amount of metal goes into solution before a dynamic equilibrium is reached, which is driven by opposing gradients in the ion concentration and the electrostatic potential.

When the zinc and copper electrodes in a Daniell cell are joined by a conductor, electrons flow via the external circuit from the zinc electrode to the copper electrode, where they combine with copper ions in solution, which are deposited on the copper electrode. The following reactions occur at the electrodes:

$$Zn \rightarrow Zn^{++} + 2e^- \text{ oxidation} \quad Cu^{++} + 2e^- \rightarrow Cu \text{ reduction}$$

Zinc loses electrons and is oxidized more readily than copper, with the result that zinc goes into solution, and copper ions accept electrons at the copper electrode and are reduced. The zinc ions in solution are neutralized by sulfate ions flowing through the porous barrier, which completes the flow of charge around the circuit. The barrier stops copper ions diffusing and depositing on the zinc electrode when no current is flowing. Zinc is the anode or negative electrode and copper is the cathode or positive electrode.

The energy required to remove an electron from zinc is less than that for copper. The available energy (as explained in Chapter 2) is minus the change in the Gibbs free energy $-\Delta G$ in the overall reaction, called a redox reaction:

$$Zn + Cu^{++} \rightarrow Zn^{++} + Cu \tag{11.14}$$

The value of ΔG depends on the concentrations; if these are one molar (1 M) at 25 °C then $\Delta G° = -213\,\text{kJ}\,\text{mol}^{-1}$. In reaction (11.14) a charge $Q = nN_A e$ is transferred, where $n = 2$ (since the ions are doubly ionized), N_A is Avogadro's number, and e is the electronic charge, so the potential difference V_B between the electrodes is related to the change in the Gibbs free energy by the formula

$$\Delta G° = QV_B = nN_A eV_B = -nFV_B \tag{11.15}$$

where F, Faraday's constant, is equal to 96 485 Coulombs per mole. Substituting the value of $\Delta G°$ gives what is called the standard potential of the Daniell cell as $V_B = 1.10\,V$.

11.14.2 The standard potential and the Nernst equation

The standard potential of a metal or compound is taken to be the voltage between the substance and a hydrogen electrode, which consists of hydrogen gas flowing over a platinum electrode that catalyses the reaction $H_2 \rightarrow 2H^+ + 2e^-$. The standard potentials for several half-reactions used in batteries are given in Table 11.1.

The dependence of the voltage of a battery on the concentration of the solutions is determined by how the Gibbs free energy depends on concentration. The result is

$$\Delta G = \Delta G^O + RT \ln Q \tag{11.16}$$

where Q, the reaction quotient, is the ratio of the concentrations of the products to that of the reactants (see Box 11.2). Substituting eqn (11.15) gives the Nernst equation relating the standard potential to the concentration of the ions in the battery:

$$V = V_B - \frac{RT}{nF} \ln Q \tag{11.17}$$

For $T = 25\,°C$ the voltage is then given by

$$V = V_B - \frac{0.059}{n} \log Q \tag{11.18}$$

For the Daniell cell, the overall reaction is given by eqn (11.14) so the potential difference V will be given by

Table 11.1 The standard potentials for several half-reactions used in batteries

Half-reaction	Standard potential
$Li^+ + e^- \rightarrow Li$	−3.04
$K^+ + e^- \rightarrow K$	−2.93
$Ca^{++} + 2e^- \rightarrow Ca$	−2.87
$Na^+ + e^- \rightarrow Na$	−2.71
$Zn^{++} + 2e^- \rightarrow Zn$	−0.76
$V^{+++} + e^- \rightarrow V^{++}$	−0.26
$2H^+ + 2e^- \rightarrow H_2$	+0.00
$Cu^{++} + 2e^- \rightarrow Cu$	+0.34
$Ag^+ + e^- \rightarrow Ag$	+0.80
$2H^+ + VO_2^+ + e^- \rightarrow VO^{++} + H_2O$	+1.00
$O_2 + 4H^+ + 4e^- \rightarrow 2H_2O$	+1.23
$Cl_2 + 2e^- \rightarrow 2Cl^-$	+1.36

$$V = V_B - \frac{0.059}{n} \log \frac{[Zn^{++}]}{[Cu^{++}]}$$

where $[Zn^{++}]$ and $[Cu^{++}]$ are the concentrations of the Zn^{++} and Cu^{++} ions. The activities of the metals (the electrodes) do not change and can be set to unity, as it is only the difference in concentrations (see eqn (11.17)) that gives rise to a change in potential. If the concentrations start at one molar then after a current I has flowed for a time t, we have

$$\frac{[Zn^{++}]}{[Cu^{++}]} = \frac{1 + It/(nF)}{1 - It/(nF)}$$

The logarithmic dependence on concentration means that the potential falls only very slowly until nearly all the Cu^{++} ions have deposited on the copper electrode, dropping only 14 mV after 0.5 mole, 38 mV after 0.9 mole, and totally discharging to $V=0$ volts after 1 mole of charge has flowed.

11.14.3 Battery types

The first battery invented by Volta in 1800 consisted of a pile of zinc and silver disks separated by cloth disks soaked in salt solution (brine). Dilute sulfuric acid can be used instead of brine as the electrolyte. The passage of current produces H_2 bubbles at the silver electrode, which form an insulating layer that causes the internal resistance of the pile to increase, a phenomenon called polarization, which limits the useful life of the battery. The reactions occurring when the electrolyte is sulfuric acid are

$$Zn \rightarrow Zn^{++} + 2e^- + 0.76V \quad \text{and} \quad 2H^+ + 2e^- \rightarrow H_2 + 0V$$

at the zinc and silver electrodes, respectively, with the overall reaction being

$$Zn + 2H^+ \rightarrow Zn^{++} + H_2 + 0.76V \tag{11.19}$$

The Volta cell reaction, eqn (11.19), is not reversed if an opposing emf greater than 0.76V is applied, since the hydrogen has escaped and is not available for the reaction to proceed. Since the overall reaction is irreversible, the Volta cell is a primary battery.

The Daniell cell was invented in 1836 and avoided the generation of hydrogen by having separate electrodes in different electrolytes. This cell and variants of it were used widely in telegraph systems.

Dry cell and alkaline batteries

Polarization can also be avoided by using an electrode rich in oxygen such as MnO_2, which accepts electrons and is reduced to Mn_2O_3. The first such battery was the Leclanché cell (1866) which used zinc and manganese dioxide as the electrodes and ammonium chloride as the (acidic) electrolyte. The first dry cell battery (1887) was made by adding plaster of Paris to the electrolyte so it became a paste. This battery became the main primary battery until its performance was improved in the 1960s using an alkaline electrolyte. The so-called alkaline battery has zinc as one electrode and manganese oxide as the other, with both in contact with a KOH paste which acts as the electrolyte for both half-cells. A carbon rod in the middle of the MnO_2 is used as one of the terminals as it resists attack by KOH. The half-cell reactions are

$$2Zn+4OH^- \rightarrow 2ZnO+2H_2O+4e^- \quad \text{and} \quad 3MnO_2+2H_2O+4e^- \rightarrow Mn_3O_4+4OH^-$$

The overall reaction is

$$2Zn+3MnO_2 \rightarrow 2ZnO+Mn_3O_4+\sim 1.5 \text{ V}$$

Lead–acid battery

For energy storage a battery's overall reaction must be reversible, i.e. the direction of the reaction is reversed if an opposing emf is applied that is larger than the cell potential. The battery can then be recharged. The first practical rechargeable battery was the lead–acid battery, invented in 1859. It has a series of lead anodes and lead oxide cathodes immersed in sulfuric acid (H_2SO_4), with each cell connected in series (Fig. 11.8). Its energy density is comparatively quite low but it can provide large currents and is still used in cars today. A lead–acid battery can be recycled several hundred times, depending on the discharge rate and depth of discharge. Those designed to produce large currents (as for starting cars) have more and thinner plates than those designed for discharge to lower levels.

On discharge, the lead anode oxidizes to lead sulfate by the half-reaction

$$Pb+H_2SO_4 \rightarrow PbSO_4+2H^++2e^-$$

and the lead oxide cathode reduces to lead sulfate by the half-reaction

$$PbO_2+2H^++2e^-+H_2SO_4 \rightarrow PbSO_4+2H_2O$$

The overall reaction is

$$Pb+PbO_2+2H_2SO_4 \rightarrow 2PbSO_4+2H_2O+\sim 2.0 \text{ V}$$

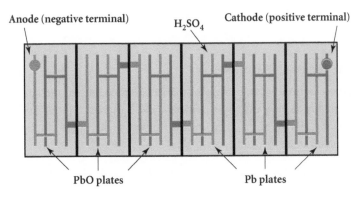

Fig. 11.8 Lead–acid battery with six cells: output voltage ~12 V.

The density and concentration of the sulfuric acid decrease during discharge. The voltage also therefore drops during discharge and, as in the Daniell cell, the fall is slow initially with a rapid drop close to full discharge (see Exercise 11.19). The specific energy density is ~20 Wh kg^{-1}.

Lead–carbon batteries are being developed, in which carbon powder is added to the spongy lead paste used in the negative electrodes to improve the cycle lifetime and reduce the charging time of the lead–acid battery. While still having a low energy density, the lead–carbon battery may provide effective cheap energy storage when size and weight are not important. (In Europe and the USA over 95% of lead–acid batteries are recycled.)

Sodium-sulfur and magnesium-antimony batteries

The Na–S battery, first developed in the 1960s, uses a solid electrolyte (beta alumina) that selectively conducts Na$^+$ ions, a liquid metal (sodium) negative electrode, and molten sulfur in porous carbon as the positive electrode. The battery operates at a temperature of ~300 °C and during discharge the sodium ions pass through the beta alumina and combine with sulfur to form sodium polysulfide. The overall reaction can be represented by

$$2Na + 3S \rightarrow Na_2S_3 + \sim 1.9 \text{ V}$$

The materials are cheap and have low densities. The theoretical specific energy is high, ~720 Wh kg^{-1} (see Exercise 11.20) and values of 150 Wh kg^{-1} have been obtained. The cycle lifetime is long and the charge efficiency (the ratio of energy from discharge to energy required to charge) is high (~90%), but economies of scale, the high temperature, and the corrosive nature of the materials make this battery most suitable for large-scale energy storage, such as for the grid, and it is still under development. A related molten-salt battery that uses beta alumina and a sodium anode but with a nickel chloride cathode, called a ZEBRA battery, has been developed that also has a high specific energy of ~90 Wh kg^{-1} and has been used to power vehicles.

A battery designed to be suitable for grid storage, which is at the R&D stage, is a magnesium–antimony liquid-metal battery. The electrolyte is a salt intermediate in density between magnesium and antimony, and in operation at 700 °C the metals and salt form three distinct layers. During discharging and charging, Mg^{++} ions pass through the electrolyte to and from the antimony. In the antimony the magnesium forms a MgSb alloy. All the constituents are readily available and relatively cheap, and the battery may enable a low-cost energy storage solution.

Nickel metal hydride (NiMH) battery

These rechargeable batteries first became available commercially in the late 1980s and have a considerably higher energy density than lead–acid batteries. The positive electrode is a hydrogen-absorbing alloy and the negative electrode is nickel oxyhydroxide, NiOOH. The discharge half reactions are

$$M + e^- + H_2O \rightarrow MH + OH^- \quad \text{and} \quad Ni(OH)_2 + OH^- \rightarrow NiOOH + H_2O + e^-$$

where M stands for an inter-metallic compound, which typically includes a mixture of rare earths, nickel, and manganese. The electrolyte is generally potassium hydroxide. The batteries are widely used in hybrid electric cars. Their specific energy density is ~60 Wh kg^{-1}.

Lithium-ion battery

A further significant advance in battery technology was made in the 1980s and 1990s with the development of the lithium-ion and lithium-ion polymer battery. Lithium has the highest electronegativity (see Table 11.1) and a low density, so is ideal for providing high energy density, but it is very reactive. A way of controlling its reactivity was found when it was discovered that lithium ions could be moved into or out of graphite without breaking up its structure—a reversible process called intercalation. Lithium can also be intercalated into certain lithium oxides that have a layered structure, such as $LiCoO_2$ or $LiMn_2O_4$.

The lithium-ion battery contains a negative electrode (anode) of usually graphite and a positive electrode (cathode) of lithium oxides. The cell voltage, ~3.7 V, is the difference in free energy between Li^+ ions in the crystal structures of the two electrode materials. (The use of a silicon rather than a graphite anode offers increased capacity.) The operation of the battery involves the transfer of lithium ions from one electrode to the other—'the rocking chair' effect—with the intercalation of the lithium controlling its reactivity. The electrodes are separated by a Li^+ conducting electrolyte, originally a lithium salt in an organic solvent but now generally a lithium salt in a solid polymer. The specific energy density of lithium-ion polymer batteries is ~150 Wh kg^{-1}.

Lithium–air and lithium–sulfur batteries

These batteries are both under development. Lithium–air batteries replace the lithium oxide cathode with a porous cathode that is permeable to oxygen from the air. One design uses a porous carbon cathode and a lithium metal anode with a polymer electrolyte. During discharge the lithium ions migrate to the porous cathode where they combine with oxygen from the air to form lithium oxide, while on charging lithium migrates to the anode and oxygen is given off from the cathode. A lower energy density, but still about twice that of lithium-ion batteries, is offered by Li–S batteries.

The low density cathode and anode of a lithium–air battery means that an energy density of about five times that of a lithium-ion battery, i.e. ~1000 Wh kg^{-1}, might be achieved. This would give a comparable energy density to that of an internal combustion engine (ICE), the higher efficiency (~90%) and higher power-to-weight ratio (~6 kW kg^{-1}) of an electric motor—compared with a thermal efficiency of ~25% and power-to-weight ratio of ~0.75 kW kg^{-1} for an ICE—offsetting the higher energy density (~13 000 Wh kg^{-1}) of the ICE fuel (petrol (gasoline) or diesel) (see Exercise 11.21).

There is considerable R&D to be undertaken, and the use of nanotechnology to increase the surface area of the electrodes and to produce materials with the right properties may help in achieving a reliable, very high energy density Li–air battery that can be cycled (discharged and charged) many times.

Flow batteries

Flow batteries store their electrical energy within their electrolytes rather than within their electrodes, so their capacity is limited only by the volume of their electrolyte containers. An example is the vanadium-redox flow battery (VRB), which has two separated electrolytes, each containing vanadium ions in different charge states in sulfuric acid. The electrolytes are separated by a semi-permeable ion-exchange membrane (IEM) which allows protons (H^+) ions to pass, and the electrodes are graphite plates. The half reactions are

$$\text{Anode: } V^{++} \rightarrow V^{+++} + e^- \qquad \text{Cathode: } 2H^+ + VO_2^+ + e^- \rightarrow VO^{++} + H_2O$$

The arrangement is similar to that of a Daniell cell in that the electrolytes are separated. Several demonstration systems have been made; for example, in 2005 a 4 MW/6 MWh VRB was installed at the Subaru Wind Farm in Japan for energy storage and wind power stabilization. However, reducing the cost of the IEM is important for effective commercialization of these batteries.

Box 11.2 Dependence of the Gibbs free energy and chemical potential on concentration

---

The Gibbs free energy $G = U + PV - TS$, and for a gas $dG = VdP - SdT$. For n moles of gas changing pressure from P_1 to P_2 at a constant temperature,

$$\Delta G = \int_{P_1}^{P_2} \frac{nRT}{P} dP = nRT \ln\left(\frac{P_2}{P_1}\right)$$

If we take P_1 to be 1 bar as our standard pressure P^0 then

$$G_2 = G^0 + nRT \ln(P_2/P^0)$$

For molecules in a solution of volume V, the partial pressures P_i of the different molecules are related to their concentrations c_i by $P_i = (n_i/V)RT = c_iRT$, and the Gibbs free energy of each compound can be written by analogy as

$$G_i = G_i^0 + nRT \ln(c_i/c^0)$$

The Gibbs free energy per mole is the chemical potential μ, so the chemical potentials of the compounds in solution are given by

$$\mu_i = \mu_i^0 + RT \ln\left(c_i/c^0\right) \tag{11.20}$$

(NB This only holds at sufficiently low concentrations that the interactions between the molecules in solution can be ignored, as in an ideal gas; when this does not hold an effective concentration, called activity, is used.)

Consider a chemical reaction

$$aA + bB \leftrightarrow cC + dD \tag{11.21}$$

This can be written in the form

$$\sum_i v_i N_i = 0 \tag{11.22}$$

where $v_A = -a$, $v_B = -b$, $v_C = c$, and $v_D = d$. If the number of moles of N_i changes from n_i to $n_i + \Delta n_i$, then from eqn (11.22) $\Delta n_i = v_i \Delta x$, where Δx is in moles. When v_A moles of A and v_B moles of B react then $\Delta x = 1$ and the resulting change ΔG in the Gibbs free energy per mole of reactants is given by

$$\Delta G = \sum_i \mu_i v_i \tag{11.23}$$

If the initial concentrations are c_A, c_B, c_C, and c_D, then using eqn (11.20) the change ΔG is given by

$$\Delta G = \sum_i v_i \left[\mu_i^0 + RT \ln\left(c_i / c^0\right) \right] \tag{11.24}$$

which can be re-expressed as

$$\Delta G = \Delta G^0 + RT \ln\left(\frac{a_C^c a_D^d}{a_A^a a_B^b} \right) \tag{11.25}$$

ΔG^0 is the change in the Gibbs free energy when the initial concentrations are standard, i.e. c^0, and a_X, which equals c_X / c^0 at sufficiently low concentrations, is the activity of X. The general expression for ΔG is therefore

$$\Delta G = \Delta G^0 + RT \ln Q \tag{11.26}$$

where the reaction quotient Q is given by

$$Q = \frac{a_C^c a_D^d}{a_A^a a_B^b} \tag{11.27}$$

and ΔG^0 by

$$\Delta G^0 = -RT \ln K \tag{11.28}$$

where K, the equilibrium constant, is the value of the reaction quotient Q when the reaction is in equilibrium.

11.14.4 **Supercapacitors**

While not a battery, a supercapacitor is an electrochemical device using electrodes and electrolytes that complements a battery in its energy storage capabilities. Electrochemical double-
layer capacitors (EDLCs) are made from carbon electrodes immersed in an electrolyte and
separated by a thin membrane (see Fig. 11.9). When a voltage is applied across the electrodes,
positive ions are attracted to the negative electrode and diffuse through the separator, which
provides electronic insulation, and form an electric double layer of charge separated by about
an ion diameter. A similar double layer forms on the other electrode. The carbon electrodes
form a porous structure that has a very large surface area ($\sim$3000 m^2 g^{-1}), which combined with
the small separation of the charge layers gives a very large capacitance. The maximum voltage
is limited by the breakdown voltage of the electrolyte, typically 1–3 V.

The two electrodes, separator, and electrolyte form a thin strip which is rolled up into a
cylindrical or rectangular shape. The capacitance C can be consider to be that of two parallel plate capacitors in series, each of area A and plate separation d immersed in an electrolyte
with dielectric constant ε, i.e.

$$C = \varepsilon \varepsilon_0 A / 2d$$

where $\varepsilon_0 = 8.85 \times 10^{-12}$ F m^{-1}. We can estimate C by taking $d = 0.5$ nm, $\varepsilon = 2$, and $A =$
3000 m^2 g^{-1}, which gives $C \approx 50$ F g^{-1}. A typical voltage is $U = 2.5$ V, so the energy stored
$E_S = CU^2 / 2 = 150$ J g^{-1} = 0.15 MJ kg^{-1} = 40 Wh kg^{-1} (More accurate estimates require evaluating the effect of image charges.) Recently (2010) a supercapacitor using graphene-based
electrodes to give a very large surface area achieved an energy storage of 85.6 Wh kg^{-1}.

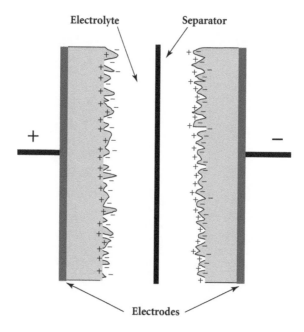

Electrolyte   Separator

+   −

Electrodes

Fig. 11.9 Schematic diagram of a supercapacitor.

Supercapacitors have self-discharge times that are shorter than batteries, typically weeks rather than months.

The maximum power P_m that a supercapacitor can deliver is when the load resistance R_L is equal to its equivalent series resistance R_s, and is given by $U^2/4R_s$. For $U=2.5V$, values of $R_s \approx 0.1\,m\Omega\,kg^{-1}$ are typical, giving $P_m \approx 10\,kW\,kg^{-1}$.

Since no charge is transferred between the electrodes and the electrolyte, supercapacitors can be cycled a very large number of times ($\sim 10^6$) and can release their stored energy quickly ($\sim 1\,s$). They are therefore ideal in a hybrid vehicle for storing the energy generated when braking electromagnetically and for providing boosts in power.

11.14.5 Specific energy and the Ragone plot

In a battery the charge capacity Q_c of an electrode is given by $Q_c = nF/(3600 \times M)\,Ah\,kg^{-1}$, where F is Faraday's constant, n is the charge state of the ion, and M is the molecular weight. For example, for a $LiMn_2O_4$ electrode in a lithium-ion battery, $n=1$ and $M=0.181\,kg\,mol^{-1}$, so $Q_c = 148\,Ah\,kg^{-1}$. The energy density E_s can be estimated by $E_s = V_p Q_c$, where V_p is the potential of the battery when half depleted. For a lithium-ion battery $V_p \sim 3.7V$ so $E_s \approx 550\,Whkg^{-1}$. This is the theoretical upper limit since the practical limit must account for the mass of the other components of the battery and also the fact that not all the lithium can be utilized. As of 2011, lithium-ion polymer batteries had an energy storage of $\sim 150\,Wh\,kg^{-1}$.

When a battery is discharged very slowly, the voltage is close to that given by the standard reduction potentials. However, when discharged quickly, there are significant voltage drops associated with the higher current through the internal resistance, with the voltage required to drive the reactions (activation potential), and with concentration gradients within the battery, which affect the voltage through the Nernst relation (eqn (11.17)). These voltage drops

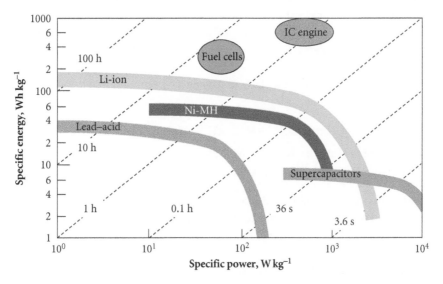

Fig. 11.10 Ragone plot of specific energy versus specific power.
Source: LBL2008.

mean that the amount of energy extracted from a battery by the time it reaches a particular voltage is less than when it is discharged slowly, and the effective battery capacity is reduced. There is also a reduction in battery capacity at high discharge rates, caused by the diffusion rate of ions being insufficient to match the current drawn.

The performance of batteries and supercapacitors can be compared on a Ragone plot (see Fig. 11.10). This is a graph of the energy stored per unit mass (specific energy) versus deliverable power per unit mass (specific power).

For a car, the specific energy is related to the range of the car, and the specific power to the acceleration. Also plotted are the corresponding values for an internal combustion (IC) engine and for fuel cells. While the weight of a battery remains the same during discharge, an IC engine's fuel is consumed so the specific energy value is for the initial mass of fuel, approximately half a kilogram of fuel per kilogram of engine. Taking the efficiency of an IC engine as 25% then gives about 1000 Wh kg^{-1} of fuel plus engine.

11.15 Electrolysis

Electrolysis is the decomposition of an ionic compound by passing an electric current through a liquid containing the compound. The process is important in the extraction of certain metals such as aluminium or sodium from their ores or oxides. The passage of a current provides the energy required to make a chemical reaction occur in the non-spontaneous direction, i.e. in the direction that results in an increase in the Gibbs free energy. The electrolysis of water is an important application in the synthesis of low-carbon fuels and as a way of storing energy through the production of hydrogen.

In pure water the fraction that is ionized is very small (H$^+$ as H$_3$O$^+$ and OH$^-$ concentrations are 10^{-7} M at STP), so the water acts as an insulator and requires the addition of an electrolyte to be easily electrolysed. The electrolyte must be more difficult to oxidize or reduce than water for the products of electrolysis to be hydrogen and oxygen. A suitable electrolyte is sulfuric acid, as it is essentially completely ionized when in aqueous solution and the sulfate ion SO$_4^{2-}$ is very hard to oxidize.

The reactions that take place at the electrodes when a current is passed through this electrolytic cell are

Cathode (negative potential): $4H^+ + 4e^- \rightarrow 2H_2$ (hydrogen gas)

Anode (positive potential): $2H_2O \rightarrow O_2$ (oxygen gas) $+ 4H^+ + 4e^-$

The overall reaction is: $2H_2O \rightarrow 2H_2 + O_2$

The concentration of the sulfuric acid remains constant and the water is electrolysed to hydrogen gas and oxygen gas. The change in the Gibbs free energy corresponds to a standard potential of 1.23 V, which is the voltage required for the electrolysis of water to occur under standard conditions (i.e. 1 molar at 25 °C).

With increasing temperature, the change in the Gibbs free energy (ΔG_e) required to electrolyse water decreases while the overall energy needed (ΔH_e) only increases very slightly. Since $\Delta H_e = T\Delta S_e + \Delta G_e$ in a reversible process, the amount of electrical work decreases with temperature and a larger amount of heat ($T\Delta S_e$) can supply the energy required. Higher temperatures also improve the reaction rate and reduce the energy losses during steam electrolysis. About twice as much heat energy is required per unit of electrical energy (NB Modern thermal power stations have an efficiency of ~50%), so the overall thermal efficiency improves with temperature. High-temperature electrolysis may be a good way to produce hydrogen if low-carbon sources of heat and electricity, such as solar or nuclear (fission or fusion) plants, are readily available.

11.16 **Fuel cells**

A fuel cell is an electrochemical device that can be used to generate electricity or store energy in the form of hydrogen. Unlike a battery, a fuel cell is fed continuously by chemicals. The chemical feed consists of hydrogen and oxygen and the fuel cell acts as a means of combining them to make water, i.e. the opposite of electrolysis. There are two electrodes (the anode and cathode), where the chemical reactions take place, and a catalyst speeds up the reactions at the electrodes. There is virtually no pollution and the only by-product is water. Also, since a fuel cell is not a heat engine operating in a closed cycle, the Carnot limit to its efficiency does not apply.

The fuel cell was invented in 1839 by William Grove but interest in it soon fell away when cheap fossil fuels became widely available. It was not until the 1920s and 1930s that significant progress was made in their development. In the early 1960s, NASA decided to use fuel cells to provide electricity for the Gemini and Apollo space capsules, which led to further improvements. With their carbon-free energy and high efficiencies, interest in fuel cells is now very high. Fuel cell development has centred on the materials used for the electrodes, the catalysts, and the choice of electrolyte. Some devices use liquid electrolytes (alkalis, molten carbonate, and phosphoric acid), while others use solid electrolytes (proton exchange membrane (PEM) and solid oxide). We concentrate here on PEM cells, whose main features are shown in Fig. 11.11, since they illustrate the physical principles involved (see Box 11.3). They are also well developed and suitable for use in cars, though not yet cost-effective compared with petrol- or diesel-fuelled engines.

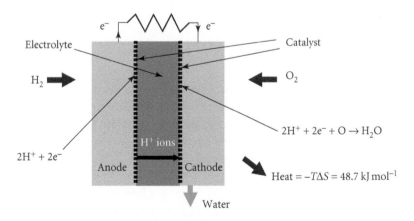

Fig. 11.11 PEM fuel cell (schematic). The electrolyte is a thin plastic membrane and the anode and cathode are porous.

Box 11.3 PEM fuel cell (PEMFC)

Hydrogen is introduced on one side of the cell and flows through a porous anode, where it dissociates into hydrogen ions and electrons on contact with a catalyst. The porous electrodes are often made of a porous carbon-impregnated cloth or paper 100–300 μm thick. The membrane acts as the electrolyte since it is permeable to hydrogen ions but is impermeable to electrons. One such membrane is perfluorosulfonic acid. This is a plastic made from polytetrafluoroethylene (PTFE) by reacting it with sulfonic acid, which adds ion clusters of $SO_3^-H^+$ to the polymer. Water is attracted to these clusters and, once the plastic has absorbed water, the hydrogen ions (but not the SO_3^- ions) become mobile. A hydrogen ion bonded to a water molecule can move from one SO_3^- ion to another. Consequently the hydrogen ions can migrate through the plastic. However, since the plastic is an insulator, with no free electrons, electron transport is blocked. The membrane is typically ~100 μm thick. The electrons flow through the external circuit, after which they combine on the catalyst with oxygen and the hydrogen ions, which have passed through the plastic electrolytic membrane, to form water. The overall reaction is

$$H_2 + \tfrac{1}{2}O_2 \rightarrow H_2O, \quad \Delta H = -285.8 \text{ kJ mol}^{-1}$$

where ΔH is the change in enthalpy. The reaction occurs at constant pressure, so $-\Delta H$ gives the energy released when hydrogen and oxygen combine plus the work done when the volume of the gases decreases (as explained in Chapter 2). The entropy of the gases decreases in this process, since the number of moles is reduced. In a reversible process the entropy of the entire system, reactants and surroundings, remains constant. As a result, an amount of heat equal to $-T\Delta S$, where ΔS is the change in the specific entropy of the gases, is transferred to the surroundings. The amount of energy available as electrical energy is minus the change in the Gibbs free energy $-\Delta G$, where

$$\Delta G = \Delta H - T\Delta S = (-285.8 + 48.7) = -237.1 \text{ kJ mol}^{-1}$$

In this reversible (ideal) process the efficiency of conversion is $\Delta G/\Delta H = 83\%$. The voltage V generated is given by equating ΔG to the total charge ΔQ that flows, multiplied by the voltage V, i.e.

$$\Delta G = \Delta Q V = n N_A e V$$

where n is the number of electrons released per mole, N_A is the number of molecules in 1 mole (Avogadro's number), and e is the charge on an electron. The product $-N_A e \equiv F$ (known as the Faraday constant) is equal to the amount of charge in one mole of electrons and has the value 9.65×10^4 coulombs per mole. Since two electrons are released per molecule of hydrogen, $n=2$, the voltage generated is

$$V = 237.1 \times 10^3 / (2 \times 9.65 \times 10^4) = 1.23 \text{ volts}$$

This is the open circuit voltage of a PEM cell, but when a current flows through an external circuit, the voltage drops as shown in Fig. 11.12.

The initial drop in voltage reduces the activation energy governing the catalysed electrochemical reactions. The rate of these reactions, and hence the current, increases as the voltage drops (activation loss). As the current rises further, the resistance to current flow within the cell causes a linear drop in voltage with increasing current (ohmic loss). Eventually, the flow of gas through the electrodes becomes the limiting factor (gas transport loss), since the formation of water occurs at such a rate as to block the reaction sites on the catalyst. The resulting curve is called a polarization curve, and these losses are often called activation, ohmic, and concentration polarization. A catalyst layer about 10 μm thick with a loading of ~0.15 mg cm^{-1} of Pt can produce ~½ A cm^{-1} at a voltage of 0.7 V.

A power of 1W is achieved when the current drawn is 1333 mA and the voltage is 0.75 V. Higher voltages are obtained by stacking a number of fuel cells in series. Oxygen for the PEM cell is obtained from the air but hydrogen has to be produced and stored in a container alongside the fuel cell system. There are a number of ways of producing and storing hydrogen, which are described in Section 11.17.

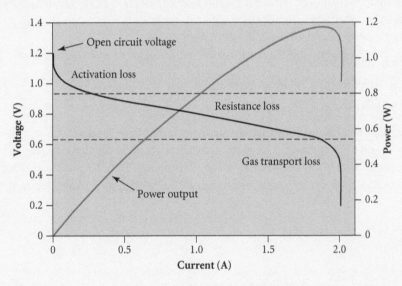

Fig. 11.12 Polarization curve of a PEM cell.
Source: USCD05.

11.17 **Storage and production of hydrogen**

As yet there is no way of storing hydrogen very compactly. It can be compressed up to 34 MPa ($\equiv$340 bar) in lightweight polycarbonate bottles, but even at that pressure there are only 31 g litre^{-1} of hydrogen. Liquefying H_2 is expensive and only increases the density to 71 g litre^{-1}. Metal hydrides such as $LaNi_5H_6$ and $NaAlH_4$ can hold 1.3 wt% and 3.7 wt% of hydrogen, respectively, which can be released using catalysts at temperatures suitable for PEM fuel cells (~80 °C). However, these densities correspond to only 13 and 37 g per kg of hydrogen. Some hydrides release hydrogen when water is added, an example being $NaBH_4$, which can hold 4 wt%.

The energy required to store and release the hydrogen needs to be considered. It takes energy to compress hydrogen to 340 bar—approximately 10% of the stored amount. Research on storing hydrogen is ongoing and there is considerable research and development on new materials for storing hydrogen: magnesium borohydride, $Mg(BH_4)_2$ holds 14.9 wt% of hydrogen and is less thermally stable than other borohydrides, decomposing at around 270 °C. Recent research suggests that additives can significantly increase the rate as well as the weight percentage of hydrogen released. Aluminium hydride, AlH_3, is another promising material, holding 10.1 wt% and releasing hydrogen at temperatures of 60–175 °C, depending on the particle size; however, it currently requires very high pressures to form AlH_3.

The aim of the US Department of Energy is to achieve 1.8 kWh kg^{-1} by 2017, corresponding to 5.5 wt% using the lower heating value (LHV) of hydrogen of 121 MJ kg^{-1}. Table 11.2 compares the energy densities of various fuels and storage devices. The energy density by volume is most relevant when size is more important than weight.

Hydrogen can be produced in a number of ways:

(a) electrolysis of water

(b) reaction of a hydrocarbon with steam (a process called reforming)

(c) from biomass.

Hydrogen can be made in (a) using energy from carbon-free sources such as hydropower, solar, wind, and nuclear power. The associated carbon dioxide emissions are low. In (b) there is no net production of CO_2 if the hydrocarbon is produced from biomass, e.g. bioethanol or biodiesel. An example of the steam-reforming reaction is with methanol using copper-based catalysts:

$$CH_3OH + H_2O \rightarrow CO_2 + 3H_2$$

Table 11.2 Energy storage capacity of various fuels and stores including weight and volume of storage containers

Fuel/store	Wh litre^{-1}	Wh kg^{-1}	Fuel/store	Wh litre^{-1}	Wh kg^{-1}
Diesel	9950	11 890	Liquid H_2	1400	1900
Petrol	8990	12 070	H_2 (34.5 MPa)	600	1800
Dry wood	2720$^\$$	5430	Hydride*	400	400
Ethanol	5910	7490	NaNiCl$^{\$\$}$	160	100
Methanol	4430	5580	Lead–acid	70	30
Coal	5500	6500	Li-ion	300	125

*300 °C; $^\$$density 0.5 kg litre^{-1}; $^{\$\$}$known as a ZEBRA battery
Source: StoreA, StoreB, StoreC.

As described in Chapter 8, hydrogen can also be produced from biomass using micro-organisms.

Hydrogen can also be made within the fuel cell itself. This is the process used in the direct methanol fuel cell (DMFC) (UCSD05). This cell contains a plastic electrolyte which is permeable to H^+ ions, as in a PEMFC, but methanol rather than hydrogen is injected into the anode. The temperature of operation is slightly higher than in a PEMFC at 120 °C, at which temperature the methanol reacts with water in the presence of a platinum catalyst to produce carbon dioxide, hydrogen ions, and electrons. At the cathode the reactions are the same as in the PEMFC, i.e.

$$CH_3OH + H_2O \rightarrow CO_2 + 6H^+ + 6e^- \quad \text{at anode}$$

$$\frac{3}{2}O_2 + 6H^+ + 6e^- \rightarrow 3H_2O \qquad \text{at cathode}$$

$$CH_3OH + \frac{3}{2}O_2 \rightarrow CO_2 + 2H_2O \qquad \text{overall reaction}$$

The DMFC is less developed than the PEMFC and has a lower operating voltage of ~0.4V, but requires neither hydrogen storage nor separate cooling to remove heat.

11.18 Outlook for fuel cells

Fuel cells have received a lot of attention throughout the world. They can provide carbon-free electricity with very low emissions and good efficiencies of ~50%. A fuel cell has no moving parts, so it is vibration-free, quiet, and reliable. There are many types of feedstock that can provide hydrogen either directly or indirectly as in the direct methanol fuel cell. Many of these supplies can be obtained using renewable sources such as hydro, wind, and solar power, or nuclear reactors. Hydrogen can be transported by pipeline or truck or can be reformed locally.

However, for the process of hydrogen production to be regarded as low-carbon it must come from renewable sources, and then the efficiency of the overall process can be an important factor. The efficiency, defined as the output divided by the input electrical energy, when using electrolysis to produce hydrogen and then a fuel cell to generate electricity, is about 25%—electrolysis ~70%, compression and transport of H_2 ~80%, and fuel cell ~45%. This can be compared with the efficiency for a battery driven electric vehicle of ~70%—transmission of electricity ~90%, battery charging ~85%, and electric drive ~90%. This difference means that the CO_2 emission reductions when using renewable energy are considerably larger for electric than for fuel cell powered cars if the hydrogen is from the electrolysis of water using electricity derived from renewable energy, as the energy is used more efficiently. But if hydrogen is produced thermochemically using solar energy or from excess low-carbon electricity, then there is very little emissions penalty.

The advantages of a hydrogen fuel cell compared with a battery powered car are the increased range and the speed of refuelling. The increased range when compared with a

Li-ion battery electric car reflects the difference in both the volumetric and gravimetric energy densities (see Table 11.2). Refuelling with hydrogen takes about 3–5 minutes, i.e. similar to refuelling with petrol or diesel (at 35 MJ per litre of petrol), so refuelling a tank of 60 litres in 3 minutes corresponds to a power of ~10 MW. Recharging a battery, though, could take several hours unless battery swaps are available.

Hydrogen storage could also be used on a very large scale. To cope with the variability of electricity from wind and solar generators, Germany is considering using huge electrolysers and fuel cells to store and deliver electricity when renewable generation is insufficient. Even though the efficiency of the system is low (~33%), it may prove to be the most cost-effective method. Some hydrogen could be mixed into the existing gas pipelines and used to generate electricity from gas-fired generators, or it could be pumped into salt caverns and then used to power fuel cell generators.

Fuel cells are now becoming more competitive with diesel generators; and they can be quieter, less polluting, and more reliable. As the cumulative global production of fuel cells increases, cost reductions are expected (see Section 12.3.3, Learning curve estimation). Considerable improvements have been made since the developments for the NASA space programme—the amount of catalyst required per square centimetre (~0.4 mg) is now over a factor of 10 smaller and the platinum catalyst is no longer the main cost. Fuel cells are now being used in portable electronics such as laptops, and in some cars and trucks, and many more uses are expected to be found over the next decade.

SUMMARY

- The emf generated by a coil of N turns and area A, rotating in a uniform magnetic field B with angular velocity ω, is given by

$$V = NBA\omega \sin \omega t$$

- Large-scale electrical generators consist of a rotor and a stator. The magnetic field is produced in the rotor by passing direct current through coils, and the rotating magnetic field due to the rotor induces a current in the stationary coils of the stator.

- Electricity is usually generated as three-phase current, which never drops to zero, delivers constant power through a resistive load, and uses less material to conduct a given quantity of power than single-phase current at the same voltage.

- The fractional loss of power along a transmission line is inversely proportional to the operating voltage.

- Transformers provide a convenient and efficient means of increasing or decreasing the voltage of a transmission line. To transform from a voltage V_1 to a voltage V_2, the ratio of the number of turns is given by

$$\frac{N_2}{N_1} = \frac{V_2}{V_1}$$

- Most high voltage transmission is AC but DC is used for very long lines and for underground and underwater transmission.

- Thyristors are used to rectify high voltage AC to DC and to convert back from DC to AC.

- Pumped storage uses cheap electricity to pump water from a low-level reservoir to an upper reservoir. The water is discharged during periods of peak demand or to compensate for the sudden loss of a generating unit on the electricity grid.

- Compressed air energy storage facilities are useful for smoothing out power fluctuations, but finding suitable sites is a problem.

- Thermal storage has been used for a long time in buildings. Large tanks of salt have been used to store solar energy very effectively.

- Flywheels are efficient devices for smoothing power fluctuations but the capital cost is significant, so applications are restricted to niche markets. Advances in lightweight materials and magnetic bearings have enabled 100 000 rpm to be achieved.

- Superconducting magnetic energy storage (SMES) is an efficient system providing very high power for short periods and for manufacturing processes where ultra-smooth power supply is important. The disadvantages are that it only operates at low temperatures and the capital cost is high.

- Lead–acid batteries are used as a back-up supply. Newer battery systems, such as Li–polymer, Na–S, and vanadium-redox flow batteries, are extending the range of large-scale applications. Li–air batteries with potentially very high specific energy densities are under development.

- Supercapacitors are useful energy stores that complement batteries.

- Fuel cells provide carbon-free electricity with very low emissions and efficiencies of around 50%, but low-carbon production of hydrogen is required, such as with high-temperature electrolysis using solar generated electricity. Fuels cells are becoming more competitive with diesel generators and are quieter, less polluting, and more reliable.

FURTHER READING

Fletcher, S. (2011). *Bottled lightning.* Hill & Wang, New York. From electric cars to smart power grids, and the role of lithium in future energy technologies.

Sullivan, J. and O'Loughlin, R. (2012). *Electric power transmission.* Nova Science, Hauppauge, NY. Discussion of electric power transmission, smart grids and related policy issues.

WEB LINKS

battery.berkeley.edu/ More detailed discussion of batteries.

www.chem1.com/acad/webtext/elchem/ec1.html Useful notes on electrochemistry.

www.chem.mun.ca/courseinfo/c1051/plw1051/15ChemicalEquilibrium.pdf Useful slides on chemical equilibrium.

www.energy.gov US Department of Energy website, appraising current programmes of research on batteries and fuel cells.

www.fuelcellsuk.org UK Hydrogen and Fuel Cell Association.

www.mpoweruk.com/index.htm Useful information on batteries.

web.missouri.edu/~puckettj/ChemistryTablesfolder/StandardReductionPotentials298K.pdf Standard reduction potentials.

www.smartgrids.eu/ Smart Grids European Technology Platform.

www.smartgrid.gov/ DOE smart grids.

www.wikipedia.org Online encyclopaedia, general information on topics in this chapter.

workspace.imperial.ac.uk/energyfutureslab/Public/Lithium%20Batteries%20-%20Going%20the%20Distance%20-%20Allan%20Paterson.pdf Lithium-ion battery technology.

www.nrel.gov/csp/troughnet/pdfs/2007/41422.pdf (DOE2007).

bestar.lbl.gov/venkat/files/batteries-for-vehicles.pdf (LBL2008).

www.pnl.gov/fuelcells/ (StoreA).

hydrogen.energy.gov/annual_review12_storage.html (StoreB).

escholarship.org/uc/item/7425173j A. Burke and M. Gardiner, Hydrogen storage (StoreC).

chemelab.ucsd.edu/fuel05/index.htm Methanol fuel cell miniaturization (UCSD05).

LIST OF MAIN SYMBOLS

φ	magnetic flux	Q	reaction quotient
A	area	R	resistance
B	magnetic field	t	time
E	energy	T	torque
G	Gibbs free energy	V	voltage
I	current, moment of inertia	θ	angle
K	equilibrium constant	μ_o	magnetic permeability of free space
N	no. of turns	ρ	resistivity
P	power	ω	angular velocity
p	pressure	Ω	angular velocity

EXERCISES

11.1 How many turns are needed for a coil consisting of cross-sectional area $0.1\,m^2$ rotating at $50\,Hz$ in a uniform magnetic field of $0.5\,T$ to generate a current of $1000\,A$?

11.2 Discuss the feasibility of replacing the brushes in a generator by a liquid metal contact.

11.3 Prove that the sum of the currents in a three-phase supply is zero.

11.4 Propose a circuit to generate three-phase electricity.

11.5 Prove that the power dissipated through a resistance by a three-phase current is independent of time.

11.6 A coil with 500 turns of cross-sectional area $2\,m^2$ rotates at 60 Hz in a uniform magnetic field of 0.3 T. What is the maximum current flowing through a load of $10\,000\,\Omega$?

11.7 A power plant transmits 2000 MW of power along a transmission line of length 10 km with a resistance of $0.02\,\Omega\,km^{-1}$. Calculate the percentage loss of power if the line is at (a) 110 kV; (b) 400 kV.

11.8 Compare the advantages and disadvantages of using copper instead of aluminium for overhead high voltage lines.

11.9 It is desired to increase the reliability of part of a grid by installing extra components. Assuming the components have a probability of failure of 0.01, what is the total probability of failure if two components are connected (a) in series; (b) in parallel?

11.10 It is desired to reduce the voltage of a transmission line from 400 kV to 12 kV. Calculate the ratio of the number of turns from the primary to the secondary side of a transformer.

11.11 A pumped storage plant has a head of 600 m and a working volume of $500\,000\,m^3$. How much power can be generated if the plant is required to operate at maximum output for 2 hours a day? Assume the efficiency of generation is 85% and the density of water is $10^3\,kg\,m^{-3}$.

11.12 A wind farm produces an average output of 250 MW. A pumped storage facility is planned to provide up to 300 MW for 5 days. The drop from the upper reservoir, whose depth is 30 m, to the power generator is 500 m. Estimate the area required for the wind farm and the area of the reservoir.

11.13 Estimate the energy stored in a compressed air energy storage chamber with a capacity of $100\,000\,m^3$, if the pressure of the air in the chamber is 50 bar. Assuming the overall efficiency of the plant is 30%, estimate the electrical power output if the air is discharged over a period of 1 hour.

11.14 Consider a ring of inner radius a and outer radius $a+t$ with a square cross section ($t \ll a$) that is rotating at an angular velocity of ω. The material has a maximum tensile stress σ_m and a density ρ. Deduce that ω_m, the maximum value of ω, is given by $\omega_m^2 = \sigma_m/(a^2\rho)$, and that the maximum kinetic energy per unit mass equals $\tfrac{1}{2}\sigma_m/\rho$.

11.15 A uniform cylindrical flywheel of radius 10 cm rotates at 100 000 rpm. Calculate the kinetic energy of the flywheel.

11.16 A solar heated thermal store is made from 30 kg of zinc, which melts at 420 °C. What is the maximum amount of energy (in MJ and in kWh) available if the maximum temperature that the zinc is heated to is 600 °C and the lowest useful temperature of the store is 150 °C? (Specific heat of zinc $= 0.39\,kJ\,kg^{-1}\,°C^{-1}$; latent heat of zinc $= 118\,kJ\,kg^{-1}$.)

11.17 The 50 MWe Andasol concentrated solar thermal power plant uses two tanks of molten salt as its thermal store. Each tank contains 28 500 tonnes of a 60%–40%

mixture of sodium and potassium nitrate which has a specific heat of $1.47\,kJ\,kg^{-1}\,°C^{-1}$. The tanks operate between 393 and 293 °C. What is the amount of energy stored in each tank in MWh (thermal)? How long can the plant operate at night if the efficiency of the 50 MWe plant is 16%?

11.18* The equilibrium constant for the water–gas shift reaction,

$$CO(g)+H_2O(g) \rightleftarrows CO_2(g)+H_2(g)$$

is approximately given by

$$K=\exp\left(-4.33+\frac{4577.8}{T(K)}\right)$$

Calculate the equilibrium concentrations of the gases at $T=750\,K$ if only CO and H_2O are presently initially, at concentrations of $0.2\,mol\,litre^{-1}$.

11.19* An approximation for the discharge reaction in a lead–acid battery is

$$Pb(s)+PbO_2(s)+2H^+(aq)+2HSO_4^-(aq) \rightarrow 2PbSO_4(s)+2H_2O(l)$$

The standard Gibbs free energy values in $kJmol^{-1}$ are: Pb(s) 0.0; PbO_2(s) −217.3; H^+(aq) 0.0; HSO_4^-(aq) −755.9; $PbSO_4$(s) −813.0; H_2O(l) −237.1

Show that the standard voltage for this cell is 1.92 V.

The standard state of a liquid (l) or solid (s) is the pure substance, so the activities (concentrations) of Pb(s), PbO_2(s), and $PbSO_4$(s) are constant and equal to 1. Assuming that the activity (concentration) of H_2O is unaffected by the dissolved sulfuric acid, calculate, using the Nernst equation (eqn (11.17)), the cell voltage when the molality of the sulfuric acid is 5, 3, 1, 0.5, 0.1, and 0.02 molal.

11.20 Calculate the theoretical specific energy density of a sodium–sulfur battery, whose overall reaction can be represented by $2Na+3S \rightarrow Na_2S_3+\sim1.9\,V$.

11.21 Calculate the specific energy of a battery that would be comparable to that of an internal combustion engine running on (a) petrol; (b) bioethanol. State your assumptions.

11.22 Write an article of 1000 words for a popular science magazine on the prospects for fuel cells. Do not assume that the readers have any prior knowledge of the subject.

Energy and society

✔ List of Topics

- ☐ Global warming
- ☐ Impact of more CO_2
- ☐ Energy-efficient technologies
- ☐ Cash flow analysis
- ☐ Learning curve estimation

- ☐ Life-cycle analysis
- ☐ Risk–cost–benefit analysis
- ☐ Designing safe systems
- ☐ Carbon abatement policies
- ☐ Strategies to reduce CO_2

→ Introduction

The amount of energy consumed per capita by any country is closely related to its standard of living. As the developing countries of the world become more industrialized and their population grows, their demand for energy increases. Most of our energy is currently produced from fossil fuels. If this does not change, the projected increase of carbon dioxide in the atmosphere is such that the world is at risk from dangerous climate changes due to increased global warming.

We begin by reviewing the environmental impact of energy production, in particular that of global warming. We then explain the importance of energy-efficient technologies before discussing the economics of energy production, which is a key factor affecting the uptake of more low carbon energy. We also show how the current and future costs of electricity can be estimated and look at how the risks and benefits of a particular technology can be assessed.

The proven reserves of fossil fuels are 46 and 59 years for conventional oil and gas, respectively, and 118 years for coal. The estimated amount of unconventional oil and shale gas approximately doubles the oil and gas reserves. The alternatives are more expensive at present, so there is little immediate economic incentive to reduce consumption of fossil fuels, although there may be reasons of energy security for a reduction. Priority therefore needs to be given to mitigating the world's dependence on fossil fuels.

We review the contribution that energy savings (i.e. reducing the demand for energy) can make before outlining some of the mitigation policies, such as carbon emissions trading, which are already being implemented. Finally, we discuss strategies to avoid the threat of damaging climate change that current projections indicate.

12.1 Environmental impact of energy production

12.1.1 Global warming

The concentration of carbon dioxide in the Earth's atmosphere has risen sharply over the last 50 years. As explained in Chapter 1, water vapour and carbon dioxide are the two main greenhouse gases that trap the infrared radiation emitted by the Earth and thereby raise the temperature of the Earth's surface. The current level of CO_2 is 390 ppm, compared with 312 ppm in 1955 and 280 ppm in the pre-industrial era before ~1750 (Fig. 12.1(a)).

Analysis of ice cores shows that over a much longer period (see Fig. 12.1(b)) there have been large fluctuations in the concentration of CO_2 in the atmosphere. These variations are mostly related to slight periodic changes in the Earth's orbit about the Sun (and hence the solar irradiance), called Milankovitch cycles after the Serbian mathematician who postulated their effect on climate in the 1920s. These cycles are linked to previous ice ages—the change in solar irradiance ($W\,m^{-2}$) triggering a change in temperature.

The end of the last ice age is thought to have been due to a rise in temperature associated with a Milankovitch cycle in which the huge ice sheets covering Canada and Europe melted. The

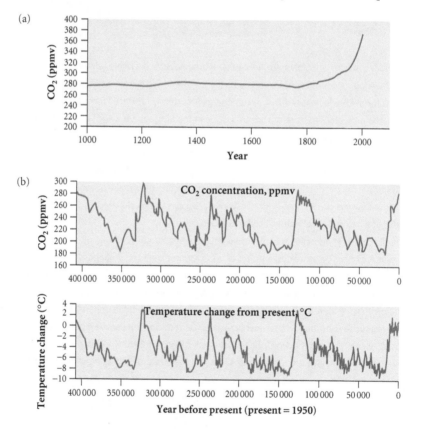

Fig. 12.1 (a) Carbon dioxide concentration over the last 1000 years. CO_2; (b) Fluctuations in CO_2 concentrations and temperature over the last 400 000 years.
Source: CO_2, Vostok.

resulting flow of fresh water into the North Atlantic disrupted the heat flow to northern latitudes and caused the Antarctic to warm and release CO_2. The increase in CO_2 concentration gave rise to an increase in the mean global temperature. The existence of a correlation between global temperature and atmospheric carbon dioxide concentration does not indicate which one caused the other, nor indeed whether both were the result of some other effect. In order to address these questions it is necessary to look closely at what is affecting the Earth's climate.

The rise in the global mean temperature of about 0.5 °C witnessed over the last 50 years is correlated with the anthropogenic (human-induced) increase in CO_2 during this period. While some of the observed temperature changes (see Fig. 1.10) are caused by variations in solar irradiance and by natural phenomena such as volcanic activity, the rise over the period 1970–2000 can only be explained if the anthropogenic change in CO_2 concentration is also taken into account. As we pointed out in Chapter 1, the predicted temperature variation due to natural causes alone over this period actually shows no rise (see Fig. 1.10). The actual temperature has been rising over the last 40 years, with 10 of the warmest years on record occurring in the period 1998–2010. It is this temperature rise over the last few decades that is often referred to as global warming.

Besides the higher temperatures there is a vast amount of corroborating evidence for global warming in recent years. The ice cover in the Arctic regions has decreased significantly. In September 2012 the extent of the ice reached a new record low, dropping below 4 million square kilometres, which is a 45% decrease compared to September conditions in the 1980s and 1990s. The Greenland ice sheet is also disappearing at a rate of ~200 cubic kilometres per year, and glaciers worldwide are in retreat. For example, in Argentina, the Uppsala glacier— once the biggest in South America—is now disappearing at the rate of 200 m per year.

The warming climate has affected coastal ice shelves in the Antarctic, but since these are afloat they do not affect the sea level when they melt. Inland in the Antarctic, nearly all of the ice is so cold and high up that it remains frozen all the time.

An analysis of ocean temperatures over the last 40 years has shown an increase of 0.5 °C in surface temperatures and 0.15 °C at greater depths, consistent with the predictions of climate models that include the anthropogenic increase in CO_2. Marine life has also been affected. The population of whales and walruses has dropped sharply off Alaska, affecting the livelihood of the Eskimos. Many species of plankton, an important part of the marine food chain, are moving to cooler waters. Large amounts of coral worldwide are dying; there has been a large loss of coral reef in the Caribbean off Puerto Rico and the Virgin Islands.

The effects of large-scale global warming could be catastrophic for many parts of the world. It could accelerate the melting of glaciers in the Andes and western China, depriving millions of people of enough water as the seasonal stores that the glaciers provide, which hold water in the winter and release it in the summer, are lost. And, once melted, there would be a significant loss of water due to evaporation. A rise in sea levels would flood low-lying regions, leading to massive movements of populations. Over the last century alone sea levels have risen ~20 cm. The predicted rise over the period 2000–2100 is about 0.4 m for a temperature rise of 3.6 °C; a rise of between ~1 °C and ~6 °C is predicted by the IPCC. A warming in excess of 3 °C could initiate the melting of the Greenland ice sheet and lead to its disappearance, raising sea levels by 7 m within 1000 years. A 1 m rise would submerge a significant fraction of Bangladesh.

Some extreme weather conditions, such as hurricanes, El Niño events, and heat waves, could be affected by global warming. For certain regions, the intensity, duration, and frequency of some of these events are predicted to increase. The change in the salinity of the oceans from ice melting could affect the ocean flow patterns. The 'Gulf Stream' carries warm water from the Tropics and the Gulf of Mexico up to the North Atlantic, where it heats the atmosphere; this in turn helps to keep Europe warm. The evaporation and freezing of the surface water in the Gulf Stream in the North Atlantic increases its salinity, and hence its density, and the water sinks and then flows south at depth. This thermohaline circulation could be slowed by a decrease in the salinity of northern waters, and that could have significant climatic effects.

There would also be an increased threat to human health from higher temperatures, particularly in tropical and subtropical regions. Food production and water supplies would be affected. Biodiversity is also likely to be irreversibly altered, with the possible extinction of some 20–30% of plant and animal species, assuming an increase in global temperature of 1.5–2.5 °C.

12.1.2 Impact of more CO_2 in the atmosphere

The amount of carbon dioxide in the atmosphere was about 3000 Gt in 2005, corresponding to a CO_2 concentration of 375 ppmv, or 8 GtCO$_2$ per ppmv CO_2 (ppmv means parts per million by volume). If a pulse of 8 Gt of carbon dioxide is released into the atmosphere the concentration of CO_2 initially rises by 1 ppmv and then decays—fairly quickly at first and then slowly over a period of a few hundred years (see Exercise 12.5* for an approximate expression for the time dependence) until after 100 years about a third of the CO_2 remains in the atmosphere. The global warming caused by an emission of 1000 Gt of CO_2 would be ~0.5 °C, most of which would be achieved after 30 years. The amount of carbon dioxide emitted into the atmosphere was ~31 Gt y^{-1} in 2010 and this is predicted to rise to ~60 Gt y^{-1} by 2050 if there is no action to reduce carbon emissions. The resultant rise in CO_2 and in global temperature would be ~75 ppm and ~0.6–1.2 °C, respectively, or a rise in temperature of ~1.5–2.1 °C since pre-industrial times.

However, there are effects that can amplify or diminish the interdependence. Examples of amplifying effects are the increase in areas of open (dark) sea, due to ice melting, and the increase in water vapour, due to increased evaporation. Water vapour is the major greenhouse gas in the atmosphere, and solar radiation is absorbed by open sea, rather than being reflected by ice, so both contribute to further warming. A rise in temperature could also lead to the release of methane frozen under the Siberian permafrost; methane could also be released from methane hydrate deposits in ocean sediments. As methane is roughly 25 times as potent a greenhouse gas as CO_2, any release of methane would further increase global warming. These are examples of positive-feedback effects. Should one of these cause global warming to reach a threshold, then we could have what has been called a tipping point, which is an effect that triggers dangerous climate change (NB Definitions of 'tipping point' vary).

Water vapour can form clouds and these reflect solar radiation, which causes a cooling. However, clouds also can cause a warming by absorbing the infrared radiation emitted by land or sea at a higher temperature than the cloud. The net result is a cooling but there is a significant uncertainty in predictions because of the effect of clouds.

An important cooling mechanism is the absorption by the oceans of heat from the atmosphere. However, their ability to do this is expected to decrease as the surface temperature of the oceans increases, because mixing of the less dense warm surface water with the colder denser deep water decreases. The solubility of CO_2 also decreases as the temperature of the water rises, reducing the ability of the ocean to act as a carbon dioxide sink. The ocean is the largest sink of CO_2 and has absorbed about a third of the ~500 Gt of carbon (~1800 Gt of CO_2) emitted from the burning of fossil fuels, land-use change, and cement production over the last two centuries. The deep, cold waters still contain pre-industrial concentrations of CO_2, and full mixing of the ocean takes over 1000 years. Hence the ocean will continue to act as a vital sink for many centuries. It is currently absorbing about 2 Gt of carbon per year.

Absorption of CO_2 by the land has depended on the relative scale of land clearance for agriculture and the uptake by re-growth of vegetation, which may be aided by higher CO_2 concentrations. However, growth could be adversely affected by drier and hotter climates. It has been estimated that the land will act as a net sink of ~1–2 Gt carbon per year, though this estimate is very uncertain.

Pollutants from fossil fuels, burning forests, and crop wastes can form aerosols that disperse worldwide. (Aerosols are tiny liquid and solid particles suspended in the air.) Many of these are damaging to health, but the overall effect of these aerosols is to reduce the rise in temperature by shading the Earth. It is expected that the level of aerosols is likely to rise less quickly than that of CO_2 because of emission controls, so the effects of CO_2 on temperature will be exacerbated. Aerosols also have a considerably shorter lifetime in the atmosphere than CO_2.

The prediction of the IPCC in 2007 was that the global temperature would rise over the period 1990 to 2100 by between 1.1 °C and 6.4 °C, the amount varying according to different assumptions, with the business-as-usual (BAU) fossil fuel intensive scenario, called A1F1, giving a rise of 4 °C. For a range of emission scenarios, the temperature rise over the next two decades is predicted to be 0.2 °C per decade. The expected impacts of different temperature rises since pre-industrial times, as given in the Stern and IPCC reports (2007), are as follows:

- 1–2 °C rise (mild impact): loss of small glaciers in the Andes, threatening the water supply for 50 million people; in parts of South Africa and South America, 20–30% less water available; up to 10 million people affected by coastal flooding; 40–60 million people in Africa exposed to malaria.

- 3 °C rise (significant impact): southern Europe: a severe drought expected every decade; 1–4 billion people suffer water shortage; 150–550 million people at risk from lack of food; up to 170 million people affected by coastal flooding; 20–50% of animal species may face extinction.

- 4 °C rise (strong impact): Mediterranean and southern Africa: up to 50% less water available; agricultural yields in Africa drop by 15–35%; up to 300 million people at risk from coastal flooding; half of Arctic tundra lost; West Antarctic ice sheet may collapse causing significant sea level rise.

- >5 °C rise (catastrophic impact): most Himalayan glaciers will disappear with loss of water for hundreds of millions of Indians and Chinese; London, New York, and Tokyo and other cities at risk from the rise in sea level; increase in ocean acidity threatens fish stocks; CO_2 release from soils and methane from melting permafrost may trigger positive feedback on the climate—a 'tipping point'.

As a result of these dire forecasts there is now an overwhelming consensus in the scientific community that decisive action must be taken now to reduce our carbon emissions if we are to reduce the risk of potentially catastrophic effects from climate change. These emissions can be reduced by introducing energy-saving measures as well as low-carbon sources of energy. We will first look at energy savings, which can come from energy efficiency and energy conservation, before looking at economic considerations that limit the adoption of low-carbon technologies.

12.2 Importance of energy-efficient technologies

The global emissions of carbon dioxide are related to population, GDP, and energy consumption through the Kaya identity:

$$CO_2 \text{ emissions} = \text{population} \times \frac{\text{GDP}}{\text{population}} \times \frac{\text{energy}}{\text{GDP}} \times \frac{CO_2 \text{ emissions}}{\text{energy}}$$

Improvements in welfare—in particular, education and access to work for women, and family planning—have meant that the global birth rate is now close to two per family, and there will be the same number of children under 15 (2 billion) in 2050 as there are now. Since there are currently correspondingly fewer older people, e.g. between 40 and 55, the population will increase and stabilize at close to 10 billion after ~2050.

Population control and wealth (GDP/population) reduction raises ethical, social, and economic concerns, so attention has focused on energy intensity (energy/GDP) and carbon efficiency (CO_2 emissions/energy). Energy intensity is improved by energy efficiency and energy demand reduction, and carbon efficiency by energy supply switching to lower-carbon sources of energy (e.g. coal to gas).

The expected future demand for energy can be very significantly reduced by making energy savings, either through improved efficiency, which refers to improvements in technology (e.g. LEDs rather than incandescent bulbs), or through conservation measures, which refers to gains arising from changes in lifestyle (e.g. taking public transport rather than driving a car). Industry, buildings, and transport are the major areas where large energy savings are needed, since they account for about a third, a third, and a quarter, respectively, of global energy-related CO_2 emissions.

In order that the world can keep on track to reduce its emissions to zero by 2100 and thereby restrict the global warming to less than ~2 °C, and thus reduce the risk of dangerous climate change, energy-related CO_2 emissions in 2050 need to be about half what they were in 2010, i.e. ~15 Gt as against ~31 Gt per annum. The IEA has estimated that emissions will approximately double by 2050 under a BAU scenario where the world remains reliant on fossil fuel energy. So, to halve our emissions by 2050 will require finding energy savings and decarbonized energy supplies equivalent to 43 $GtCO_2 y^{-1}$ or ~10 000 GWe (assuming an equivalence of ~0.5 kgCO$_2$/kWh). About 17% and 19% of the energy savings could come from improvements in the building and industry sectors, respectively, but an even greater reduction might be possible in the transport sector (see Fig. 12.2).

The IEA BLUE map scenario (IEA-ETP-2008/2010) is an optimistic and aggressive scenario that now looks unlikely to be achieved given the slow pace of change and that much of the present infrastructure will still be in use in 2050. The scenario assumes rapid global adoption of a wide-ranging series of carbon-reduction measures. On the supply side, CCS of both coal and gas plants, together with improved efficiency, generation III and IV nuclear, solar PV and CSTP, and wind all contribute to the decarbonization of the power sector. Biomass and biofuels could contribute in the industrial, air, and shipping sectors, but emissions associated with change of land use need to be avoided.

We will now look at the emissions reductions that might be possible in the industry, buildings, and transport sectors. However, these gains can be difficult to realize in practice, partly because of a lack of capital in the developing world, while in the developed world the rebound effect can negate the savings. This refers to a situation in which money saved through efficiency improvements is spent on activities or energy-consuming devices whose associated emissions (e_a) offset those saved through improved efficiency (e_s); $e_a > e_s$ is referred to as the 'Jevons paradox'. For example, improving the efficiency of a ship's engines would result in a reduction in emissions provided the total mileage remained the same. But the emissions could rise if the reduction in the cost of transport opened up new markets that increased the ship's total mileage significantly. Historically there have been many instances of the rebound effect. There can also be hidden costs to improving efficiency, e.g. the cost of redecoration after installing cavity wall

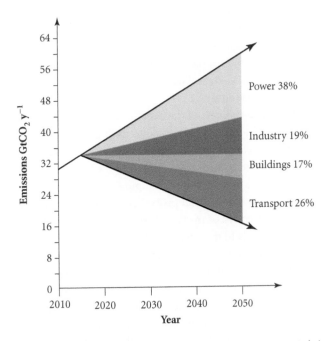

Fig. 12.2 Sectoral emissions reductions in the IEA 450 ppm or BLUE map scenario (schematic).
Source: IEA-ETP-2008/2010.

insulation, and the disruption to production while installing new technology. As a result it has been argued (Keay2011) that many predicted savings may well be too optimistic. Improving energy efficiency helps meet the demand for energy, but to reduce emissions it needs to be part of a low-carbon strategy that includes regulation and a sufficiently high carbon price.

12.2.1 Emissions reduction in the industry, building, and transport sectors

The buildings sector provides the largest potential for emissions reduction, estimated by the IPCC (fourth assessment report) to be ~6 Gt CO_2 y^{-1} at a cost of less than \$100 per tonne of CO_2, while the corresponding reductions for the industry and transport sectors are about 4 and 2 Gt per annum, respectively. In the buildings sector about 2 Gt per year could be provided by 2020 at a net cost savings. In 2010 our emissions were ~31 Gt y^{-1}.

New buildings can achieve the largest savings in energy by minimizing energy loss and maximizing passive energy inputs: the passive house (passivhaus) principle. Passive energy gains are solar heat and heat from internal sources, such as appliances and people (~100 W per person). Electricity consumption can be reduced by, for example, improving the efficiency of domestic appliances, using natural lighting and changing bulbs to LEDs. Heat loss is reduced by using insulation with a very low U-value (see Section 2.3) and triple glazing, and by ensuring air-tightness. Ventilation with heat recovery then provides fresh air. The ground under the building can be used as a thermal store to even out the seasonal demand for heating (or cooling), and internal thermal mass will help reduce high temperatures in summer. The remaining heating and cooling demand is then very small and can be provided very effectively by heat pumps. These techniques, which will also need to be applied at greater cost to existing buildings, can reduce the energy demand per square metre of floor space in a building from ~150 kWh m^{-2} y^{-1}, a typical energy demand for a UK home built in ~2000, to ~15 kWh m^{-2} y^{-1} (1 kWh $m^{-2} \equiv 317$ Btu ft^{-2}), with a 50–80% reduction more typical. Zero energy use is possible through incorporating renewable sources of energy. How the challenge of reducing energy demand has been met in a hot climate is illustrated in the case study carried out at Shinawatra University in Thailand (see Case study 12.1).

However, a general lack of awareness of the benefits and entrenched practices are hindering progress. There is a need for higher appliance standards, building codes, and energy efficiency requirements. Higher fossil fuel costs would also provide an incentive. It is particularly important that new buildings are built to high standards since they last for many decades. How this might be achieved at relatively low cost in the developing world is discussed in Case study 12.1.

In the industry sector, CCS in industrial processes and more efficient systems help on the demand side. The energy intensity of many industrial processes is at least 50% greater than the theoretical minimum and is higher than what is possible with available technology. The efficiency of electric motor driven systems can also be improved. Furnaces and process heaters can be further optimized to reduce energy consumption. Globally 43% of emissions in industry is associated with steel and cement production, and demand for these is expected to double by 2050. While it is possible with efficiency improvements plus CCS to reduce cement emissions by half by 2050, this will only be possible for steel emissions if demand is reduced through, for example, reusing components, making objects lighter, substituting other materials, and extending the life of objects by making individual components replaceable.

Case Study 12.1 Energy-efficient building at Shinawatra University, Thailand

Thailand is a hot and humid country which has undergone rapid urbanization in the last few decades, and most people now live in high-density conurbations. Air conditioning has become the standard way to keep cool, unlike in the traditional Thai wooden house, which has natural ventilation and shady areas to escape the sun. The situation is exacerbated by the fact that urban buildings are typically constructed with reinforced concrete, brick, plaster, and concrete roof tiles, which act as heat sinks. Air conditioning and other energy consuming devices for modern lifestyles have led to a massive rise in energy consumption in recent years. Given the general lack of skilled workers in the construction industry and the need for high quality heat insulating materials to be imported, a new approach to construction is needed that makes best use of the natural resources and significantly reduces energy demand.

A novel solution to the problem has been developed at Shinawatra University in Bangkok for an energy-efficient building that incorporates a number of different functions. It is a concept that also works on a small scale, owing to its modular nature, and provides a model for other countries with similar climates and resource limitations. The challenge was to use indigenous materials and a largely unskilled labour force to create an attractive, multifunctional environment that significantly reduced the energy usage for cooling purposes. The design minimizes the amount of hard-top surface on the outside of the building and uses natural vegetation and water resources to create a comfortable microclimate in three separate zones: a lecture theatre, a laboratory, and a canteen. The specific requirements of each zone (lighting, temperature, humidity, and wind speed) are satisfied in such a way as to minimize the overall energy consumption of the building. Pond cooling is used to reduce the cost of air conditioning in the lecture theatre and the computer centre in the laboratory, the dining area is cooled by rerouting cool air streams, the heat generated in the canteen is vented through the upper openings of the building using the stratification effect, the large thermal mass of concrete is exploited to provide thermal stability, and glazing material is chosen that allows the penetration of the visible part of the spectrum but not the infrared. The net result is that the lighting and air-conditioning systems use only about one-sixth of the electrical energy of typical buildings. The capital cost is somewhat larger than that for a typical building but this is soon paid off by savings in running costs. Thus, the concept makes economic sense, reduces the carbon footprint, and significantly improves the quality of life of the people who use the building.

In the transport sector, the immediate gains will be achieved by improving the efficiency of petrol and diesel ICEs. Improvements in the last couple of decades have tended to be used to increase the size, weight, and power of vehicles, rather than their fuel economy. Regulations to ensure that cars have good fuel economy and hence low emissions per km will be required, and a recent study by the UK Environment Agency concluded that a limit of $60\,\mathrm{gCO_2}$ per km for cars will be needed. Ultimately, the introduction of electric vehicles coupled with a decarbonized electricity supply or, if cheap hydrogen made from low-carbon energy is available, cars powered by fuel cells, will be needed.

For electric vehicles, the range and the cost of batteries are the present limitation, though plug-in hybrid electric vehicles (PHEV) could provide a useful bridging technology. PHEVs and EVs could be cost-competitive with conventional cars on a life-cycle analysis as fossil fuel costs rise; however, there is a tendency for people to be influenced more by the immediate cost than by the long-term fuel cost savings. PHEVs could enable a significant fraction of many people's daily car journeys to be made just using the battery supply of electricity, with longer journeys using the ICE to generate electricity. In Europe, 50% of trips are less than 10 km and 80% less than 25 km, while in the United States, ~60% are less than 50 km and 80% less than 100 km. PHEVs are an extension of the hybrids that are currently available, which use an ICE to generate electricity for an electric motor. One such hybrid uses an ICE with an Atkinson cycle, which has a longer exhaust stroke than an Otto cycle, which improves the motor's efficiency but reduces its power. Hybrids also take advantage of the storage of braking energy in the car's battery (regenerative braking) or in supercapacitors that electric motors afford. While these features improve efficiency, the most significant carbon savings would arise from the decarbonization of the electricity grid. Currently there are about 1 billion cars worldwide. Their average fuel economy is ~27, ~37, and ~47 mpg (US gallon) in the USA, China, and Europe or Japan, respectively. There are plans in the USA and Europe for a target of around 60 mpg (or $100 \, gCO_2 \, km^{-1}$) by 2020–2025. This target would produce savings of about $70 \, gCO_2 \, km^{-1}$ and, estimating the average annual mileage as 15 000 km, an annual global reduction in emissions of $\sim 1 \, GtCO_2$. With the present fuel economy of ~35 mpg, a similar reduction would be achieved by reducing the average annual mileage by 6000 km.

An additional benefit of battery-driven cars is that the batteries can be used to provide grid storage, which will become increasingly important as the percentage of renewable power increases. The network, though, would need strengthening to homes (or to garages if batteries were swapped). However, the price of batteries, currently about $500 per kWh, needs to decrease, and their energy density and the number of recharge cycles (from low to full charge) to increase, before PHEVs and EVs will have a large fraction of the market.

Urban planning can also have a significant effect through the building of communities that are close to bus and train stations and have cycle lanes so that cars are needed much less. Furthermore, the use of video conferencing can reduce travel. One counterproductive result of energy (and hence cost) savings is the rebound effect. For consumers, it appears that this has the effect of making the share of their budgets spent on energy roughly constant, regardless of the cost of energy.

Large as these energy savings are, we see from Fig. 12.2 that a significant amount of low-carbon energy will be needed to keep within ~450 ppm CO_2. Carbon dioxide reduction in the power sector is mostly obtained from introducing more low-carbon sources, and the amount required is about 3500 GWe, approximately equivalent to 14 Gt y^{-1} of CO_2, which is ~33% of the total reduction needed. The accessible potentials by 2050 of the main low-carbon sources, which have been described in earlier chapters, are shown in Table 12.1.

While the accessible potentials of low-carbon sources by 2050 total more than 3500 GWe, the challenge will be to ensure that they are actually deployed. There are two main reasons why the low-carbon technologies described in earlier chapters are not more widespread: the

Table 12.1 Accessible potential by 2050 of low-carbon sources of energy

Source	Potential (GWe)	Source	Potential (GWe)
CCS	~1000	Solar PV	~500
Hydro	~500	Solar CSTP	~500
Marine*	~100	Biomass	~200
Geothermal	~100	Nuclear	~1000
Wind	~800	Total	~4700

*tidal stream and wave

deployment of those close to being competitive with fossil fuels is partly limited by manufacturing capability, and for the rest it is largely a question of their cost. In Section 12.3 we summarize some of the main economic considerations that determine the cost of energy.

12.3 Economics of energy production

For all energy technologies we can break down the money involved into three parts. First, capital is required to build the plant. Second, money is needed to operate and maintain (O&M) the plant, and, finally, there is the revenue obtained from selling the energy. A simple measure of the competitiveness of a plant is the time required to pay back the cost of manufacture and installation, allowing for the annual cost of O&M. However, this payback time does not quantify the return on the investment. To calculate this we must take account of the fact that the value of revenue received in the future is worth less than if it were received today. For example, £100 invested today at 5% interest would be worth £105 after one year; conversely, the value of £105 of revenue in a year from now would have a present value of £100.

12.3.1 Discounted cash flow analysis

The translation of future money to its present value is called discounted cash flow analysis and the interest rate used is called the discount rate R. For N successive equal payments A, the present value V_P is given by

$$V_P = A[1-(1+R)^{-N}]/R \tag{12.1}$$

Derivation 12.1 Discount rate and present value

An amount of Q pounds invested today would be worth $Q(1+R)^n$ pounds after n years if the interest (discount) rate is R. So if A pounds are received in n years time, its present value is $A/(1+R)^n$. The present value V_P of a series of annual payments A made over N years is therefore

$$V_P = A/(1+R) + A/(1+R)^2 + A/(1+R)^3 + \cdots + A/(1+R)^N$$

Put $x=1/(1+R)$; then

$$V_p = Ax(1+x+x^2+\cdots x^{N-1})$$

Let $S=1+x+x^2+\cdots x^{N-1}$; then

$$xS = x+x^2+\cdots x^N = S-1+x^N$$

So $S = \left(1-x^N\right)/(1-x)$ and

$$V_p = Ax(1-x^N)/(1-x)$$
$$= A[1-(1+R)^{-N}]/R$$

Discounting is particularly important when revenue is expected over many years, as from a wind turbine scheme. In Example 12.1 we consider a simple case for a wind turbine where we ignore O&M costs and only take into account the capital cost of manufacture and installation and the annual revenue.

EXAMPLE 12.1

Calculate the present value of the revenue generated over a 30-year lifetime from a 3 MW turbine. The capital cost C_{capital} of the turbine is \$2 000 000 and the discount rate R is 6%. The cost of electricity charged by the wind turbine operator is 4¢ per kWh. The capacity factor of the turbine is ¼.

At a capacity factor of ¼ the energy produced per year E will be

$$E = 24 \times 365 \times \tfrac{1}{4} \times 3 \times 10^3 = 6.57 \times 10^6 \text{ kWh}$$

At 4¢ per kWh the revenue per year A is $(0.04 \times 6.57 \times 10^6) = \$263\,000$.

Substituting in eqn (12.1) with $N=30$ gives the present value V_p of the revenue as

$$V_p = 263[1-(1.06)^{-30}]/0.06 = \$3\,620\,000$$

This amount is less than half of what we would have calculated had we neglected the effect of the interest rate.

We are now in a position to calculate the cost of producing energy by this wind turbine and the rate of return on the capital invested to build it. If we subtract the capital cost from the present value of the revenue we get the net present value V_{NP}:

$$V_{\text{NP}} = V_p - C_{\text{capital}} \tag{12.2}$$

In this example $C_{capital}=\$2\,000\,000$ and $V_P=\$3\,620\,000$, so $V_{NP}=\$1\,620\,000$. As V_{NP} is positive the rate of return is greater than the discount rate, which is generally a requirement for investment. The rate of return R_{return} is given by finding the discount rate R_{return} that makes V_{NP} zero, i.e. when

$$C_{capital} = A[1-(1+R_{return})^{-N}]/R_{return} \tag{12.3}$$

In the above example, with the revenue per annum $A=\$263\,000$ and $N=30$, this is when R_{return} is 12.8%.

We can calculate the cost of energy C_{cost} by finding what annual revenue A_{cost} at the given discount rate R would make V_{NP} zero. This revenue A_{cost} is given by

$$C_{capital} = A_{cost}[1-(1+R)^{-N}]/R \tag{12.4}$$

The cost of energy C_{energy} is then obtained by dividing A_{cost} by the energy produced per year E, i.e.

$$C_{energy} = A_{cost}/E \tag{12.5}$$

In our example, the cost of energy is 2.2¢ per kWh.

Equations (12.4) and (12.5) show that the cost of energy is effectively the cost, per unit of energy generated, of tying up capital in the project for the time that the energy is generated. The difference between the revenue per unit and this cost is the profit per unit.

In Example 12.1 we ignored the time taken to build the wind turbine. To compare costs and revenues that occur at different times we measure them all in terms of their value at a particular time. For revenues we have chosen the time to be when the turbine starts to generate. So, if it takes two years to manufacture and install a wind turbine, then we need to discount the capital to the time the turbine starts to produce energy. This discounted amount will be larger than the sum of the annual capital payments, since these would accrue interest if invested elsewhere; e.g. $\$1000$ spent two years ago with an interest rate of 10% would be worth $\$1000\times(1.1)^2=\$1210$ at the start of production. For equal annual capital payments of $C_{capital}/N$ over N years we need to increase the capital sum $C_{capital}$ in eqns (12.2), (12.3), and (12.4) by the factor $F_{discount}$ given by

$$F_{discount} = (1+R)\left[(1+R)^N -1\right]/NR \tag{12.6}$$

A long construction time costs money and increases the cost of energy. From eqn (12.6) the effect of increasing the construction time by one year is to increase $C_{capital}$ by approximately $(1+R/2)$. We will illustrate the importance of construction time by considering its effect on the cost of energy from a nuclear power plant.

EXAMPLE 12.2

A 1 GWe nuclear power plant cost $2500 million to build. The plant has a capacity factor of 85% and a lifetime of 30 years. Assume a discount rate of 10% and annual costs of fuel and of O&M of $5 per MWh and $10 per MWh, respectively.

Calculate the cost of electricity when the construction period is (a) 5 years; (b) 4 years. By how much are these costs reduced when (c) the construction cost is reduced by 25%; (d) the discount rate is reduced to 8%?

At a capacity factor of 85% the energy produced per year E will be

$$E=24\times365\times0.85\times10^9=7.45\times10^6\,\text{MWh}$$

(a) From eqn (12.6) for a construction period of 5 years, $F_{discount}=1.343$, so $C_{capital}$ becomes 3.358×10^9.

Substituting in eqn (12.4) with $N=30$ gives the annual revenue required to repay the capital $A_{costcapital}$ as

$$A_{costcapital}=3.358\times10^9\times0.1/[1-(1.1)^{-30}]=\$356\times10^6$$

There is an additional cost of $15 per MWh for fuel and O&M so the total annual cost is

$$A_{cost}=\$(112+356)\times10^6$$

The cost of electricity $C_{electricity}$ is therefore given by

$$C_{electricity}=A_{cost}/E=6.28\text{¢ per kWh}$$

(b) From eqn (12.6) for a construction period of 4 years, $F_{discount}=1.276$. So $C_{capital}$ becomes 3.191×10^9. $A_{costcapital}$ is therefore 338×10^6. The annual cost is

$$A_{cost}=\$(112+338)\times10^6$$

The cost of electricity is therefore $C_{electricity}=A_{cost}/E=6.04\text{¢ per kWh}$

(c) Cutting construction costs by 25% reduces $A_{costcapital}$ by 25% and the cost of electricity becomes 5.09¢ per kWh for a 5-year construction period and 4.91¢ per kWh for a 4-year one.

(d) Repeating the calculations for (a) and (b) with a discount rate of 8% gives the cost of electricity as 5.28¢ and 5.13¢ per kWh, respectively.

These simple examples illustrate some of the key points, but an actual economic analysis is much more complicated. In practice, a particular uncertainty is over future revenues, and projections would be made using different assumptions on revenue. The actual discount rate is difficult to predict—one estimate would be the long-term interest rate (e.g. 30-year rate on bonds) minus the expected rate of inflation over the period of the project. But, as the example above shows, the discount rate directly affects the return.

The relative cost of alternative ways of producing power can also affect the profitability; for instance a significant drop in the price of gas could have a large effect. Renewable energy has an additional value in saving a utility power company from using other fuels;

these direct savings are called avoided costs. Their use will also reduce the requirement for conventional generating capacity. The desire to reduce global warming can also result in requirements on utilities to use a certain percentage of renewable energy, such as from wind power.

12.3.2 Costs of electricity generating technologies (2010)

Table 12.2 shows cost estimates for some fossil fuel and renewable technologies made by the European Academies Science Advisory Board in 2010—a discount rate of 10% over 25 years and a conversion rate of $1 \equiv €0.755$ were assumed. Data were obtained from the US Department of Energy (2010) and the location for the solar technologies was taken to be Arizona, where the intensity of solar energy (DNI) is $2500 \, kWh \, m^{-2} y^{-1}$. Onshore wind can already be seen to be competitive with a mid-load coal-fired plant, though for the solar technologies either a subsidy such as a feed-in tariff or a carbon price is required; for CSTP a price of €80–100 per tonne of CO_2 would be needed. (The PV estimate does not take into account recent price decreases.)

In Table 12.2 the lifetime is taken as 25 years, i.e. $N=25$; the capital cost ($C_{capital}$) covers the engineering, procurement, and construction costs; the fixed operation and maintenance costs ($O\&M_f$) are taken as a fraction of the capital cost, while the fuel costs (Fuel) and the variable O&M costs ($O\&M_v$) are dependent on the amount of electricity. The levelized electricity cost (LEC) is given by

$$LEC = \left(annuity \times C_{capital} + O\&M_f\right) / (8760 \times CF) + O\&M_v + fuel$$

where CF is the capacity factor and 8760 is the number of hours in a year. The annuity is the fraction of the capital cost that gives the annual revenue required to make the present value of

Table 12.2 Cost estimates for some fossil fuel and renewable technologies

Technology	LEC (c/kWh)	Assumed capacity (MW)	Capital (€/kW)	Assumed capacity factor	Fuel (c/kWh)	$O\&M_f$ (€/kWy)	$O\&M_v$ (c/kWh)
CSTP* no storage	17.9	100	3542	0.28	0	48	0
Coal base-load	6.9	650	2391	0.90	2.9	27	0.3
Coal mid-load	9.0	650	2391	0.57	2.9	27	0.3
Gas mid-load	6.1	540	738	0.40	3.2	11	0.3
Wind onshore	8.5	100	1841	0.30	0	21	0
Wind offshore	15.3	400	4511	0.40	0	40	0
Photovoltaics*	21.2	150	3590	0.22	0	13	0

*Arizona

†1 c = 1 euro cent

Source: EASAC2011.

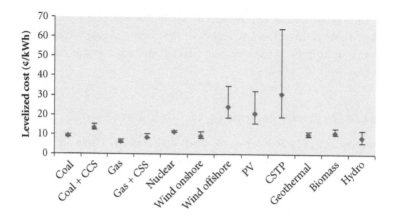

Fig. 12.3 Regional variation of levelized costs for 2016 in the USA.
Source: EIA2011.

the total revenue from N years of operation, when the discount rate is R, equal to the capital cost. From eqn (12.4), the annuity equals $A_{cost}/C_{capital}$, and is given by

$$\text{Annuity} = R/[1-(1+R)^{-N}]$$

For comparison, Fig. 12.3 shows estimates made by the Energy Information Administration (EIA2011) of the regional variation in the levelized costs of different generating technologies in the USA for 2016.

As expected, the LECs for wind and solar depend strongly on the resource intensity. For the marine technologies, wave and tidal, Mott MacDonald (MOTT2011) have estimated: wave (fixed) 37p per kWh, wave (floating) 60p per kWh, tidal stream 29p per kWh, and tidal barrage 52p per kWh. We can see that, of the technologies with a large accessible potential by 2050, only onshore wind, nuclear, and hydro are currently cost-competitive.

One of the challenges to a government or a regulator that wants a particular source of energy to be used (e.g. nuclear or wind) is to devise a pricing regime that guarantees a certain rate of return but still maintains competition among utilities. Estimates of future costs are also very important in planning an energy strategy. Some of the technologies that we have discussed are relatively young and prices are dropping. In order to predict what the costs will be in the future, a technique called learning curve estimation is useful.

12.3.3 Learning curve estimation

It is common experience from a broad range of technologies that costs fall as production increases. In particular, costs decrease roughly linearly with cumulative production when plotted on a log/log scale. The learning rate for a technology is the percentage reduction in costs for a doubling in cumulative production; typically the learning rate is between 10 and 30% for industrial products and tends to be larger in the developing stages. It is the global cumulative production that matters, not the production in a particular country. Although it appears rather simplistic, learning curves have been shown to provide good cost estimates—the technique was first noted in the construction of aircraft in 1930s. The tendency for the slope of a learning curve to decrease with time as a technology becomes more mature needs to be allowed for when extrapolating into the future. To work out the time when the cost will have fallen to a particular value requires assumptions about the rate of production.

As an example, the learning curve for global onshore wind turbines is shown in Fig. 12.4. The learning rate between 1984 and 2011 was 8.6%. There was a rise in the price of turbines in 2004, mainly due to an increased cost of materials and a strong demand, which reflected a lack of manufacturing capability. Over-supply can also distort a learning curve, temporarily causing the price to drop. Another example of a learning curve is given for photovoltaic panel production in Chapter 7 (see Fig. 7.18).

12.3.4 Life-cycle analysis (LCA) and external costs

Besides the costs involved in manufacture, installation, and operation there are related costs that can affect the cost of electricity. If there were a tax on the emission of carbon then that would increase the cost of electricity from fossil fuel power stations. There is a cost in reducing other emissions that are damaging, such as SO_2 from coal which gives rise to acid rain. Renewable energy has substantial environmental benefits and these can be given a monetary value from the amount of CO_2 and other emissions saved. The external cost (mainly environmental) of coal-fired generation has been estimated to be 4c (euro cents) per kWh compared to 0.2c per kWh for wind energy. If this external cost were included then the price of electricity in Europe from coal would roughly double.

In the combustion of municipal solid waste (MSW), although the fuel is mainly organic, the combustion is not carbon-neutral because some of the material is derived from fossil fuels (typically 20–40%). The analysis of all the emissions involved in producing energy from a particular source is called a life-cycle analysis. It calculates the amount of CO_2 (and other gaseous emissions) per kWh of energy produced. While the burning of agricultural wastes gives less than 30 g per kWh, that of MSW, also referred to as energy from waste (EfW), gives ~360 g

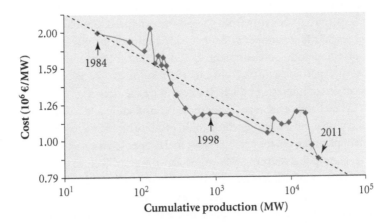

Fig. 12.4 Learning curve for global onshore wind. The dashed line shows a learning rate of 8.6%.
Source: Bloomberg.

per kWh compared with ~1000 g per kWh from coal and ~450 g per kWh from a natural gas CCGT power plant (see also Section 1.3.1).

There are costs and benefits as well as risks associated with energy production and we now look at how these can be analysed. The perception of risk and the evaluation of the risks associated with a technology are also very important and these are also considered.

12.4 Cost-benefit analysis and risk assessment

Cost–benefit analysis (CBA) is the basic method used when trying to decide whether to embark on a given project. The first step is to identify the benefits of the project and to assign a monetary value to them. The cost of the project is then calculated and, if the benefits outweigh the cost, then the project is cost-effective. CBA is usually used for calculating the costs and benefits to society. In terms of these a tax has no net effect, whereas external costs such as pollution and global warming should be included. When the project also involves an assessment of the risk of fatalities or injuries, the analysis is called risk–cost–benefit analysis.

Assigning a cost to a human life is something that we are reluctant to do. A more acceptable approach is to assign a cost to avoiding a human fatality or injury. We can estimate this cost by looking at situations where people accept higher risks for more pay, e.g. on oil rigs. We can then determine the value people put on accepting risks and hence the cost they place on reducing risks. These costs may well be controversial and difficult to estimate, but it is important to recognize and try to evaluate them.

A simple example of CBA is the building of a motorway. The main cost is buying the land and building and maintaining the road. The benefits include the economic improvements and the journey time savings that the motorway brings. These can be given a value by comparison with previous similar projects. The new motorway may also reduce the number of car accidents on journeys between places linked by it.

The benefits and risks of a project may well affect different groups of people. In the siting of a nuclear power station, the local population is at the highest risk from an accident at the site, while a much larger group benefits from the power produced. Should the local population receive cheaper electricity to compensate for the proximity of the nuclear power plant? Who should decide on whether the nuclear plant should go ahead? Should it just be the local population, a panel of experts, scientists from other disciplines, or an even wider involvement? Experts may have vested interests in the technology they are assessing.

There needs to be a balance between a widespread consultation and a consequently slow decision process, and a small review panel which could produce a much quicker decision. The latter helps government planning on, for example, national energy security, as well as on the international issue of tackling global warming, while the former is more likely to gain public confidence.

The public perception of risk is often different from that of experts who focus on the probability of an accident. Risks are perceived to be greater when they are involuntary, uncontrollable, and potentially catastrophic. These concerns contribute to what has been called the dread factor. Conversely, risks are more acceptable if they are voluntary, controllable, and limited. Figure 12.5 shows a plot of perceived fatalities associated with various activities against actual

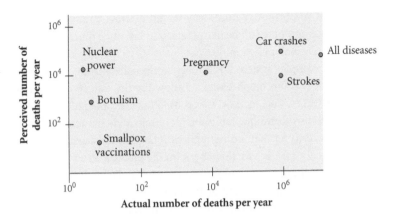

Fig. 12.5 Perceived risks versus actual risks.
Source: Psychol.

fatalities. Nuclear power is an example of an industry with a high dread factor; the actual number of deaths per year due to nuclear power is low compared with much more familiar incidents such as fatal car crashes.

There is also a tendency to think that if something can occur then it will, which increases the perceived danger of risks with a very low probability. Conversely, there is a reduction in the perceived danger from risks with a higher probability. Awareness of these differences is important when considering any project.

Public involvement and trust in the utility companies and government agencies involved are key factors that can help speed the approval of a project. It is important that the public are aware of the benefits that the project will bring, both to them and to others. It is also important not to assume that the general public cannot understand technological arguments—if you believe a project to be safe then you should be able to explain why you think so. This raises the question of how you make a system reliable and safe. We now consider methods used in designing safe systems.

12.5 Designing safe systems

One of the ways that a system can be made safer is by adding redundancy. If we have a key component, e.g. a relay, which has a probability of failure of q, then by adding another relay in parallel, so that both would need to fail for the system to fail, the reliability (i.e. the probability of not failing) is improved from $(1-q)$ to $(1-q^2)$ (see Exercise 11.9). For example, if $q=0.1$ the reliability improves from 90% to 99%. However, it is important that both components are independent. The reliability would be lower, for example, if both components were from the same batch using the same type of bad resistor, or if the main power supply and the back-up cables ran through the same duct, because they would both fail if there was a fire in the duct. These are examples of common-mode or common-cause failures.

The introduction of new technology can initially result in unexpected failures since experience of the new technology is lacking. This can lead to a very conservative approach.

For example, the most important requirement for an aircraft engine is reliability, and an engine designed 40 years ago may still be in operation today since it has a proven track record (an 'if it works then don't fix it' philosophy).

In designing a large, complex system it is important that safety is incorporated in the design right from the start—it is often much easier to allow for the failure of a component at the design stage than afterwards. One method that is used to assess the risk of failure is probabilistic risk assessment. In this technique faults are identified and characterized by the probability of failure and the severity of their consequences. The possible causes of any fault are then identified (fault-tree analysis). As a further check on the identification of all significant faults, the consequences of the failure of any component are analysed (event-tree analysis).

An ideal system is fail-safe in the event of any fault. For example, if the pilot light goes out in a boiler, the gas supply is automatically turned off by the action of a bi-metallic strip that closes off the supply valve when it is cold. The mechanism is passively activated, as it requires no mechanical or operator intervention. The overflow on a bath is an example of an inherently safe design as it relies just on a physical law, in this case gravity, to work.

An illustration of an inherently safe passive device is the mechanism used to operate the control rods in a pebble bed nuclear reactor. In this type of reactor the control rods are held out of the core by an electromagnet whose current flows through a resistor attached to the core. If the core temperature rises, then the current flow through the electromagnet drops since the resistance of the resistor increases with temperature. This causes the rods to be released when the temperature rises too far, and they drop under gravity and shut down the reactor.

A further degree of protection in a system is to incorporate a number of units that will carry out the same function should one fail: a fault-tolerant system. These units can either be in parallel, when they give redundancy, or in series, when another unit or units (defence in depth) come into operation should the first unit fail. An example of the latter is an uninterruptible power supply (UPS); this is a power distribution unit connected to the main power supply that includes a large battery that provides power if the main supply is cut. Another example is the outer containment building in a nuclear reactor plant.

It is important to look at the system as a whole and not just at its components. Incorrect interactions between subsystems can lead to accidents, e.g. if the software controlling a unit detects a fault, the programmed response should take into account the state of other connected units. Software must also do exactly what is expected of it—called safety critical software. For some operations controlled by computer, it may be required that the operation can be also carried out manually, providing defence in depth.

One of the difficulties with probabilistic risk assessment is to take into account feedback between subsystems and, in particular, the effect of the management's attitude on safety and the safety culture of the workforce. The overall safety culture of an industry has been identified as a very important factor in reducing accidents. In France, a safety culture was encouraged 150 years ago in the explosives industry by a law requiring the manufacturer and his family to live on the premises!

The desire to save time and be more productive and efficient can lead to procedures being adjusted and safety being compromised. In the Bhopal chemical factory disaster, cost-cutting, workforce reductions, and other actions had reduced the safety margins over a period of time and led to the worst industrial accident on record. During the Three Mile Island nuclear reactor incident, the operators had incorrect and insufficient information on the state of the nuclear reactor core—a signal was sent from one valve to say that it had received power to close it, but not that it had closed. The lamp indicating the true position had been eliminated during construction to save time.

It is important that a 'best practice culture' is established in power plants, where comments on safety are encouraged. As those who have watched Homer Simpson in his control room will know, repetitive operations need to be made interesting to reduce boredom. It is also important that both scientists and engineers are on the staff to increase the chances that all aspects are considered in unexpected situations. A consideration of human behaviour under stress also needs to be made (human factors).

One way of dealing with possible conflicts between safety and profitability is to appoint a regulator: an independent authority that can enforce safety requirements, maintenance schedules, and inspections. Preventive maintenance is a well-established way of reducing the chance of an accident—replacement of components in a car after fixed distances is an example of this. All plant modifications should be justified beforehand by a formal safety case: a thorough explanation of physical processes should be given by experts in the field and challenged by independent experts appointed by the regulator. They then judge whether or not the safety case is convincing.

Independent regulation enables the approval of a plant design to be separated from planning consent. This can help in speeding up the planning process. Public confidence would be helped by getting internationally agreed safety criteria for the plant, as would allowing the public to be involved in the decision-making process.

Even if a power plant is accepted as safe and desirable on environmental grounds, the fact remains that most of the alternatives to fossil fuels are currently uneconomic. As many countries realize the importance of reducing CO_2 emissions, what policies have been proposed to abate carbon emissions?

12.6 Carbon abatement policies and issues

At the first UN international Earth Summit, held in Rio de Janeiro, Brazil, in 1992, the United Nations Framework Convention on Climate Change (UNFCCC) was adopted. Its provisions on tackling global warming were strengthened in 1997 in Kyoto, Japan, where an amendment was negotiated, called the Kyoto Protocol. Under this protocol, industrialized countries agreed that they would reduce their emissions of greenhouse gases by 5.2% compared with those of 1990 for the period 2008–2012. If a given country's emissions were not falling in line with this reduction, then that country would be allowed to engage in emissions trading. The greenhouse gases specified are carbon dioxide, methane, sulfur hexafluoride, HFCs, and CFCs. The protocol also acknowledged that the bulk of the increased levels of CO_2 in the atmosphere had been produced by the developed nations. As a result India, China, and other developing countries were not required to reduce emissions.

The protocol came into force in February 2005. Over 160 countries have ratified the agreement, with the notable exception of the United States, which is concerned that the reduction in emissions would be damaging to its economy. The American government also felt that the protocol should have included targets for developing countries as well as developed countries. In 2003 China emitted an estimated 3.5 Gt of CO_2, compared with 5.8 Gt by the United States, but by 2010 China had increased its emissions to 8.95 Gt whereas those of the United States had decreased to 5.25 Gt, though China's per capita emissions are still 2.5 times less than those of the USA (see Table 12.3). (NB Imported and exported goods have associated carbon emissions and these need to be taken into account to give an accurate picture of a particular country's emissions.) The six highest emitting countries contribute 70% to the global emissions of CO_2 and, although the reduction agreed to under the Kyoto Protocol looks like it will be achieved, the annual emissions are still increasing at an alarming rate.

It was hoped that the UNIPCC meeting in Copenhagen in 2009 would produce a protocol to follow the Kyoto Protocol which ends in 2012, thereby establishing a further international commitment to reduce emissions. However, the Copenhagen Accord failed to produce any agreement on emissions targets, and only resolved that on the basis of the scientific evidence for global warming the world should strive to keep the temperature rise since pre-industrial times to below 2 °C. Since that meeting there has been little progress, in part because of the global economic crisis in 2008.

Table 12.3 Carbon dioxide emissions in 2010 (Mtonne CO_2) and per capita emissions 1990–2010 (tonne CO_2 per person) (includes cement production, ~8% of global total)

Country	CO_2 emissions 2010 (Mt)	Per capita CO_2 emissions (t per person)			Change since 1990 (%)	
		1990	2000	2010	CO_2	Population
Annex 1						
United States	5250	19.7	20.8	16.9	5	23
EU-27	4050	9.2	8.5	8.1	−7	6
Russian Federation	1750	16.5	11.3	12.2	−28	−4
Japan	1160	9.5	10.1	9.2	0	4
Australia	400	16.0	18.6	18.0	46	30
Canada	540	16.2	17.9	15.8	20	23
Non-Annex 1						
China	8950	2.2	2.9	6.8	257	17
India	1840	0.8	1.0	1.5	180	40
South Korea	590	5.9	9.7	12.3	134	12
Indonesia	470	0.9	1.4	1.9	194	30
Brazil	430	1.5	2.0	2.2	96	30
Mexico	430	3.7	3.8	3.8	39	35
Saudi Arabia	430	10.2	12.9	15.6	159	70

Source: PBL2011.

But it should be understood that there are real difficulties for many countries in reaching a common accord. In a developing country there can be millions of people who need energy to improve their standard of living, and their governments are understandably reluctant to spend large sums on projects such as building homes that require little energy to run and that can even save money over the lifetime of the project, but that reduce the amount for immediate help to the poor. Balancing the responsibility for the present generation against that for future generations is a particular challenge.

One means of help is to subsidize the cost of fossil fuels. Many governments give subsidies on the grounds that it is making energy more accessible to the poor, which is important for raising their standard of living. It has been estimated that these subsidies run at between $150 billion and $250 billion each year, far in excess of the subsidies for low-carbon energy, which were estimated at $33 billion per year in 2004. However, it is argued that removing these subsidies and redirecting the money just to the poor would be much more cost-effective than a fossil fuel subsidy which benefits all. Such subsidies also make it harder for low-carbon technologies to compete.

Actions like insulating homes have a negative cost over the lifetime of the project and many only have a cost that a relatively small carbon price would cover. The cost neutral carbon price is given by the price of CO_2 per tonne that, when multiplied by the lifetime savings in CO_2 in tonnes, equals the cost of the project. It has been estimated that up to ~27 Gt CO_2 y^{-1} could be abated at a carbon price of less than €40 per tonne. At negative or zero cost an abatement of ~7 Gt y^{-1} could be achieved by improvements in insulation, vehicles' fuel efficiency, lighting systems, air conditioning, water heating, switching to sugarcane biofuel, and to nuclear power; while at a positive cost of less than €40 per tonne, actions could include the introduction of CCS, wind power, co-firing with biomass, solar power, and avoiding deforestation. (Note that ~43 Gt y^{-1} is the estimated reduction by 2050 to be on track for 450 ppm CO_2.)

The Stern Report concluded that taking action and spending ~1% of GDP annually by 2050 on mitigation could save ~5–20% of global GDP arising from the costs of climate change under a BAU scenario. However, there are pressures, particularly in an economic downturn, to invest in schemes that benefit the economy in the short term rather than in the long term. The developing countries are at a disadvantage since they are generally warmer than developed regions, so further warming would affect them more, particularly from poor agricultural yields and flooding. The heat waves that occurred in Europe in 2003 when some 35 000 died will also become more frequent.

The cost of climate change arises in the future and is therefore subject to discounting (see Section 12.3.1); however its present value is very dependent on the discount rate, which is difficult to guess. Nevertheless, there are strong ethical, environmental, as well as economic reasons why the world should act, and we will now look at some of the carbon abatement policies that are in place.

12.6.1 Emissions trading

Each country that has emissions reduction obligations under the Kyoto Protocol has set limits on the amount of greenhouse gases that it can emit, but many countries have limits above their current production. The surplus amount can be purchased by other countries, which allows them not to reduce their emissions if they so wish. Countries also receive credit for any

CO_2 sinks that they develop, such as creating a forest, or for work on carbon abatement in developing countries. Abatement tends to occur where costs are lowest and results in transfers from rich to poor, but there are lower reductions in industrialized countries. Trading appeals to industry as there is the possibility of profits. Caps on emissions are also easier to agree than taxes, but tight caps are required to ensure scarcity and a high carbon price. A long-term trading period is also needed to reduce market risk.

Setting up the scheme has proved difficult since it requires records and monitoring of emissions. An important concern over manufactured items is the assigning of carbon emissions in the Kyoto Protocol to the country producing the goods rather than to the country importing them. A scheme for emissions trading has been set up by the European Union. However, an initial over-allocation of credits resulted in a surplus and as a result a low carbon price; a re-launched scheme with tighter allocations also had difficulties caused by the global economic downturn in 2008, which caused production to drop and created a consequent surplus of credits. Introducing a minimum carbon price might be an effective way to make the scheme more robust and give confidence to producers.

12.6.2 Carbon tax

A simpler scheme to set up than emissions trading is to impose a tax (collected by the government) on all fuels that emit CO_2. The tax is based on the amount of carbon emitted, so coal would have a higher tax per kWh than gas. This scheme, which was introduced in Sweden, Finland, the Netherlands, and Norway in the 1990s, provides an inducement for everyone to reduce emissions and applies to transport, domestic consumers, and industrial consumers. It also demonstrates the importance attached to reducing global warming by the government concerned. It is more costly for countries with less efficient energy usage, which was a reason why the United States was in favour of emissions trading rather than a carbon tax. However, there is an uncertainty in the tax required for a particular environmental outcome, i.e. emissions reduction, and a tax can affect the less well off adversely.

12.6.3 Regulations

Regulations can be an effective way to reduce emissions. Examples would be building standards, where the minimum amount of insulation or the use of renewable energy supply is specified, and fuel efficiency (mpg or km per litre) requirements on vehicles. The usual market forces provide an incentive to meet the standard as cost effectively as possible, but there is no market incentive to innovate and go beyond the standard. So regulation alone is unlikely to find the most cost-effective way of reducing the emissions, though in conjunction with price mechanisms it can be very useful. It can encourage the deployment of—and research into reducing the costs of—efficient technologies.

12.6.4 Feed-in tariffs (FITs) and tradable green certificates (TGCs)

To help reach their commitment by 2020 to produce 20% of energy demand from renewable sources, EU countries have introduced a number of incentive schemes. These can be classified as aimed at either price or quantity, and by whether they focus on investment or on generation. The two main mechanisms are feed-in tariffs (FITs), which are price and generation based, and tradable green certificates (TGCs), which are quantity and generation based.

A feed-in tariff is a guaranteed price that a producer of renewable electricity will receive, and the price is set so that a reasonable profit can be made. The extra cost of production over that for conventional generation, which is currently mainly fossil fuel based, is shared by all the consumers of electricity in the province or country. The level of tariff can be based on the levelized cost and so will be different for different technologies. The value can depend on the size of the plant so as to encourage both centralized and distributed generators. It can also be revised to take into account decreasing costs arising from the learning effect. FITs are used in 20 of the 27 countries in the EU and have been successful in promoting renewable energy. The long-term period of the tariff reduces risk for investors and the price can be set to give a good return. Also, the set-up costs of the scheme are relatively low.

A TGC scheme requires electricity suppliers to obtain a specified fraction of their energy from generators using renewable energy sources. The obligation sets up a market in tradable green certificates. The number of TGCs per MWh of generated electricity can vary between technologies in order to avoid only the current cheapest technology being promoted. Investment in a wide range of new technologies is important for stimulating innovation, building up expertise, and avoiding prejudgements as to the best technology. Suppliers have to buy a certain number of TGCs to show that they have obtained the specified fraction of renewable energy. This they can do by buying TGCs directly from the renewable generator or on the open market. Renewable generators of electricity will earn revenue from selling both electricity and TGCs, which will give them extra income. The idea is that, as the price of TGCs rises, developers will be encouraged to build more renewable energy generators. Market forces will generate competition and favour the most economic renewable sources. Furthermore, the increase in production will help reduce prices as the technology improves (see Section 12.2.2). However, in practice, the price volatility of TGCs has deterred investors, and feed-in tariffs have achieved a larger deployment of renewable electricity generators at lower cost than TGC schemes. The UK introduced a FIT scheme in 2010 to encourage deployment of small-scale, low-carbon generation to complement its existing TGC scheme of Renewable Obligation Certificates (ROCs).

The need for a reduction in CO_2 emissions is now well established and widely recognized. But the amount of reduction required is very large, and will need a portfolio of measures. In an attempt to make the challenge appear less daunting, Pacala and Socolov in 2004 proposed dividing up the total required into seven equal sized parts, each one providing an increasing reduction with time (see Fig. 12.2 for an illustration of unequal sized wedges). Each action was called a stabilization wedge, and at that time it was thought to be enough to hold emissions constant until ~2050 and then reduce them to zero, ending up with a level close to ~500 ppm. Now the scientific consensus is that the level of CO_2 needs to be below 450 ppm in order to reduce the threat of dangerous climate change significantly, and that emissions must start to fall very soon.

12.7 Strategies for lowering CO₂ emissions

The IEA BLUE map scenario, mentioned in Sections 3.7 and 12.2, is a strategy to achieve stabilization at about 450 ppm. Although it is now (2012) looking almost impossible to achieve, it illustrates very clearly the global challenge that we face. As shown in Fig. 12.2, it would require reducing our emissions by 2050 to about a half of what they were in 2010 and to zero by 2100. We can roughly estimate the increase in CO_2 from this course of action by noting that the emissions at the start of 2010 were ~31 $GtCO_2 \, y^{-1}$, or about 8 Gt y^{-1} of carbon, and assume they rise to 34 Gt y^{-1} by 2015 and then fall linearly to zero by 2100. The amount emitted would be ~1600 Gt, of which about a third would remain in the atmosphere, causing the concentration of CO_2 to rise to ~465 ppm and the global temperature to rise by a further ~0.8 °C relative to that in 2010.

The contributions to the emission reductions in this strategy are shown in Fig. 12.6. This shows the same reduction as Fig. 12.2, but with the contributions from the different technologies rather than sectors shown. The electricity generation (in $TWh \, y^{-1}$) in 2050 assumed from renewables (IEA2011) is: CSP ~2500, PV ~2500, geothermal ~1000, wind ~5000, hydro ~5000, biomass and waste ~2000, and nuclear ~10 000 (300 GWe continuous output generates ~2500 $TWh \, y^{-1}$). Fuel switching refers to changing to low-carbon fuels (e.g. from petrol (gasoline) to electricity from low-carbon sources or from coal to gas). An illustration of carbon capture and storage (CCS) in industry would be CCS added to efficient coal gasifiers to replace natural gas and provide hydrogen for process heat. But, as discussed in Section 3.6, there are uncertainties about the deployment of CCS. Substituting biofuels for fossil fuels would be an example of fuel switching in industry. Switching about 1000 GWe of power from coal-fired to natural gas-fired plants would avoid about 4 Gt of CO_2 emissions each year, since the amount of carbon from gas per kWh is about half that from coal (0.45 compared to 1.0 $kgCO_2$/kWh).

In the scenario shown in Figs 12.2 and 12.6, end-use reduction plays a very major role and corresponds to about one-half of the total of ~43 $GtCO_2 \, y^{-1}$ required by 2050. To

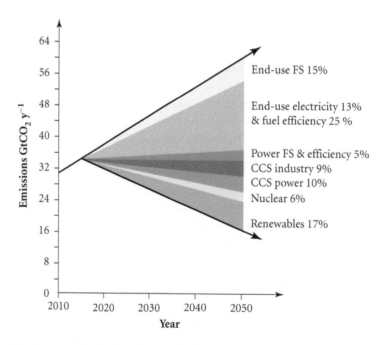

Fig. 12.6 Emissions reductions by technology (FS: fuel switching) in BLUE map scenario (schematic). *Source:* IEA-ETP-2008/2010.

achieve the total, CCS is expected to play a significant part, together with additional low-carbon generation of ~2300 GWe continuous, of which an extra 600 GWe from nuclear power is envisaged.

A recent study (UofCamb2011) on the practical limits to reducing energy demand considered end-use efficiency improvements in transport (cars, trucks, planes, ships, and trains), industry (furnace, driven, and steam systems), and buildings (space heating and cooling, hot water systems, domestic appliances, and lighting). The report concluded that 73% of global energy could be saved through practical design changes reducing demand, with the largest savings of 83% in buildings, mainly in heating and cooling spaces and in appliances. Savings in transport were 68%, largely from cars, where reductions in the mass and rolling resistance were significant, and in industry 62%, with similar percentages in all three areas considered. This report also concluded that the global energy savings could be increased to 85% with conversion efficiency improvements. These are upper limits rather than best estimates, but suggest that the aggressive target of 38% efficiency improvements in the IEA BLUE map scenario over and above the assumed 0.9% per annum gain in the BAU scenario (making a total of ~74% by 2050) is nonetheless possible.

For comparison, in the BP 2030 projection (BP2030), where only some relatively small reductions over BAU are predicted to occur, the 2030 estimate of ~39 GtCO$_2$ y^{-1} for annual emissions in 2030 is 27% larger than the 2010 value of 31 GtCO$_2$ y^{-1}, compared with the IEA BAU estimate of 44 GtCO$_2$ y^{-1}. A strategy has been developed by the State of California to meet an 80% emissions reduction target by 2050 compared to 1990 levels, which illustrates many aspects of the challenge, and is described in the case study below.

Case Study 12.2 California emissions reduction plan 2050

The State of California plans an 80% reduction in its emissions by 2050 from 1990 levels. The population is expected to increase from 37 million in 2005 to about 55 million by 2050. Even with moderate economic growth and BAU efficiency gains, the demand for energy is predicted to double by 2050. Emissions will need to fall from ~13 tCO$_2$ per person to ~1.6 tCO$_2$ per person. A report on these plans (California2011) identified that about a 60% reduction in emissions is possible if known technology is rapidly deployed: in particular, aggressive efficiency measures and electrification (to reduce and avoid fossil fuel use wherever possible), decarbonizing the electricity supply while doubling electricity production, and developing zero-emissions load balancing. The remaining 20% reduction would require decarbonizing the remaining fossil fuel use in transportation and heating.

Under efficiency and electrification, the committee recommended that (i) by 2040 all new buildings should use 80% less energy and the remaining buildings be retrofitted or replaced (though this would be very expensive); (ii) 70% of space and water heating should be shifted from gas to electric heat pumps; (iii) energy intensity in industry should be reduced by efficiency gains and electrification; and (iv) new cars should be predominantly (70%) PHEVs or EVs and the fuel efficiency of the remaining cars should increase to 58 mpg.

The report supports developing and deploying several low-carbon technologies since no single approach can solve the problem on its own. While nuclear can provide reliable base-load generation, further expansion in California is dependent on finding a satisfactory way of dealing with nuclear waste. Furthermore, the Fukushima accident will cause a re-examination of nuclear safety precautions. California has large renewable energy resources that provide energy security as well low-carbon power, but their variability means that, unless zero-emission load balancing (ZELB; the matching of supply and demand) can be achieved, there would be significant emissions from gas-fired back-up generators (spinning reserve). ZELB might be possible through the use of smart grids and energy storage, and decreasing the cost of large-scale storage would make a major impact. CCS could also help but relies on large-scale underground storage that is not yet developed. CCS with biofuels is attractive since it gives negative emissions and would allow the continued use of some fossil fuels. But increasing biofuel production risks adverse effects on food production and water availability and increased demand for fertilizer.

Hydrocarbon fuels have very high energy density (see Table 11.2) and will be needed in several processes where neither electrification nor CCS is practicable, such as in heavy-goods vehicles and ships. The sustainable supply of biofuels in 2050 would probably only meet about half the demand in these areas. To achieve 80% reduction would require innovation in multiple low-emission technologies: making carbon-neutral fuel from sunlight and advances in nuclear power were identified as possible game changers, CCU with the photocatalytic conversion of CO_2 plus renewable hydrogen to hydrocarbons and the development of controlled fusion being two possibilities.

The energy supply system chosen was an approximately equal mix of renewables, nuclear power, and CCS, with half the load balancing requirement provided by ZELB technologies and half using gas, and half of the remaining fuel requirement by biofuels. Only by using efficiency and electrification together, largely decarbonizing the electricity supply and remaining fuel, could a 60% reduction be achieved, with the efficiency measures making the largest single contribution, 20%.

The build rates required are ~5 nuclear, 6 CCS, 15 wind, and 15 solar farms every 15 years and about 40 000 distributed PV systems a year. This programme would require aggressive and fast permitting processes. The remaining emissions would nearly all come from fossil fuel use in transportation and heating and could most easily be reduced if more sustainable biomass were available, though CCS could provide an alternative source of low-carbon fuel.

Emission savings can also be made by enhancing natural carbon sinks. Planting some 250 million hectares of tropical forest, or a slightly larger area of plantations, would avoid about 3.5 Gt of CO_2 per year. When land is tilled for replanting, about half of the stored carbon in the soil is lost, mainly through aeration. Extending conservation tillage, in which seeds are planted without re-ploughing the land, to all the world's cropland, coupled with using ground cover crops that help prevent soil erosion, would avoid a further 3.5 Gt of CO_2 per year.

While Table 12.1 suggests that ~3500 GWe of low-carbon supply is possible, it will require substantial support through subsidies to help build up production, and consequently to drive down costs through learning, but also through funds for R&D on these technologies. It is only by reducing the costs of alternative technologies below that of fossil fuels and by developing the necessary energy storage and smart and super grids for its deployment that we can be independent of fossil fuels and reach zero emissions.

However, whether the world can manage to achieve this goal without emitting a dangerous amount of CO_2 is far from clear. As we may well exceed our emission targets, we will have to work on adaptation as well as on mitigation, and should also carry out research on geo-engineering as a means of reducing global warming. One idea to reduce the mean temperature is to create aerosols that will artificially increase the albedo. However, such action is likely to cause significant changes in precipitation and temperature in some regions and will not reduce the acidification of the oceans. Carbon dioxide capture from the air would be ideal if it could be carried out affordably at the required scale. An interesting idea, which could conceivably make a significant contribution, is to enhance the weathering of magnesium silicate rocks such as olivine by mining it and spreading it finely in mainly humid regions. The weathering transforms CO_2 into bicarbonate and rain washes this into rivers that remove it to the oceans, where it eventually precipitates as carbonate. But many geo-engineering schemes affect large regions and there are governance issues that will have to be resolved.

For adaptation, an example would be altering buildings so that they stay cool during the hotter summers that are expected from climate change. Another would be building flood defences to guard against the increased risk of flooding in certain parts of the world. There is already a need to adapt to a certain level of climate change from the emissions we have already made, and we must endeavour to limit future emissions to keep us below 450 ppm of CO_2 if at all possible.

12.8 Conclusions

There are several established ways by which we can reduce our CO_2 emissions. Electricity decarbonization through wind, solar, nuclear, and CCS on the supply side; energy savings in the industry, building, and transport sectors on the demand side; and the introduction of smart and super grids are all methods that could make a significant difference in the near future. However, these all require a serious commitment by governments and a determination by the public to curb CO_2 emissions immediately. The Kyoto Protocol was at least a step in the right direction but needs all nations to be involved, and the rate of CO_2 abatement needs to be speeded up very considerably.

If the world carries on as it has (i.e. Business As Usual) then there is a serious threat of dangerous climate change within 100 years. People will need to understand and support a significant shift in our energy supply and use. Limits and quotas on CO_2 emissions need to be agreed among nations. Carbon taxes would bring home the seriousness of the situation to individuals and drive changes in behaviour, greater efficiency, and increased private investment in R&D. But concerns over competiveness, for example that taxes and the increased prices will drive industry abroad and thereby give other countries an advantage, are a serious barrier. (Border carbon adjustments may make a difference. These are import fees levied on goods made in countries without a carbon tax by countries with a carbon tax.). Unless there is a significant improvement then we may even need to consider rationing carbon-emitting energy sources such as petrol.

Successful reduction in CO_2 emissions will require a diverse approach and strong leadership. Fusion and breeder reactors could enable us to provide low-carbon base-load power after ~2050 and it is important that research in these as well as in more established areas, in particular in energy saving technology, in solar, and in wind power, is strongly supported. But all technologies that we have described in this book that can significantly reduce our carbon emissions should be explored. That way we can hope to be successful in sufficiently reducing the total amount of carbon emitted to ensure a good quality of life for future generations.

 SUMMARY

- There is strong evidence that anthropogenic emissions of CO_2 have caused a significant global temperature rise of about 0.5 °C over the last 50 years. The IPCC have predicted under a business-as-usual scenario a rise of between 1.1 and 6.4 °C in the period 1990–2100, with a risk of catastrophic climate change.

- To reduce the risk of dangerous climate change will require limiting CO_2 levels in the atmosphere to about 450 ppm. But this limit is looking increasingly unlikely since it would require the world to start within a few years to reduce carbon emissions to zero by ~2100.

- Measures to limit anthropogenic contributions to global warming have been endorsed by the vast majority of nations in the Kyoto Protocol, but attempts to establish a new protocol establishing a further international commitment to reduce emissions have so far failed. In the Copenhagen Accord of 2009 no emissions targets were agreed, only that on the basis of the scientific evidence for global warming the world should strive to keep the total temperature rise since pre-industrial times below 2 °C.

- End-use energy savings as well as low-carbon sources of energy are essential if we are to provide the energy needed to raise the standard of living in developing countries and avoid excessive global warming. But the predicted reductions in emissions are likely to be hard to achieve, in particular because of lack of capital and the rebound effect.

- Raising awareness and changing social attitudes about global warming coupled with strong political leadership to introduce the necessary measures is crucial.

- Economics is a key factor in limiting the uptake of more low carbon energy. To address this, carbon emissions trading, carbon taxes, and feed-in tariffs have been introduced, but concerns over competiveness are a serious barrier.

- While cost reductions through 'learning' from increased production are very important, strong funding of R&D of new technologies for generation and storage is vital.

- Cost–benefit analysis and probabilistic risk assessment of projects together with public involvement are important in developing energy strategies.

- Many different carbon mitigation strategies are required in parallel to hold the level of CO_2 as low as possible. Electricity decarbonization through wind, solar, nuclear, and CCS on the supply side; energy savings in the industry, building, and transport sectors on the demand side; and the introduction of smart and super grids are all methods that could make a significant difference soon.

FURTHER READING

Borowitz, S. (1999). *Farewell fossil fuels*. Plenum, New York. Good book on the need to reduce our use of fossil fuels.

Botkin, D.B. and Keller, E.A. (2003). *Environmental science*. Wiley, New York. Interesting chapter on climate and global warming.

Boyle, G. (ed.) (2004). *Renewable energy*. Oxford University Press, Oxford. Interesting comments on renewable resources.

Cassedy, E.S. and Grossman, P.Z. (1998). *Introduction to energy*. Cambridge University Press, Cambridge. Informative section on risk and decision-making.

Jaccard, M. (2005). *Sustainable fossil fuels*. Cambridge University Press, Cambridge. Interesting book on using fossil fuels cleanly with carbon capture.

Jarman, M. (2007). *Climate change*. Pluto, London. Good discussion of the impact of climate change on the world's poor.

Jeffs, E. (2012). *Greener energy systems: energy production technologies with minimum environmental impact*. CRC Press, Boca Raton, FL. Critical review of the greener energy systems which need to replace coal.

Lichtenstein, S. et al. (1978). Judged frequency of lethal events. *J. Exp. Psychol.* **4**, 551 (Psychol).

Lovins, A. and Rocky Mountain Institute (2011). *Reinventing fire*. Chelsea Green Publishing, White River Junction, VT. Energy-efficient technologies in transportation, buildings, industry, and electricity.

MacKay, D.J.C. (2009). *Sustainable energy—without the hot air*. UIT, Cambridge Available from *www.withouthotair.com*. Very readable introduction to alternative energy and energy policy.

Mazo, J. (2010). *Climate conflict*. IISS, London. An important aspect of climate change.

Pacala, S. and Socolow, R. (2004). Stabilization wedges. *Science* **304**, 968.

Ramage, J. (1997). *Energy: a guidebook*. Oxford University Press, Oxford. Good survey of energy sources and options.

Remme, U. (2011). *Energy technology perspectives: BLUE map scenario*. International Energy Agency. Available from ebookbrowse.com/re/remm?page=4. (IEA 2011).

Richter, B. (2010). *Beyond smoke and mirrors*. Cambridge University Press, Cambridge. Good non-technical account of climate change science and on how we might reduce our reliance on fossil fuels.

Shepherd, W. and Shepherd, D.W. (2003). *Energy studies*. Imperial College Press, London. Useful overview of various energy sources.

Sorensen, B. (2004). *Renewable energy*, 3rd edn. Academic, New York. Good advanced reference book.

Thompson, E.R. (2012). *Energy policy and issues*. Nova Science, Hauppauge, NY. Critical analysis of energy policies and obstacles to progress.

Twidell, J. and Weir, T. (2006). *Renewable energy resources*. Taylor & Francis, London. Good discussion of institutional and economic factors.

Walker, G. and King, D. (2008). *The hot topic*. Bloomsbury, London. Clear discussion of the problems of global warming and of possible solutions.

WEB LINKS

www.climatetechnology.gov/stratplan/final/CCTP-StratPlan-Ch4.htm USA end-use efficiency strategic plan.

www.earth-policy.org/Indicators L. Buckley (CO_2).

www.hm-treasury.gov.uk/d/Executive_Summary.pdf Summary of the Stern review on the economics of climate change, 2006.

www.ipcc.ch/pdf/assessment-report/ar4/syr/ar4_syr.pdf AR-4 2007 Climate change mitigation in the building sector.

www.newscientist.com/topic/climate-change Good articles on climate change.

www.rmi.org/Knowledge-Center/Library/E05–16_EnergyEndUseEfficiency Interesting paper on end-use efficiency.

go.bloomberg.com/multimedia/wind-innovations-drive-down-costs-stock-prices/ (Bloomberg).

bp.com/energyoutlook 2011 (BP2030).

www.ccst.us/publications/2011/2011energy.pdf California's energy future: the view to 2050 (California2011).

www.easac.eu/fileadmin/Reports/Easac_CSP_Web-Final.pdf (EASAC2011).

www.eia.gov/oiaf/aeo/electricity_generation.html (EIA2011).

www.iea.org/techno/etp/index.asp Energy Technology Perspectives 2008 and 2010 (IEA-ETP-2008/2010).

www.oxfordenergy.org/wpcms/wp-content/uploads/2011/12/SP_24.pdf Interesting article arguing that not too much should be expected from energy efficiency (Keay2011).

hmccc.s3.amazonaws.com/Renewables%20Review/MML%20final%20report%20for%20CCC%209%20may%202011.pdf Marine costs (MOTT2011).

www.pbl.nl/en J.G.J. Olivier, G. Janssens-Maenhout, J.A.H.W. Peters and J. Wilson (2011), Long-term trend in global CO_2 emissions, 2011 report. PBL/JRC, The Hague (PBL2011).

pubs.acs.org/doi/abs/10.1021/es102641n Reducing energy demand: what are the practical limits? (UofCamb2011).

www.nature.com/nature/journal/v399/n6735/full/399429a0.html J.R. Petit et al. (1999). Climate and atmospheric history of the past 420 000 years from the Vostok ice core in Antarctica. Nature **399**, 429 (Vostok).

? EXERCISES

12.1 Discuss the evidence for an enhanced greenhouse effect occurring over the last 100 years.

12.2 Explain three possible global catastrophes if nothing is done to stop global warming.

12.3 Discuss what can be deduced from the observation in the climate record that the levels of CO_2 are correlated with the global temperature over a period of time.

12.4 Show that a concentration of 375 ppm (by volume) corresponds to ~3000 Gt of CO_2 in the atmosphere.

12.5* The additional concentration $\rho_{CO_2}(t)$ of carbon dioxide above the pre-industrial level of 280 ppm is approximately given by

$$\rho_{CO_2}(t) = C_{CO_2}\left\{\int_0^t E_{CO_2}(t')\left[a_0 + \sum_{i=1}^{3} a_i \exp\left\{-\left(\frac{t-t'}{\tau_i}\right)\right\}\right]dt'\right\}$$

where t is the time since anthropogenic emissions began, $C_{CO_2} \cong 0.128$ ppm Gt^{-1} is the conversion of $GtCO_2$ emitted to ppm, $E_{CO_2}(t')$ is the emission rate $(Gt\,y^{-1})$ at time t', $a_0 = 0.152$, $a_1 = 0.253$, $a_2 = 0.279$, $a_3 = 0.316$, $\tau_1 = 171$ y, $\tau_2 = 18.0$ y, $\tau_3 = 2.57$ y (the Bern Model).

An approximate expression for the emission rate in a BAU scenario is $E_{CO_2} = ct'^2$ with $c = 0.13$ Gty^{-1} and $t' = 120$ y corresponding to the year 2010. Calculate (a) the total anthropogenic carbon emissions in Gt of carbon by 2010; (b) the CO_2 emission rate in 2010; (c) the concentrations of CO_2 in 2050 and 2100 in this BAU scenario.

12.6 Give two examples of effects that could amplify the rise in global temperatures arising from an increased level of CO_2 in the atmosphere.

12.7 Show from eqn (12.6) that the cost of electricity rises approximately by a factor $(1 + R/2)$ per year of construction time, where R is the discount rate.

12.8 Consider a wind farm consisting of 20 2-MW turbines with a capacity factor of 0.3. The price per kWh is 4¢ and the cost of each turbine is $1 500 000. Neglect other costs and take the discount rate to be 8%. Calculate for a 25-year lifetime of the wind farm: (a) the present value of the revenue V_P; (b) the net present value V_{NP}; (c) the rate of return; (d) the cost of energy.

12.9 A 4GWe nuclear power plant cost $8000 million to build. The capacity factor of the plant is 90% and its lifetime is 40 years. Assume a discount rate of 7% and an annual cost of O&M and fuel of $12 per MWh. Calculate the cost of electricity for a construction period of (a) 4 years; (b) 5 years.

12.10* Decommissioning the plant described in Exercise 12.9 will cost $1000 million. Work out what annual payment would be required to have a value of $1000 million after 40 years. Hence calculate the cost of electricity allowing for the decommissioning costs.

12.11 Explain why the effect of inflation at I% is to reduce the discount rate from R% to $(R-I)$%.

12.12 Using the data in Table 12.2, recalculate the levelized cost for: (a) CSTP for a lifetime of 35 years, discount rate of 5%, and capacity factor of 0.5; (b) PV for a lifetime of 50 years and a discount rate of 5%.

12.13* Show that a learning curve can be expressed as

$$C=aP^b$$

where C is the cost per kWh, P is the cumulative production in TWh, and a and b are constants. Relate b to the learning rate.

12.14 A person is willing to accept $500 per year less in salary for a reduction in the risk of a fatality from 10 in 100 000 to 5 in 100 000 per year. What is the implied value of a life (called the value of a statistical life)?

12.15 Is it necessary for a government to adopt a BANANA (Build Absolutely Nothing Anywhere Near Anyone) policy to counter the NIMBY (Not In My Back Yard) attitude that a project can engender?

12.16 What type of safety device is: (a) an electrical fuse box; (b) an airbag in a car; (c) a written procedure; (d) a parachute?

12.17 Discuss the relative merits of regulation, carbon emissions trading, carbon taxes, and feed-in tariffs.

12.18 Show that a reduction of about $1\,\text{Gt}\,\text{y}^{-1}$ of carbon could be achieved by: (a) improving fuel efficiency from 30 mpg to 60 mpg for an annual mileage of 10 000 miles and assuming there are 2×10^9 cars globally; (b) substituting about 1400 GWe of coal-based generating capacity with gas-based power (assume a plant efficiency of 50% and that a plant of 33% efficiency produces 1 kg of CO_2 per kWh from coal and 0.5 kg of CO_2 per kWh from gas); (c) placing wind turbines over about 40 million hectares; (d) planting about 400 million hectares of land with cellulose-based crops.

12.19 Explain why energy savings are crucial in combating global warming.

12.20 Successive energy conversions result in 100 units of input fuel energy delivering 10 units of energy output. Explain why end-use efficiency improvements are particularly important.

12.21 What are the main actions that should be taken to reduce carbon emissions in the building and transport sectors?

12.22 Is it essential to electrify all road and rail transport to reduce carbon emissions?

12.23 How does one balance the responsibility for future generations with the (energy) needs of the present?

12.24 Discuss whether the California emission reduction plan for 2050 could be adopted globally and whether there are places where it would need to be modified.

12.25 Write a critical account of the prospects of geo-engineering as a solution to global warming.

12.26 Comment critically on the following statements:

(a) The risks associated with burning fossil fuels are much greater than those associated with nuclear power.

(b) The 'hydrogen economy' does not help with the problems of global warming.

(c) It is essential to develop carbon capture technology on a large scale to tackle global warming.

(d) Money should be spent on mitigating the effects of global warming, e.g. by building dykes or relocating people away from low-lying lands, rather than on preventing global warming.

(e) The development of low-cost solar power and of energy storage are the key challenges in providing sustainable low-carbon energy.

Numerical answers to exercises

Chapter 1

1.2 73 m

1.4 2×10^4 N

1.9 ~10 billion

1.10 (a) 3.5; (b) 10.7 in 1000

1.13 (a) 948 Btu; (b) 1.06 kW

Chapter 2

2.1 11 W

2.2 5×10^3 s

2.4 1.05 W m^{-2} K^{-1}

2.5 (a) 21%; (b) 22%; (c) 17%

2.6 7920 Btu

2.7 (a) 3620; (b) 1517

2.8 21 640 kWh

2.9 11.2 m

2.10 0.59 m

2.12 16 °C

2.13 $Re \approx 21\,000$

2.14 14.9 W

2.19 0.75 km; 4 bar

2.22 ~34 y

Chapter 3

3.2 (a) 75.4 y; (b) 36.2 y; (c) 44.3 y

3.6 15 kJ

3.7 861 kJ

3.9 (a) 0.56; (b) 778 kJ kg^{-1};
(c) 342 kJ kg^{-1}; (d) 1015 kJ kg^{-1}

3.10 (a) 20 kJ kg^{-1}; (b) 3702 kJ kg^{-1}; (c) 1882 kJ kg^{-1};
(d) 0.50

3.11 854 K; 47%

Chapter 4

4.3 0.44 m s^{-1}

4.6 10^4

4.10 6.67 : 1

4.14 1.07×10^6 N

4.15 221 m s$^{-1} \equiv$ 795 km h^{-1}

4.16 62.8 MW

Chapter 5

5.2 9.8 MW

5.3 $R = 1/2$; $\eta = 0.92$

5.6 0.86

5.9 43.5

5.14 (a) 44% increase; (b) 44% decrease

5.15 20%; 10 m

5.16 1.6 MW

5.17 5.2 GW

5.20 $a = 1/2$; $b = 1/2$

5.21 12.5 m s^{-1}

5.24 10 m s^{-1}

5.25 25.6 MW

Chapter 6

6.1 9.85 MW

6.4 43 tonne-weight

6.8 3.45 m, 9.9°; 1.77 m, 2.6°;
1.19 m, 0.1°; 0.89 m, −1.2°

6.11 $N = 3.6 \times 10^8$

6.12 97–131 Pa

6.13 1.68 MW

6.14 (a) 0.028 MW ha^{-1};
(b) 0.036 MW ha^{-1}

6.19 (a) 4.8c per kWh, 3.0c per kWh;
(b) 8.5c per kWh, 5.4c per kWh

Chapter 7

7.1 42%

7.2 (a) 300 W m^{-2}; (b) 520 W m^{-2}

7.4 (a) 5777 K

7.11 (a) 84.6 mW, 3.61 Ω;
(b) 83.0 mW

7.12 ~9 m^2

7.13 13.63 V; 14.15 V; 14.45 V; 14.66 V; 14.83 V

7.14 ~300 ha

7.15 (a) 1.25 eV and 0.69 eV;
(b) 38.2 mW; (c) 38.2%

7.17 61.9%

7.19 26.7% versus 18.0%

7.21 1.66 MW

7.22 0.28

7.23 €3500

7.24 (a) 23.8c per kWh;
(b) 29.3c per kWh

7.25 0.43 MW

7.26 941 m

Chapter 8

8.1 1.2×10^5 ha

8.2 ~600 litres

8.3 260 Mt of CO_2

8.4 40.4 MJ

8.5 ~22 Mt

8.8 110 Mt of CO_2

8.9 $\sim 2 \times 10^4$ km^{-2}

8.11 (a) 5.3 Mt; (b) 10.5 Mt

8.12 (a) 5×10^5 ha; (b) 7.9 Mt, 2.15 Mt;
(c) 2.3×10^7 sq km

8.13 0.75 kg

8.14 (a) 62 mpg; (b) 38 mpg

8.15 (a) 38 300 km^2; (b) 44 Mt

8.17 ~19 million ha

Chapter 9

9.1 ~209 MeV

9.2 0.25 t

9.3 ~100 kt

9.4 48 kt

9.11 (b) 1.76 m

9.12. ~32 h

9.13 (a) 0.015 mSv; (b) 1414 m

9.14 1.5 ms

9.16 67 s

9.17 1.6%

9.18 6.1%

9.19 1.7×10^4 TBq

9.20 (a) 6.8×10^7 m^{-2} s^{-1};
(b) 400 μS h^{-1}

9.21 19 y; 27.4 y

9.22 (a) ~12 m^3; (b) ~320 m^3

Chapter 10

10.1 2400 kg

10.3 1.45×10^{20}

10.4 12.5 keV; 0.08 mm; 1.9×10^6

10.6 6.1 mm

10.9 128 rad s^{-1}

10.10 8×10^7 K

10.11 ~2500 km

Chapter 11

11.1 64

11.6 11.3A

11.7 (a) 3.3%; (b) 0.25%

11.9 (a) 0.0199; (b) 0.0001

11.10 33 : 1

11.11 347 MW

11.12 (a) 1.0 km^2 (efficiency 0.9);
(b) ~100 km^2

11.13 19.56×10^{11} J; 163 MW

11.15 2.74 MJ

11.16 2.45 kWh

11.17 7.5 h

11.18 [CO_2] 0.142; [H_2] 0.142; [CO] 0.058; [H_2O]
0.058 mol litre^{-1}

11.19 2.00, 1.98, 1.92, 1.88, 1.80, 1.72 V

11.20 720 Wh

Chapter 12

12.5 (a) 511 Gt; (b) 31.2 and 41.6 Gt y^{-1};
(c) 454 and 558 ppmv

12.8 (a) $\$44.9 \times 10^6$; (b) $\$14.9 \times 10^6$
(c) 13.4%; (d) 2.7¢ per kWh

12.9 (a) 3.46¢ per kWh;
(b) 3.54¢ per kWh

12.10 (a) 3.48¢ per kWh;
(b) 3.56¢ per kWh

12.12 (a) 13.7c per kWh;
(b) 10.9c per kWh

12.14 $\$10 \times 10^6$

Index